Information Innovation Technology in Smart Cities

Leila Ismail · Liren Zhang
Editors

Information Innovation Technology in Smart Cities

Springer

Editors
Leila Ismail
College of Information Technology
United Arab Emirates University
Al Ain
United Arab Emirates

Liren Zhang
College of Physics and Electronic Science
Shandong Normal University
Jinan
China

ISBN 978-981-10-1740-7 ISBN 978-981-10-1741-4 (eBook)
https://doi.org/10.1007/978-981-10-1741-4

Library of Congress Control Number: 2017949487

Printed on acid-free paper

This Springer imprint is published by Springer Nature
The registered company is Springer Nature Singapore Pte Ltd.
The registered company address is: 152 Beach Road, #21-01/04 Gateway East, Singapore 189721, Singapore

To you my great Dad Fayez Ismail whose love, dedication, and aims for perfection inspired me in whatever I do. As always, you were keeping me company during the late nights working on this book. As I look up to the sky, I can hear you saying "I am proud of you my lovely girl".

Leila

To my daughter Haiyan, innovation director at Microsoft Research Cambridge for her valuable advices to make my work productive.

Liren

Preface

This vision of Smart Cities has emerged an excitement among talented research and development teams around the world to create smart applications, smart infrastructure and smart collaborative environments and to conduct experiments for the building blocks of the Smart City of tomorrow. This book reports some highlights of current R&D works here, but a lot of future works still remain to be done in order to build the Smart City of tomorrow; Smart Cities are complex and multi-dimensional.

Goals of this Book

This book describes Smart Cities and the information technologies that will provide better living conditions in the cities of tomorrow. It brings together research findings from several countries across the globe, from academia, industry and government. It addresses a number of crucial topics in state of the arts of technologies and solutions related to smart cities, including big data and cloud computing, collaborative platforms, internet of things, internet of people, communication infrastructures, smart health, sustainable development and energy management.

Information Innovation Technology in Smart Cities is essential reading for researchers working on intelligence and information communication systems, big data, Internet of Things, Cyber Security, and cyber-physical energy systems. It is invaluable resource for advanced students exploring these areas.

Acknowledgement

It is with great pleasure that we acknowledge the efforts of many people who have contributed to the development of this book. First and foremost, we thank the contributors who wrote the chapters and responded to numerous comments, reviews and suggestions from the editors and reviewers. We also would like to thank Dr. Michael Perrone from IBM T. J. Watson Research Center, USA, Dr. Driss Guerchi from the Canadian University, Dubai and Dr. Fady Al Najjar from the UAE University, for their help in reviews, despite their busy schedule. Finally, we gratefully acknowledge the expert help from Springer Nature during the preparation of the manuscript. We have been fortunate to work with an outstanding team at Springer Nature: Dr. Huang Mengchu, the Associate Editor of Springer Nature, and Mr. Sivajothi Ganesarathinam, the project coordinator, Mr. Parvathidevi Krishnan, and Mr. Parimelazhagan Thirumani were everything we could want from a publisher team: professional, efficient and a great delight to work with.

Introduction

Prof. Dr. Leila Ismail and Prof. Dr. Liren Zhang

Keywords: Smart Cities, Smart Environments, Smart Applications, Big Data Analytics, Internet of Things (IoT), Energy Efficiency, Sustainable Development, Cloud Computing, High-Performance Computing, Scalability, High Availability

Smart Cities' main goal is to increase resource efficiency and enhance the quality of living of citizens, driven by better services and less impact on the environment. Resource efficiency involves the deployment of smart environments to reduce the cost and increase the return values from the deployed resources driven by autonomous computing and insightful predictive actions in order to achieve a predefined objective. Figure 1 provides an overview of Smart City digital-enabled components. To enhance the quality of life, it is crucial for citizens able to effectively access to data statistics and Big Data analytics via broadband communication infrastructure and how these facilities can be applied in their daily life, including security and safety, transportation/mobility, healthy lifestyle, and for being involved in the decision-making process. In a Smart City, all components of the city's digital ecosystem—consisting of an agile smart infrastructure, information systems, sensors, and citizens—collaborate, thus allowing new classes of smart applications and smart collaborative environments to emerge, for instance, to enhance the health lifestyle by using applications based on social networking between citizens with common lifestyle interests and to use Big Data analytics to drive insightful predictive actions based on predefined objectives for resource efficiency and quality health lifestyle. Digital technologies have begun to blanket our cities, forming the backbone of a large, intelligent infrastructure [1].

L. Ismail
College of Information Technology, UAE University, Al Ain, United Arab Emirates
e-mail: leila@uaeu.ac.ae

L. Zhang
College of Physics and Electronic Science, Shandong Normal University, Jinan, China
e-mail: lirenzhang@sdnu.edu.cn

Fig. 1 Overview of digital-enabled smart cities

The global urban population is expected to grow tremendously by the next few years. This creates sustainability challenges in resources, such as energy and water, health threats due to air and water pollution, economic instabilities creating unemployment, social violence and crimes, and traffic congestion. Creating Smart Habitats and Smart Living Cities, by incorporating digital and information technologies, enabled by collaborative smart environments, based on sensors, Internet of Things (IoT) and Big Data Mining and Analytics [2], and Cloud computing technologies [3] as a supporting backbone, can address most of those sustainability problems, enhance citizens' involvements in decision-making by providing context-aware views of the city operations, enhance existing services, and create new ones. As collaboration is taking place in Smart Cities, cybersecurity and privacy are a crucial issue that should be considered.

Furthermore, [4] energy consumption in cities is significant—and growing rapidly. Half of the global population lives in urban areas. However, town and city dwellers consume 80% of all commercial energy produced globally. And as a result, cities offer a huge resource for energy efficiency improvement. This huge consumption of energy which adds to it the exponential growth of population gives insights to research works on ways to reduce energy consumption at different levels of our habitats.

We believe that building Senseable Smart Cities will have a tremendous impact on the evolution of the information technology, as it involves building smart autonomous infrastructures and smart collaborative environments across enterprises and organizations interacting with citizens to solve humans and social problems. We can clearly see, in the process of building the Smart City, the revolutionary impacts of large-scale resource sharing and virtualization within enterprises and organizations, such as governmental, educational, or health organizations, and the new information and communication technologies required to enable secure, reliable, high-performance, highly available, and efficient resource sharing on large-scale collaborative smart environments.

As different enterprises, organizations, and research and development centers in academia and industry are working on system components of a Smart City, a standard has to be established in order to ensure consistent outcomes and interoperability between the system components from the different suppliers whether hardware or software. The International Organization and Standardization (ISO) published standards, ISO 37120:2014, which describe performance indicators for city services and quality of life [5]. As an attempt to progress in standardization effort for Smart Cities, standards organizations, such as IEC (International Electrotechnical Commission), ISO, and ITU (International Telecommunication Union), met on July 13, 2016, during the World Smart City Forum for a kickoff collaboration on defining standards for Smart Cities [6, 7]. The UK Department of Business, Innovation, and Skills (BIS) has commissioned BSI to develop a standards strategy for smart cities in the UK.

References

1. Ratti C (2016) Smart cities—data can lead to behavioral change. Available at: http://www.siemens.com/innovation/en/home/pictures-of-the-future/infrastructure-and-finance/smart-cities-interview-carlo-ratti.html. Last accessed on 27 Sept 2016
2. Ismail L, Masud MM, Khan L (2014a) FSBD: a framework for scheduling of big data mining in cloud computing. In: 2014 International congress on big data. doi:10.1109/BigData.Congress.2014.81, IEEE
3. Ismail L, Khan L (2014b) Implementation and performance evaluation of a scheduling algorithm for divisible-load parallel applications in a cloud computing environment. Softw Pract Experience. doi:10.1002/spe.2258. Available at https://www.iso.org/obp/ui/#iso:std:iso:37120:ed-1:v1:en. Last accessed 28 Sept 2016
4. IEA (2008) Jollands N, Kenihan S, Wescott W (2008) Promoting energy efficiency—best practices in cities—a pilot study. IEA/SLT/EC(2008)3
5. ISO (2014) International Organization and Standardization. Sustainable development of communities—indicators for city services and quality of life (ISO 37120:2014)
6. IEC (2016) International Electrotechnical Commission. Significant milestone for smart city development. Available at http://www.iec.ch/newslog/2016/nr2416.htm. Last accessed 28 Sept 2016
7. BSI (2016) The British Standards Institution. Smart city standards and publications. Available at http://www.bsigroup.com/en-GB/smart-cities/Smart-Cities-Standards-and-Publication/. Last accessed 28 Sept 2016
8. Senseable City Laboratory. Available at: http://senseable.mit.edu/. Last accessed 27 Sept 2016

Contents

About the Editors

Prof. Dr. Leila Ismail is an Associate Professor at the College of Information Technology (CIT) of the United Arab Emirates University (UAEU), UAE, joined in 2005, and is the founder and director of the High Performance, Grid & Cloud Computing Research Laboratory at CIT, and the UAEU supercomputer, with state-of-the-art architecture and technology.

Dr. Ismail completed higher studies (DEA) in distributed systems at the Joseph Fourrier University (Grenoble I)/ENSIMAG Engineering School in France, and earned Ph.D. in distributed systems from Computer and Software Engineering/Distributed Systems from the National Polytechnic Institute of Grenoble, France, in September 2000 with very-honorable degree.

Dr. Ismail has a vast industrial and academic experience; at Sun Microsystems R&D Center, worked on the design and implementation of highly available distributed systems; and participated in the deposit of a US patent in the domain. She served in teaching at Grenoble I, France, Assistant Professor at the American University of Beirut, and has been serving as an Adjunct Professor at the Digital Ecosystems and Business Intelligence Institute Curtin University, Australia.

Dr. Ismail current research interests include performance analysis in distributed systems, energy efficiency and management, green computing, resource management and scheduling problems in distributed systems with emphasis in clouds, middleware, High Performance Computing, Big Data analytics and software security in distributed systems. She won the IBM Shared University Research Award and the IBM Faculty

Award, very competitive world-wide, won funding for major projects as a PI/Co-PI and the funded project by UAE/NRF was top ranked by external anonymous reviewers.

Dr. Ismail has international collaborations and publishing her research results in prestigious journals and international conferences. She is an Associate Editor of the International Journal of Parallel, Emergent and Distributed Systems, and an Editorial Board member of the International Journal of Engineering and Applied Sciences. She served as chair, co-chair and track chair for many IEEE international conferences, including being a General Chair for the IEEE DEST 2009 and a General Chair, Technical Program Chair and Organizing Committee Chair for the 11th International Conference on Innovations in Information Technology 2015 (IIT'15) for which Dr. Ismail obtained the IEEE Computer Society (HQs) Technical Sponsorship.

Prof. Dr. Liren Zhang received his M.Eng. (1988) from the University of South Australia and Ph. D. (1993) from the University of Adelaide, Australia, all in electronics and computer systems engineering. He is currently with the College of Physics and Electronics, Shandong Normal University, China as full professor. Dr. Zhang was Senior Lecturer with Monash University, Australia (1990–1995), Associate Professor with Nanyang Technological University, Singapore (1995–2007), Professor of Systems Engineering with University of South Australia (2006–2012) and Professor in Network Engineering with UAE University (2009–2016).

Dr. Zhang has vast experience as an engineer, academia and researcher in the field of network modeling and performance analysis, teletraffic engineering and network characterization, optical communications and networking, network resource management and quality of services (QoS) control, wireless/mobile communications and networking, switching and routing algorithm, ad-hoc networks, electro-optical sensor networks, and information security. He has published more than 150 research papers in international journals.

Currently, Prof. Zhang's research interests include biomedical imaging using ultra wideband penetrating radar sensor; Extraction of biometric signatures for forensic identification and Multi-layer security and airborne SAR signal processing, imagery and tracking of ground moving targets.

Part I
Software Design and Development of ICT-Based Smart Applications

Unified Interface for People with Disabilities (UI-PWD) at Smart City (Design and Implementation)

Ghassan Kbar, Syed Hammad Mian and Mustufa Haider Abidi

1 Introduction

A smart city can be defined as the city which employs digital technologies to improve the quality of life of the citizens, including the PWDs such as the health care [1]. According to WHO [2], the population of the disabled people has been increasing worldwide and the older people posited a high percentage of the overall population in certain counties such as Japan (87%), China (67%). These data from the WHO and the other active agencies have encouraged and motivated the researchers to develop assistive technology (AT) based solutions such as the daily mobile service, etc. [3]. According to Yang et al. [4], the AT solutions for the PWDs should be developed in such a way that they can respond to their needs in real time, especially in case of the emergency needs. The support of the intelligent healthcare system such as the University of Rochester's Smart Medical Home [5], and the use of mobile computing for continuous measurement and analysis of the health data of the PWDs [6] represent examples of the relevant AT solutions deployed at smart cities. Actually, two main issues, including the right design methodology and choosing the right technologies need to be considered when designing an AT solution.

The WHO [7] had defined AT as any product, instrument, equipment, or technology which is adopted or specially designed to improve the functioning of the disabled person. According to Hersh and Johnson [8], AT is a generic or umbrella term which encompasses technologies, equipment, devices, apparatus, services, systems, processes and environmental modifications used by the PWDs and elderly

G. Kbar (✉)
Riyadh Techno Valley, King Saud University, Riyadh, Saudi Arabia
e-mail: ghaskibar@hotmail.com

S.H. Mian · M.H. Abidi
FARCAMT CHAIR, Advanced Manufacturing Institute,
King Saud University, Riyadh, Saudi Arabia

L. Ismail and L. Zhang (eds.), *Information Innovation Technology in Smart Cities*,
https://doi.org/10.1007/978-981-10-1741-4_1

people to encounter the social, infra-structural and other obstacles. The different design methodologies have been covered in the literature. According to Magnier et al. [9], the worst part in designing for the disabled people is that the designers have to depend on their experience as a user which is very different from the disabled users. As discussed by Hersh [10], the AT design process should at least pursue the following basic principles: user-centered design (UCD), with end-users involved throughout the design and development process [11], iterative, multi-criteria approaches (including the function, form, ease of use, usability, accessibility, performance, reliability, safety and environmental factors), ease of upgrading, repair and maintenance, robust design, modular software architecture, etc. The UCD has been considered as the best design practice, especially in the systems which involve Human Computer Interface (HCI) or software. The UCD is a design philosophy, which design tools or the processes from the viewpoint of the users, i.e., how will it be perceived by the user. There are a number of design tools such as user interview, questionnaires, focus groups, survey, usability evaluation, etc. that are utilized in UCD [9]. Although, there have been a number of different approaches to the UCD, but they all utilized the following four principles that are: user involvement, user information for the design process, prototypes for user evaluation, and iterative design to spot out issues early on and rectify them [12, 13]. According to Chagas et al. [14], the primary reasons for difficulties in designing and producing AT's are the kinds and degrees of disabilities and individual characteristics associated with the users (physical, psychological, cultural, and environmental, etc.).

Presently, many ICT based solutions assist PWDs to carry out their daily tasks with or without the guidance. For example, visual and tactile signaling devices, text enlargement program, speech synthesizer, Braille display, etc. In addition to this, there also exist special devices such as the special keyboards, with overlays for pictograms, environmental control units with infrared signaling, speech recognition devices, and standard and special switches, etc. [15]. Recently, as a result of advances in technologies, the computer based devices have attracted a lot of attention, especially among the PWDs [16]. According to Kim et al. [17], a unified system can provide multi-modal feedback because it includes the requirements of the PWDs as well as the innumerable communication possibilities. One of the several instances of the multi-modal system can be a universal smart solution based on Ambient Intelligence as developed by Kbar et al. [18]. Their interface addressed the needs of people with different impairment conditions as well as comprised of an intelligent intervention based on the behavior of the disabled person. The Internet of Things (IoT) has been regarded as one of the effective solutions to enable technology in the spectra of the Smart City [19]. The smart health care system in the smart cities is very crucial for the elderly and the PWDs because it permits periodic measurement and analysis of their health data [6]. Hussain. et al. [20], developed an intelligent system with real-time monitoring and personalized health care of the elderly and the PWDs at their home. This system allowed them to interact with each other and the system as well as provided them emergency support and the virtual community.

In this work, the design considerations for the PWDs at the workplace have been discussed. It would help to identify the right technologies to be supported as explained in Sect. 2. In Sect. 3, an AT review and assessment of technologies relevant to smart city has been presented. In Sect. 4, the design methodology for AT solution for PWDs is covered, and in Sect. 5 the PWD project development is shown.

2 Design Considerations for PWDs at the Workplace

In order to develop an AT solutions applicable in smart cities, it is very important to devise a proper development process with relevant features as well as the solutions employing the right technological features. These solutions should aim to address all the necessary needs of the PWDs and provide them with smart solutions. The requirements crucial to the process of designing and implementing an AT solution for the PWDs may include the user involvement, identification of the performance criteria for validating the AT solutions, determination of the methods for assessing the AT solution outcomes, and usability and adaptability of the AT solutions. Moreover, there have been other features related to the technologies such as the accessibility and comprehensive solution to support the different impairment conditions, universality and modality, adaptability and context aware, heterogeneity with mobility and navigability and innovative solutions, etc.

2.1 Technology's Features or Criteria

The following technological features have been identified and should be incorporated in the development of an efficient AT solutions for the PWDs.

2.1.1 Universality Based Solution

The requirement of the universality and multi-modality in the AT systems are crucial to support as many users as possible. According to the Antona [21], the universal access principles should be used in designing products and the technologies instead of designing ATs for a particular group of the PWDs. The universal design is effective and efficient in terms of the saving cost, user satisfaction, convenience, etc. The system developed by Karpov and Ronzhin [22] utilized *different technologies to provide the universal access. They employed voice, video, sign language synthesis, and haptic avatar, to get the input and provide output from/to the users.* They also used a smart control system to monitor the occupants for special events such as when they fall or sneeze. In addition to that, the system was context aware, adaptive, and employed a great deal of intelligence to accommodate different users in different environments and conditions.

2.1.2 Comprehensive and Multimodality Based Solution

To support a comprehensive solution for people in the workplace, including the PWDs, the AT system should consider multimodal HCI. The multimodal HCI provides a flexible system to incorporate a variety of users with different impairment conditions. It is a complex system which uses different technologies including the audio-visual to allow smart user interaction and support of communication [23]. It should also support different interfaces (such as attentive and wearable, enactive, and perceptual), support computer input devices such as keyboards and pointing devices as well as support different human senses, such as the vision (body, facial, gaze, gesture), audio, haptic, smell and taste [24].

In addition to the multimodal feature, comprehensive model architecture [18] addressed the needs for normal and PWDs at the workplace through the support of relevant services. Furthermore, this model also supported tracking user location for the behavior analysis and intervention.

2.1.3 User Interface Adaptation and Context Aware Based Solution

The advancement in technologies and the support of ubiquitous computing have encouraged the designers and organizations to develop a system which can improve the customized access to a wide range of devices and services through context-aware systems [25]. The usefulness of the context-aware computing technology can be realized in [26]. A context aware system can link location, user identity and environmental resources to a mobile system [27]. It can be defined as a set of information that characterizes the situation of an entity [28], where the environment is a triplet of entities namely < Object; person; event >. Similarly, the context can be described as a user interface adapted to the context of use through a new triplet namely < user; platform; environment > [29]. To support context aware solution at the workplace, the AT system should also address issues related to the user profile, health condition, work and social activities, and user's location.

2.1.4 Heterogeneity, Mobility & Navigability Based Solution

The heterogeneity of the people, the technology, work environment, etc. contribute significantly towards the difficulties in measuring the effectiveness of the AT [30]. A high degree of participant heterogeneity creates difficulty in measuring outcomes [31]. It means that the outcome of the specific AT may not be applicable to many participants, if the participants in the study have a high level of variation. Heterogeneity in the work environment is related to accessibility which requires multiple technologies, as well as mobility and navigation that are needed for different people including the PWDs. The access and mobility are the important dimensions of the quality of life. For the PWDs, every movement or task is often filled with problems, including many barriers [32]. Mobility impaired people make a heterogeneous user group with a wide range of diverse requirements relating to the navigation [33]. The navigation is an important mobile activity and can be described as the key for maintaining the mobility and independence. However, many older people find increasing difficulties with it due to declines in their perceptual, cognitive and motor abilities [34].

The navigation system can be referred to as the multi-modal because of the presence of the different modes of transportation. Völkel and Weber [33], carried out the multimodal interpretation of the geographical data in the navigation of the mobility impaired users. Many AT which have been developed so far, primarily focused on helping the PWDs in unknown environments. The support of indoor navigation is also very crucial for the PWDs at the workplace as it can guide them to reach their destination quickly. The outdoor navigation is less important for the PWDs at the workplace because they get assistance most of the time by caregiver.

2.1.5 Innovative Based Solution

Design for user-centered innovation (DUCI) as proposed by Zaina and Alvaro [35], guide the software development process to integrate the business innovation and user needs. This approach helped in producing a quality requirement through the development of the business-focused skills. In addition to the support of software design process for the HCI, it allowed the designers to observe solutions by focusing on the quality of the interaction as well as the contribution to the user life. The User centered design (UCD) would engage users in the development process in order to create products relevant to them, which can satisfy their needs [36]. The Lean Startup that combines the principles of agile software development and new product development develops prototypes at an early stage in order to conform market assumptions as well as acquiring customer feedback more quickly, assisting entrepreneur to eliminate incorrect market assumptions, etc. [37]. The provision of the innovative solution for the PWDs at the workplace is important in order to provide relevant solution, i.e., practical and can satisfy the needs satisfactorily. However, it doesn't have to be profitable because its primary purpose is to support a humanity mission and not to generate profit or selling oriented marketing solutions.

The different criteria (Ft1–Ft5) for the assessing technologies of the AT solutions at the workplace can be summarized in the Table 1. The assessment criteria C1–C13 that are associated with Ft1–Ft5 and used to assess research papers and technological solutions are defined in [38].

Table 1 Matching the users and expert requirements

Criteria for assessing technologies for AT solutions at the workplace				
Ft1	Ft2	Ft3	Ft4	Ft5
Universality based solution	Comprehensive and multimodality based solution	User interface adaptation and context aware	Heterogeneity, mobility & navigability	Innovative based solutions
Support of technologies related to C1–C13 + support of partial VI, HI, SI, MI, VMI, SMI, HMI	Support of technologies related to C1–C13 relevant for partial VI, HI, SI, MI, VMI, SMI, HMI at the work environment	Support adaptability and context aware for PWDs based on user profile, environment, location, health, date and time, social, health, work, and editor utility	Support of different Technologies related to C1–C13 that can be relevant for different PWDs conditions and interface as well as support of mobility and navigability	Developed prototype that can be improved toward products relevant for marketing solutions

3 Related Work for AT Review and Assessment for Smart City

To improve the quality of life for PWDs, it is very important to provide assistance in a smart way [4]. The smart cities are such idea where digital technologies are employed to improve the quality and the performance of the urban lifestyle. Therefore, the AT based solutions implemented in the smart cities can overcome visual, mobility, and cognitive problems. Back in 1996, Allen [39] presented the concept of the smart city, where they integrated the control system with ATs to execute smart home applications for the PWDs. The comprehensive survey on features and future perspectives of the smart homes have emphasized on the use of ATs in smart home [40]. For example, Doukas et al. [41] developed intelligent platforms involving agents, context-aware and location-based services for the elderly and the PWDs in the smart city. Similarly, Hussain et al. [20], developed an AT application to monitoring the health status of the PWDs and elderly people in the Smart city.

There are many AT research solutions can be suitable applications in the smart cities, such as, the haptic and auditory maps developed by the Rice et al [42] which allowed the blind people to interact with the system through force feedback devices and listen to audio signals. Similarly, the voice recognition system based on GPS assisted the blind people while walking [43]. This system guided them through the voice, and it also detected obstacles using the SONAR unit. The mobile and cheap visual and haptic sensory feedback device introduced by the Israr et al. [44] can also be used by the visually impaired to determine the obstacles in the environment. Hawley et al. [45] also formulated a system of augmentative and alternative communication with voice input and voice output. This developed system could recognize disordered speech in order to process it and convert it into fluent synthetic speech. Similarly, Peixoto et al. [46] invented a wheelchair controlled through the humming of the motor impaired user. Rodriguez-Sanchez et al. [47] devised a smart phone navigation application for the blind people. This system provided instructions through audio and tactile feedback. It also made routine phone usage easier for the user by splitting the screen and displaying the basic features like information, destination, help, etc. Therefore, it can be concluded here, that the ATs play a significant role in the designing of the smart cities and homes.

4 Design Methodology for PWD

A user-driven process focusing on the technologies as well as the quality of life has been proposed by Federici and Scherer [48]. The objective was to develop a proper solution based on the AT and addressed the needs of the PWDs. This objective was accomplished through application of the Assistive Technology Assessment (ATA) process. The implementation of the ATA was based on the users seeking a solution, where professional team would interact with the users to get their requirements. Subsequently, the professionals matched the users' needs to suitable solution and users validated the solution provided by the professional team. Finally, the users adopted the solution and provided with appropriate training. According to the authors, three factors, the accessibility, universality, and sustainability are crucial in designing any system for the PWDs. The universal AT with multi-modal input and multimedia output interface proposed by Karpov and Ronzhin [22], represent one of the supreme examples in the present scenario. The sustainability is related to how technology can adapt over time to a person's changing needs.

In this work, an AT proper based solution (AT-PBS) is used to improve the functional capabilities of the PWDs, improve the information and communication means of the PWDs, and improve the service support accessibilities. Two design models including the ATA model [48] and the Human Activity Assistive Technology (HAAT) model [49] as shown in Fig. 1, have been adopted. In phase 1, the objective was to define the requirements and the needs of the PWDs through the users, experts and the existing solutions. The existing solutions have also been used to identify the research gap. In phase 2, the professional team of the project identified the matching solution of the PWD needs and then the interaction of the technological solution with PWDs to achieve full satisfaction. This was carried out through the environmental assessment process, matching assessment and suggested relevant solutions as described in the ATA. It has also included the technology assessment process to take into account the system technology issues, in addition to other issues for activities specifications, end user issues, and design issues as described in the HAAT model. Finally, in the Phase 3, the proposed solution would be tested with the PWDs to ensure their adoption of the solution. This has been in accordance with the Phase 3 of the ATA model, and assistive technology assessment of the HAAT model.

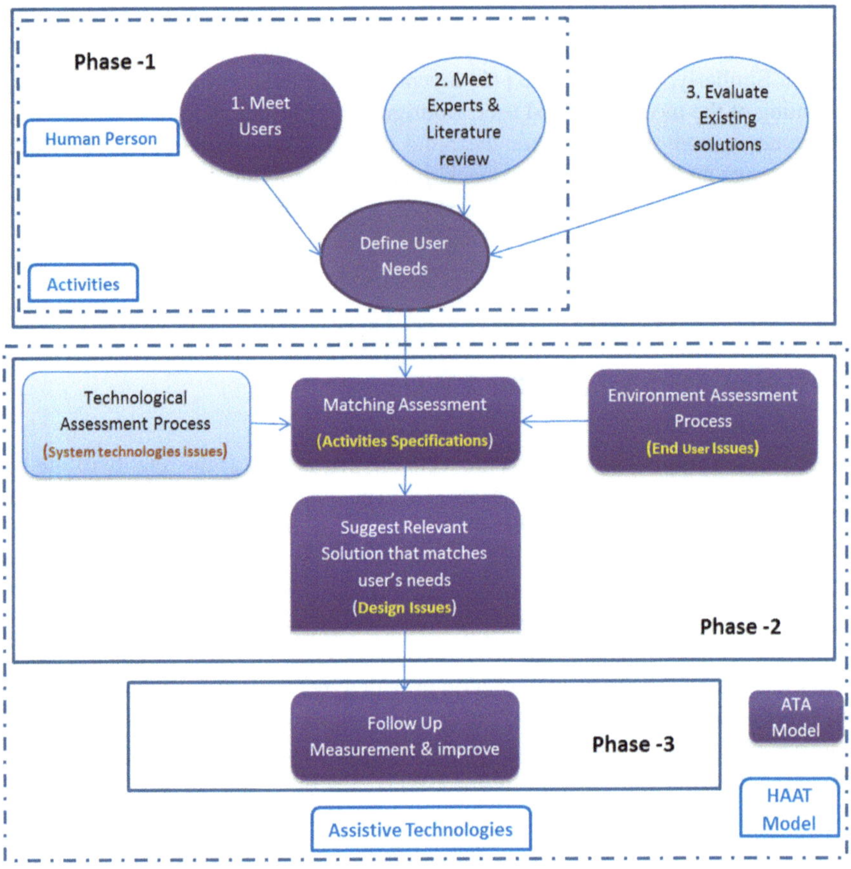

Fig. 1 Combined design models of assistive technology assessment (ATA) and human activity assistive technology (HAAT)

5 PWD Project Development

5.1 A Collection of PWD User Requirements: Identify the Requirements and Needs for PWDs (Phase-1)

In order to identify the needs of the PWDs in the university campus (Step 1), the surveying of the PWD stakeholders is crucial for UCD. Since, the PWDs might not be aware of the different AT solutions, there is a need to verify the relevant technologies based on the expert opinion and the available literature review as described in Step 2. Moreover, there is a need to assess existing research and the commercial technology solutions. It identifies the criteria associated with the best practices as well as determined the appropriate tools that can be used for voice recognition in the majority of the PWDs as discussed in the Step 3.

Step1: Survey PWDs to know their needs

To develop an effective and efficient product or service, it is always important to identify the needs and requirements of the customers.

In this work, a survey of the PWD students has been conducted to identify their needs as per the especially designed questionnaire. The results of the survey revealed the following requirements for designing AT based-solution:

- Accessible system: To access any information related to curriculum while walking
- Mobile searchable system: To search for location using Location based System (LBS)
- Supported communication system: To communicate with others and the teachers
- Auto-searchable system: To find and talk to any person that matches their profile
- Universal Interface system: To support different impairment condition of the PWDs
- Comprehensive solution: Accessing information, communicating with others, support of editing tools, and tracking user for intervention and guidance.

Step2: Identify the needs of PWDs through Experts and literature review

Based on the literature review and the issues highlighted in the introduction section, the following needs have been identified for the PWDs: AT system or application should be user friendly, it should be mobile, it has to be accurate and flexible to meet the user needs, it should provide a good help and support documentation, it should be comprehensive in nature and it should be interoperable

Step3: Survey of existing technological solutions for PWDs Gap Analysis

Many research and commercially available assistive solutions worked on the smartphone platforms. These solutions help the PWDs to perform their daily activities in a better way. Kbar et al. [50] evaluated ten research/commercial products (refer to Table 1 in [50] according to the design criteria defined in Table 2 (where (x) means a technology is relevant for a particular impairment). This would help the designers to identify the relevant criteria that supports AT based on best practices.

Based on their analysis, Table 1of Kbar et al. [50] concluded that 4 out of 10 products resulted in a good score above 75% and met the specific conditions of the PWD users. It was also clear from their analysis that 3 out of 10 products had a good score for the majority of the design criteria and therefore they met the specific conditions in terms of usability and maintainability. It can be observed that the innovative technologies assist researchers and the developers to provide intelligent solutions for the PWDs and efficiently interact with the environment.

Conclusions of Section 5.1

The current research lacks a complete unified smart solution relevant for the PWDs at the workplace or the university environment. The research and commercially available solutions have mainly focused on a particular type of PWD group, i.e., they have not considered the wider target group. These kinds of systems are wanting for a person suffering from more than one type of disability. Therefore, there is a need of a comprehensive solution, which can provide services to a wide

Table 2 Technologies essential in an assistive system for different impairment conditions

Type of interface	Type of assistive technology	Hearing impairment	Motor impairment	Speech impairment	Visual impairment
Input to system	1. Voice recognition	x	x		x
	2. Voice to text	x	x		
	3. Keyboard & mouse	x		x	
	4. Gesture control	x	x	x	x
Output from system	5. Text to voice		x	x	x
	6. Display screen	x	x	x	
	7. Vibration or flashing LED	x	x	x	x
Mobile input & output system	8. Mobile platform such as laptop and smartphone	x	x	x	x
	9. Tracking Location and behavior using RFID sensors, and WIFI	x	x	x	x
	10. Flexible adaptable interface	x	x	x	x

range of PWD group rather than focusing on one group. In this work, a comprehensive and smart solution is developed and implemented in an academic work environment.

Based on the analysis carried out in the step 1–3, an efficient smart and the context aware in the smart cities should consider specific criteria to suit the majority of PWD conditions. This can be achieved through designing a universal interface that supports the needs for occupational therapy and rehabilitation (Medical model) [51], and empowers the disabled people through user-centered solution (Social model) [52]. The following requirements, including the mandatory and the enhancement (information and support accessibilities) should be considered while designing the AT. The mandatory requirements for better functional capabilities, AT-PBS should include the following features:

- Speech to text
- Text to speech and synthesis
- Speech/Voice Command Control
- Display enhancement through magnification and zooming
- Speech volume control
- Keyboard, mouse and touch screen.

The enhancements to support full information accessibility, AT-PBS should include the following characteristics:

- Auto Emergency Response
- Comprehensive support of activities at smart cities, including smart help comprising of the auto-searchable system, profile setup, noting and communication.

The enhancements to have better service support accessibilities, AT-PBS should include the following features:

- Mobility support
- Tracking, behavior analysis and intervention
- Adaptable interface according to user profile
- Design for all by supporting universality
- Overcome existing environmental and social barriers by supporting usability solution adapted to user's conditions.

5.2 Design Solution for PWDs at Smart City

The standards for designing user interface have been addressed by [53]. These standards make interaction with system easy and user experience more friendly and effective. They identified the following requirements associated with usability:

- The interface has to be tested by the users to obtain their feedback.
- The design has to be evaluated by relevant experts. The use of formal guidelines, checklists or questions should be employed.
- To support usage logging, where useful information would be recorded automatically by the server or the software.

The proposed design is based on a flexible dynamic interface that adapts to different impairment conditions in order to satisfy the needs of the PWD users. This design has been validated by the expert of PWD and will be tested by the PWDs to get their feedback for further design optimization.

Note that this design supports eleven groups with different impairment conditions through the adaptable unified interface. This design can adjust the interface environment setup according to user group conditions as well as the working environment conditions associated with floor location, weather and day conditions, etc. The proposed solution would mainly be relevant for PWD in smart cities involving both the desktop and the smart phones.

This new integrated interface mainly comprises of the two features, namely: SMARTHELP and SMARTEDIT. The SMARTHELP provides assistive and communication services to the PWDs in the workplace. It aids the PWD in getting help information about the locations in the workplace building such as the locations of offices in various departments, directions to these departments, building information like toilets, emergency exits, lifts, etc. This interface also enables the PWDs to make a call using Voice over IP (VOIP) to the colleagues and other people in the building for help or intervention. In addition, it offers Auto Emergency Response facility to call for immediate help from an attendant. In this system, a tracking network using Ekahau Real-time Location System (RTLS) has also been implemented. It works with RFID tags and communicates through Wi-Fi network in the building [54]. When a PWD user enters the building, they are given a traceable tag. This tag is continuously monitored through the RTLS. The tracking system is deployed on 802.11b/g (2.4 GHz) WLAN technology using the building Wi-Fi network. The Wi-Fi tags transmit like any Wi-Fi client (e.g., a laptop) where the Wireless Local Area Network (WLAN) Access Points work in reception mode to measure tag received signal strength. Six APs have been deployed on the ground floor to monitor the movement of the PWDs and tracking and logging their locations (for the campus lay out, refer to [50]. When the user carrying a tag enters a signal field of a particular access point, signal measurements are taken via radio communication and delivered to the RTLS software controller, which updates and stores the information in a Tracking and Database server located inside the building. The APs send signal strength packages to the tracking server so the algorithm can determine the movement and store user location on the Database server. A behavior and reminder algorithm has also been developed. This algorithm can read the location data of PWD from Database server and analyze the behavior based on the scheduled events to determine if PWD execute the event according to setup or not. The behavior algorithm has been running on the server and sends an alert message to PWDs as well as their caregivers to inform them about missing events. The behavior result can also be logged on the server for further analysis by dependence and the experts. The SMARTEDIT feature which can facilitate different PWDs is a multimodal interface that provides a capability of writing and editing a document using voice recognition.

5.3 Development Solution

The developed solution in this work is based on the unified user interface [50] and exhibited the following characteristics and benefits to the PWDs:

- User can login using the three different methods—normal (username and password), RFID and voice recognition. New user can also sign in using the Join US button.
- Font size can also be changed as per the requirement.
- Interface can also be controlled or driven using the voice command.
- User can customize his/her interfaces using the profile and the environment set up. Environment set up helps the PWDs to set up input/output interfaces. User profile requests for the user name, age, email, etc. depending on the user disability.
- The two main features SMARTEDIT and SMARTHELP provide help to different kinds of impaired persons through one unified interface.
- The SMARTEDIT feature is a multimodal interface that provides a capability of writing and editing a document using voice recognition, to facilitate different PWDs.

- SMARTHELP provides assistive and communication services to a PWD user at workplace. It helps in getting relevant information about locations in the workplace, enables the call using Voice over IP (VOIP) and offers Auto Emergency Response facility to call for immediate help. It also allows the user to run profile set up, get help (or any information), communicate, etc. User can conduct search either using the search list, the keyword search or search by location.
- In the communication window, user can edit event, make call using the Voice Over IP to track location or set personal reminder or set up the intervention using the behavior algorithm.
- The proposed design interface also meets the standard guidelines and ISO 13407.

5.3.1 Unified Interface and Prototype

A unified interface for PWDs with different impairment conditions has been designed and implemented. User can change the font size at the first page by selecting the button "Change Font" or by speaking the word "eight". Three modes to log in into the system have been supported. These are: normal login using username and password, RFID login, and voice recognition login. User will be able to customize his/her profile by defining his/her visual and hearing capabilities. Then user will be asked to change their login details. Then user can login in the system. At this point, the user can choose one of the four available modules, namely, user profile, environment setup, smart editor, and smart help. The user can also change his/her user's profile setup and/or environment setup. All the above mentioned steps are depicted in Fig. 2.

6 Discussion and Comparison

It is clear that the implemented interface in our project is adaptable to PWD users' requirements. This smart flexible interface that can run on laptop, mobile and tablets, support many important features needed for PWDs at the workplace. This smart interface program interacts with MySQL database, and RFID WIFI tracking system. Now, the PWD users are able to use voice recognition to drive and control the interface, RFID and voice recognition to login on the system, and VOIP, email, and SMS to interact with others. In addition, this unified interface supports continuous monitoring of the PWD's movement. This has been achieved through the behavior and intervention algorithm, which alert PWDs and their caregivers for their scheduled events.

Fig. 2 Different modules of UI-PWD

7 Conclusions

Recently, ATs have become very common due to the reduction in cost of smart phones, and computer accessories and attracted researchers to develop intelligent AT solutions relevant for smart cities. Additionally, the developments in HCI and ICTs have given impetus to these technologies. Still, there is a need of a comprehensive solution that is capable to fulfill the needs of different kinds of impaired persons living and working in smart cities. Based on literature review and assessment of commercially available solutions, it can be concluded that there are available solutions, but they only focus on one type of impairment and provides lesser flexibility to the users. The developed system in this work contains two main features SMARTEDIT, and SMARTHELP. This system can provide help to various kinds of impaired persons through one unified interface. The multiple options provide flexibility to perform many tasks normally by the PWDs. The system not only focuses on the daily activities, but also provides smart context aware options for activities that are required in work environment of the smart cities. The implemented interface is adaptable to PWD users' needs and provide smart flexible interface that support many important features needed for the PWDs at the workplace. Therefore, this smart comprehensive solution can offer many opportunities to the PWDs to live their life in a better way and work effectively and productively through the support of smart AT solutions relevant to the smart cities' environment.

Acknowledgements This Project was funded by The National Plan for Science Technology and Innovation (MARRIFAH), King Abdulaziz City for Science and Technology, Kingdom of Saudi Arabia, Award Number (12-ELE3220-02).

References

1. Hollands RG (2008) Will the real smart city please stand up? City 12(3):303–320. doi: https://doi.org/0.1080/13604810802479126
2. World Health Organization, Disability and Health (2014). Available from http://www.who.int/mediacentre/factsheets/fs352/en/. Accessed on 26 July, 2015
3. Ghaffarianhoseini A, Dahlan ND, Berardi U, Ghaffarianhoseini A, Makaremi N (2013) The essence of future smart houses: from embedding ICT to adapting to sustainability principles. Renew Sustain Energy Rev 24:593–607
4. Yang L, Li W, GeY FuX, Gravina R, Fortino G (2014) People-centric service form health of wheelchair users in smart cities. Springer, Internet of Things Based on Smart Objects
5. Almudevar A, Leibovici A, Horwitz C (2005) Electronic motion monitoring in the assessment of non-cognitive symptoms of dementia. Int Psychogeriatr. Cambridge University Press, New York, NY USA
6. Pawar P, Jones V, VanBeijnum B-JF, Hermens H (2012) A framework for the comparison of mobile patient monitoring systems. J. Biomed. Inf. 45:544–556
7. World Health Organization (WHO) (2001) International classification of functioning, disability and health. Switzerland, Geneva
8. Hersh MA, Johnson MA (2008) On modelling assistive technology systems part 1: modelling framework. Technol Disabil 20(3):193–215

9. Magnier C, Thomann G, Villeneuve F (2012) Seventeen projects carried out by students designing for and with disabled children: identifying designers' difficulties during the whole design process. Assistive Technol 24(4):273–285
10. Hersh MA (2010) The Design and evaluation of assistive technology products and devices part 3: outcomes of assistive product use. In: JH Stone, M Blouin (eds). International encyclopedia of rehabilitation. Available online: http://cirrie.buffalo.edu/encyclopedia/en/article/312/
11. Dvir D, Raz T, Shenhar AJ (2003) An empirical analysis of the relationship between project planning and project success. Int J Project Manage 21:82–95
12. den Buurman R (1997) User-centred design of smart products. Ergonomics 40(10):1159–1169
13. Gould JD, Lewis C (1985) Designing for usability: key principles and what designers think. Commun ACM 28:300–311
14. Chagas BA, Fuks H, de Souza CS.(2015) Lessons learned in the design of configurable assistive technology with smart devices. In: Proceedings of the 5th international symposium, IS-EUD 2015, Madrid, Spain, May 26–29, pp 180–185. doi:https://doi.org/10.1007/978-3-319-18425-8_13
15. Foley A, Ferri BA (2012) Technology for people, not disabilities: ensuring access and inclusion. J Res Spec Educ Needs 12(4):192–200
16. Shih CT, Hsu SC (2013) Measuring the satisfaction of implementing assistive input devices for disabled people with InLinPreRa. Adv Mater Res 774–776:1967–1970
17. Kim J, Huo X, Minocha J, Holbrook J, Laumann A, Ghovanloo M (2012) Evaluation of a smartphone platform as a wireless interface between tongue drive system and electric-powered wheelchairs. IEEE Trans Biomed Eng 59(6)
18. Kbar G, Aly S, ElSharawy I, Bhatia A, Alhasan N, Enriquez R (2015a) Smart help at the workplace for persons with disabilities (Shw-Pwd). In: ICIES 2015: XIII international conference on intelligent environments and systems, Paris, France, January 23–24
19. ITU (2015) Report on internet of things: executive summary. Available via https://www.itu.int/osg/spu/publications/internetofthings/InternetofThings_summary.pdf. Accessed on 1 Mar 2016
20. Hussain A, Wenbi R, da Silva AL, Nadher M, Mudhish M (2015) Health and emergency-care platform for the elderly and disabled people in the smart city. J Syst Softw 110:253–263
21. Antona M, Ntoa S, Adami I, Stephanidis C (2009) User requirements elicitation for universal access. In: Stephanidis C (ed) The universal access handbook, CRC Press
22. Karpov A, Ronzhin A. (2014) A Universal Assistive Technology with Multimodal Input and Multimedia Output Interfaces. UAHCI/HCII, Springer International Publishing Switzerland, Part I, LNCS 8513, p. 369–378
23. Cohen PR, McGee DR (2004) Tangible multimodal interfaces for safety-critical applications. Commun ACM 47(1):1–46
24. Jaimes A, Sebe N (2007) Multimodal human–computer interaction: a survey. Comput Vis Image Underst 108:116–134
25. Korn O (2014) Context-aware assistive systems for augmented work. A framework using gamification and projection, Thesis, Institut für Visualisierung und Interaktive Systeme, Universität Stuttgart, Available Online: http://www.motioneap.de/wp-content/uploads/2014/05/Context-Aware-Assistive-Systems-for-Augmented-Work.-A-Framework-Using-Gamification-and-Projection_WEB.pdf. Accessed on 8 Mar 2016
26. Riahi I, Moussa F (2014) A formal approach for modeling context-aware Human-computer system. Comput Electr Eng 44:241–261
27. Schilit B, Adams N, Want R (1994) Context-aware computing applications. In: proceedings of the 1994 first workshop on mobile computing systems and applications, WMCSA'94. IEEE Computer Society; pp 85–90
28. Dey A, Sohn T, Streng S, Kodama J (2006) iCAP: interactive prototyping of context-aware applications. In: Fishkin K, Schiele B, Nixon P, Quigley A, (eds) Pervasive computing. Lecture notes in computer science, vol. 3968. Berlin, Heidelberg: Springer. pp 254–271

29. Calvary G, Demeure A, Coutaz J, Dassi O (2004) Adaptation des interfaces homme-machine à leur contexte d'usage plasticité des ihm. Revue d'Intelligence Artificielle 18(4):577–606
30. DeJonge D, Scherer MJ, Rodger S (2007) Assistive technology in the workplace. Elsevier Health Sciences Medical, p 253
31. Fuhrer MJ (2001) Assistive technology outcomes research: challenges met and yet unmet. J Am Phys Med Rehabil 80:528–535
32. Matthews H, Beale L, Picton P, Briggs D (2003) Modelling Access with GIS in Urban Systems (MAGUS): capturing the experiences of wheelchair users. Area 35(1):34–45
33. Völkel T, Weber G (2007) A new approach for pedestrian navigation for mobility impaired users based on multimodal annotation of geographical data. In: Stephanidis C (ed) Universal access in HCI, Part II, HCII, LNCS 4555, pp 575–584, 2007; Springer-Verlag Berlin Heidelberg
34. Kirasic KC (2000) Ageing and spatial behaviour in the elderly adult. In: Kitchin R, Freundschuh S (eds) Cognitive mapping: past, present and future (Chapter 10), vol 4. Routledge Frontiers of Cognitive Science, Routledge
35. Zaina LAM, Álvaro A (2015) A design methodology for user-centered innovation in the software development area. J Syst Softw 110:155–177
36. Rubin J, Chisnell D (2008) Handbook of Usability Testing: How to Plan, Design, and Conduct Effective Tests. Wiley, Indianápolis, Indiana
37. Ries E (2011) The lean startup crown business. Available at: http://theleanstartup.com/principles. Accessed on 10 Mar 2016
38. Kbar G, Bhatia A, Abidi MH, Alsharawy I (2016) Assistive technologies for hearing, and speaking impaired people: a survey. Disabil Rehabil Assistive Technol. doi:10.3109/17483107.2015.1129456
39. Allen B (1996) An integrated approach to smart house for people with disabilities. Med Eng Phys 18(3):203–206
40. Chan M, Campo E, Estève D, Fourniols JV (2009) Smart homes—current features and future perspectives. Maturitas 64:90–97
41. Doukas C, Metsis V, Becker E, Le Z, Makedon F, Maglogiannis I (2015) Digital cities of the future: Extending @home assistive technologies for the elderly and the disabled. Telematics Inform 28:176–190
42. Rice M, Jacobson RD, Golledge RG, Jones D (2005) Design Considerations for haptic and auditory map interfaces. Cartography Geogr Inf Sci 32(4):381–391
43. So-In C, Arch-in S, Phaudphu C, Rujirakul K, Weeramongkonlert N (2012) A new mobile phone system architecture for the navigational travelling blind. In: 9th international joint conference on computer science and software engineering, Bangkok, pp 54–59
44. Israr A, Bau O, Kim SC, Poupyrev I (2012) Tactile feedback on flat surfaces for the visually impaired. In proceedings of CHI conference, USA
45. Hawley MS, Cunningham SP, Green PD, Enderby P, Palmer R, Sehgal S, O'Neill P (2013) A Voice-Input Voice-Output communication aid for people with severe speech impairment. IEEE Trans Neural Syst Rehabil Eng 21(1):23–31
46. Peixoto N, Nik HG, Charkhkar H (2013) Voice controlled wheelchairs: Fine control by humming. Comput Methods Programs Biomed 112:156–165
47. Rodriguez-Sanchez MC, Moreno-Alvarez MA, Martin E, Borromeo S, Hernandez-Tamame JA (2014) Accessible smartphones for blind users: a case study for a way finding system. Expert Syst Appl 41:7210–7222
48. Federici S, Scherer MJ (2012) The assistive technology assessment model and basic definitions. In: Federci S, Borsci S, Mele ML (eds) Environmental evaluation of a rehabilitation aid interaction under the framework of the ideal model of assistive technology assessment process. In Human-computer interaction, Part I, HCII 2013, LNCS 8004, pp 203–210
49. Cook AM, Hussey SM (2002) Assistive technology: principles and practice. 2nd edn, Mosby

50. Kbar G, Bhatia A, Abidi MH (2015b) Smart unified interface for people with disabilities at the work place. In: 11th international conference on innovations in information technology (IIT'15), Dubai, UAE, 1–3 Nov
51. WHO (1980) International Classification of impairments. Disabil Handicaps, World Health Organization, Geneva, Switzerland
52. Barnes C (1994) Disabled people in Britain and discrimination: a case for anti-discrimination legislation. Hurst & Co., London
53. JISC Guide (2015) Graphical user interface design: developing usable and accessible collections. Available at: http://www.jiscdigitalmedia.ac.uk/guide/graphical-user-interface-design-developing-usable-and-accessible-collection. Accessed on 25 Jan 2016
54. Ekahau (2015) Available at: http://www.ekahau.com/. Accessed on 5 July 2015

A Software Model Supporting Smart Learning

Shamsa Abdulla Al Mazrouei, Nafla Saeed Al Derei and Boumediene Belkhouche

1 Introduction

Smart devices have grafted themselves into our bodies and our environment at large, thus extending individual and collective capabilities [1, 2, 3, 4]. Connectivity and accessibility to resources provide users with sophisticated forms of play, communication, collaboration, and knowledge acquisition. In this context, learning in cyberspace (i.e., cyber-learning) has become a pervasive and dominant activity [5]. Cyber-learning is a critical component of cyber-cities, which provide the fundamental infrastructure for smarter cities. Cyber-learning supports constructivism by affording learners independence, exploration, self-discovery, and knowledge construction. Novel models are being proposed to support and enhance the current educational models by integrating technology, learning, and playing, resulting in what is generically termed "smart learning". Smart learning provides capabilities to support the learning process through guidance, customization, independent knowledge construction, interactivity, and accessibility [6]. Game-based learning is one example of smart learning, which integrates gameplay and explicit learning outcomes. Smart learning aims to have specific, measurable, attainable, relevant, and time-limited learning contents.

A premise motivating the design of educational games is the support of an independent learning process that is effective in easily transferring the acquired skills into the real word [7]. That is, knowledge acquisition is relevant and

S.A. Al Mazrouei · N.S. Al Derei · B. Belkhouche (✉)
College of Information Technology, United Arab Emirates University, Al Ain, UAE
e-mail: b.belkhouche@uaeu.ac.ae

S.A. Al Mazrouei
e-mail: 200807204@uaeu.ac.ae

N.S. Al Derei
e-mail: 200812303@uaeu.ac.ae

L. Ismail and L. Zhang (eds.), *Information Innovation Technology in Smart Cities*,
https://doi.org/10.1007/978-981-10-1741-4_2

comprehensive. It covers learning contents and outcomes from various subjects, such as information technology, engineering, sciences, languages, history, and many others. Players (i.e., learners) are engaged in learning by being challenged to perform a variety of cognitive tasks, such as information collection, analysis, decision-making, reasoning, problem-solving, pattern recognition, and other physical and intellectual activities. As a simple example, learning the alphabet by children involves interacting explicitly with shapes, sounds, and images, and implicitly with meanings, relationships, and composition. Learners engage in the knowledge acquisition process through pattern recognition, meaning formation, concept association, and concept composition. These basic cognitive tasks, coupled with independent progress, constitute one aspect of what we define as *smart learning*.

Even though traditional education and GBL share learning as their main goal, GBL is a radical departure from tradition. As such, it presents researchers with the grand challenge of how to design a computational model of the learning process that captures effectively the cognitive tasks identified by cognitive scientists. In the traditional education context, teachers, mentors, peers, and students act together as facilitators of learning, while GBL seeks the introduction of an unknown context based mostly on assumptions from traditional education. Thus, another major challenge is to disentangle GBL from these assumptions.

Nevertheless, despite these challenges, researchers have been proposing frameworks designed to support the game based learning process [8]. However, software models based on these frameworks, which would form the basis for designing educational games have not been adequately addressed. Indeed, new software object oriented models are needed to facilitate the implementation of games that support GBL. Also, completely new models of the learning process are needed to integrate technologies and learning sciences [9, 10].

The rest of the paper is structured as follows: Sect. 2 presents a literature review of previous works related to GBL framework design. In Sect. 3, we elaborate a game-based learning software model and describe its architecture. Section 4 describes the implementation of our proposed software model. Section 5 presents the results of our experimental evaluation. Section 6 summarizes the tasks we have completed in this research.

2 Literature Review

There is a wide agreement among researchers that games can effectively improve education and can be helpful and useful for teaching complex concepts and skills [11]. Several comprehensive meta-analyses demonstrate the positive impact of GBL on learning [12, 13, 14, 15]. A study found out that 70% of high school students left school because they were not interested or motivated to continue their education [11]. Moreover, a pan-European study surveyed 500 teachers showed that motivation was increased when computer games were included into the educational process [11]. The work in [16] posits that students are motivated through: competition and goals, as players feel personal attachment to a goal, rules, choice, fantasy, and challenges [17]. Yet, others see that GBL research has been based on claims that cannot be substantiated [12]. The disagreement stems from the lack of balanced integration of concepts from pedagogy, instruction, technology, and content into an operational software model. Without this balance, GBL development becomes ad hoc, and the resulting product will not meet the expectations of game makers, students and educators [12].

Proposed frameworks to guide GBL development deal with design at highly abstract conceptual level. In general, these frameworks are far removed from software design. An early model was elaborated by Kiili in [18]. Kiili's Experiential Gaming Model tries to integrate game design, flow theory, and experiential learning theory. Game challenges should commensurate with the player's skill levels to keep the player engaged with learning activities [18]. Another model is the Game Object Model (GOM) [19], which is a naïve attempt at developing an object-oriented model of GBL. The decomposition of the model into pseudo objects and the use of notation and terminology inconsistent with object-oriented design highlight the wide gap between GBL researchers from the educational side and GBL software developers. By its reliance on highly abstract cognitive tasks, the GOM proposal fails to address modeling issues. A recent proposal is the synthesis of a GBL framework developed in [7, 8]. This framework identifies learning, instruction, and assessment as the three major dimensions of GBL all linked with game elements [8]. Each dimension contains the relevant components that contribute to learning, instruction, or assessment. Specifically, the dimensions are:

1. Learning consisting of learning objectives, goals, and learning content.
2. Instruction consisting of games elements (context, pedagogy, learner specifics, and representation) and instructional design.
3. Assessment consisting of feedback and debriefing.

In order to have an effective educational design of games, the alignment between learning, instruction and assessment aspects should be achieved. The learning objectives, the player goals and the learning content are defined in the learning column. The game learning cycle imbedded in the instruction column consists of user feedback, user behavior, user engagement and user learning. While designing the instructional design, designers must take into consideration that the player actions must be followed by sufficient feedback in order to engage the player in the game for better effective learning. Finally, the assessment column contains two elements which are: debriefing element by which the player makes a connection between the experience gained by playing the game and the real life experience, and the system feedback element which represents displaying the score to the player.

This framework can be used as a design aid that can guide new GBL design and improve existing game designs by identifying their structures and components and refining them to meet the requirements. Moreover, this framework can be used as an evaluation tool as one of its elements is the assessment concept, which helps in evaluating the player after playing the game [8].

3 Proposed Software Model

None of the proposed frameworks addresses GBL from a software design perspective. These frameworks are described in highly conceptual terms that do not provide any guidance for software design and implementation. For example, the concept of learning remains undefined, and even though we have an intuitive idea about it, we still need an operational definition to guide its software implementation. Hence, we elaborate an analysis and refinement of the framework in [8] in order to derive an object oriented software model, which is then used as the basis for the implementation, thus bridging the gap between informal design and implementation. Thus, our main concern is to refine the framework by expressing it as a software design so that implementability issues can be directly addressed. As an example, Fig. 5 illustrates how we decompose the concept of learning into concrete tasks. Our resulting design will lend itself easily to implementation using any object-oriented approach. The proposed software model consists of four perspectives: system architecture, model-view-controller architecture, structural design (class diagram), and the behavioral design (state chart). We present each one of them in the following sections.

3.1 Three Tier Architecture

We mapped the game-based learning framework into a three-tier system architecture (See Fig. 1) which consists of: (1) the presentation layer that represents the user interface to translate the tasks and the results to visual objects the user can understand; (2) the business logic layer that executes commands and calculations, and makes logical evaluations; and (3) the data layer where the information is stored and retrieved from a data model repository.

When the learner plays an educational game, he/she interacts with the concepts and the objects of the game. Selecting the right concept and associating them with the objects reflects the learning goal and the user behavior and helps the player to gain decision making skills.

One way to support user engagement is interactivity. There are three forms of interactivity: the first is the physical interactivity which represents the player's behavior while interacting with the game. The second is the visual interactivity which reflects the player responses to the visual objects of the game. And the third is the sound interactivity which is represented by the player's reactions to sound triggered by game objects, and the system feedback provided by the game (Fig. 2).

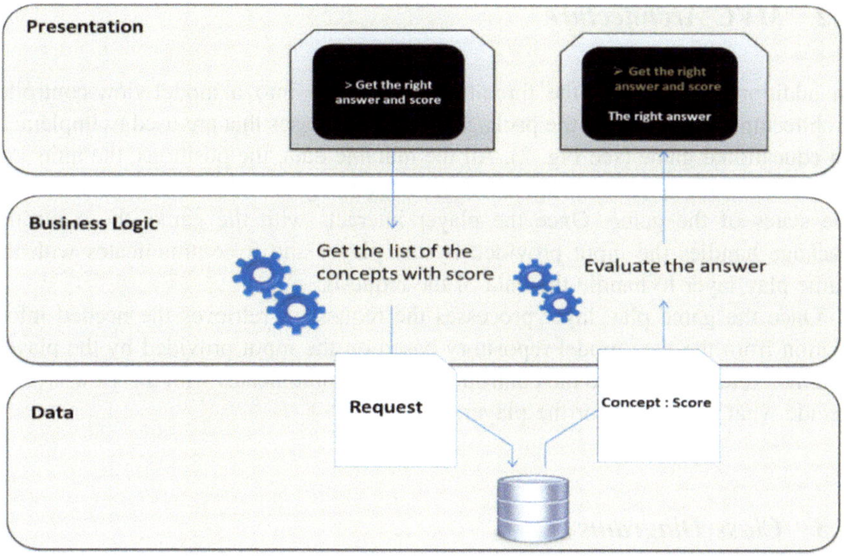

Fig. 1 System architecture

Fig. 2 MVC architecture

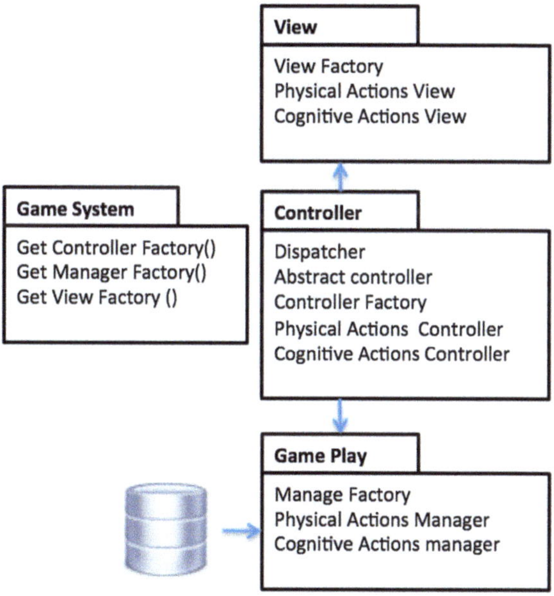

3.2 MVC Architecture

In addition, we mapped the three-tier architecture into a model-view-controller architecture, which shows the packages and their classes that are used to implement an educational game (see Fig. 2). All the mutable data, the positions, the units and the status of the objects are included in the game system that checks and changes the states of the game. Once the player interacts with the game, the controller package handles the input provided by the player and it communicates with the game play layer to handle the data of the requests.

Once the game play layer processes the request, it retrieves the needed information from the data model repository based on the input provided by the player, and then returns it back to the controller which communicates with the view layer to decide what to display for the player.

3.3 Class Diagrams

For the refinement of the MVC architecture we developed a class diagram. For example one interesting class diagram is shown in Fig. 3. This class diagram models the player, which consists of three classes, one representing the player and the other two representing the cognitive and physical actions that the player is capable of performing. The player class consists of the activities that he/she can perform while playing, and descriptive attributes. Examples of descriptive attributes of the player may consist of: memory, acquired knowledge, position, sprite, speed and direction.

The physical actions class represents the physical actions available to the player which are carried out by the avatar that' is under the control of the player. The

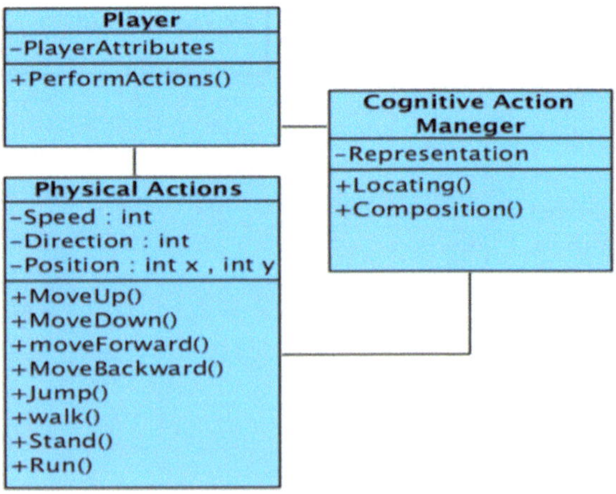

Fig. 3 Player class diagram

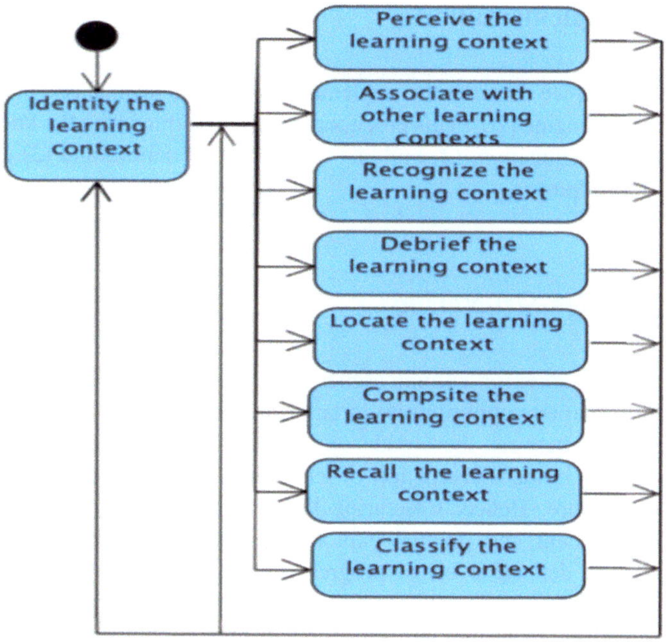

Fig. 4 Player state chart

cognitive actions contribute to knowledge acquisition (i.e., learning) through concept formation elaborated from basic building blocks. These blocks are individual atomic concepts that can be composed to form more complex blocks, which, in turn, can themselves be used for further composition, thus, increasing the complexity of the knowledge being acquired.

3.4 State Chart

For the player to learn while playing an educational game, he must perform cognitive activities by interacting with objects in the game in order to acquire knowledge. The following state chart (Fig. 4) illustrates the states of one object of the game, which is the player. For an object to move from one state to another it must carry out relevant actions during the interaction with the other objects, thus resulting in a state change. Basic cognitive actions that are provided consists of: perception, association, recognition, debriefing, locating, composition, recalling, and classification. Physical actions are the basic movements, such as: select, jump, walk, run, and many others.

4 Game Implementation

As an illustration, we consider the implementation of the main class "player" using the game maker engine [20]. This process requires: (1) structuring the knowledge to support gradual acquisition; (2) building an interaction matrix; and (3) elaborating the main object "player".

4.1 Knowledge Structuring

The literature on games for learning argues that the multiplicity of exploration paths gives players (learners) opportunities to tailor their learning experiences. However, there is a no link between these paths and explicit learning content structuring to guide the learner in the construction of learning paths by "freely" navigating the underlying structure. Hence, structuring the learning content as a concept map captures independent and gradual exploration, whereby links define paths and concept ordering defines levels of progressive evolution from basic to advanced concepts. Viewed as small-scale ontologies, concept maps help learners organize their knowledge and their thinking, and stimulate their cognitive skills. Even though the underlying representation is highly structured, the learner's exploratory activities are spontaneous and guided by his/her actions and decisions. Figure 5 shows a concept map that structures the learning content of Arabic for KG students.

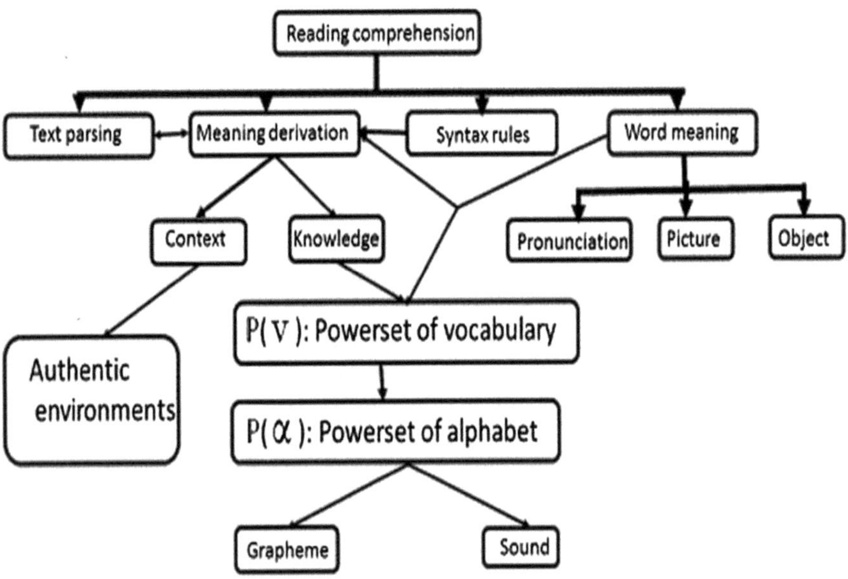

Fig. 5 Knowledge structuring using a concept map

4.2 Interaction Matrix

The interaction matrix captures all the interactions among objects within the game world. Our player interacts with objects through physical and cognitive actions, as mentioned earlier. Table 1 provides a generic summary of the interactions between the player and the various objects.

Note that the set ca belongs to ‖ (Cognitive Actions) (i.e., *ca* ⊙ ‖ (Cognitive Actions)), the set *pa* belongs to ‖ (Physical Actions) (i.e., *pa* ⊙ ‖ (Physical Actions)), and the objects O_i and O_j are members of the set {O: O is an object in the game world), where ‖ is the power set construct. The value in a given matrix cell indicates the nature of the interactions. In Table 1, X stands for no interaction, *ca* stands for a set of cognitive interactions, and *pa* stands for a set of physical interactions. Instantiating O_i, O_j, *ca*, and *pa* in the table will result in a complete description of all the possible interactions among the objects and the player.

We need to associate meanings with the actions involving the player and the generic objects. For example: what is the meaning of the cognitive action "perceive"? Through playing, the player acquires knowledge about the instantiated objects in terms of the cognitive actions. As for the meaning of "perceive" as a cognitive action, we define it as "the ability to see and recognize an object". Hence, perceiving the object means, the ability to see the object and knowing everything related to it. In object-oriented terms, given that any object is endowed with attributes and services to reveal itself, this ability amounts for the player having full access to the object of interest. The information that the player perceives about a given object are the characteristics and the effects of methods of that object. For example, Fig. 6 depicts a simple object "SPARROW", which, when discovered by the player, will expose knowledge about itself, and even sing. Table 2 summarizes some of the cognitive activities of the player and the type of questions he may be able to answer once having acquired new knowledge.

Agent technology was used to implement the game architecture we defined in the previous section by having two agents, the first one is to represent the player and the second one is the assistant agent to help in game calculations, evaluations and management. The actual game introduces alphabets and a picture as objects, and the player must construct the right word that is related to the picture using the displayed alphabets. In this way the game supports two of the knowledge acquisition activities: identification by which the player identifies the context and association by which the player relates the context to the right object.

Table 1 Generic interaction matrix

	Player	O_i	O_j
Player	X	*ca*	*pa*
O_i	*ca*	X	X
O_j	*pa*	X	X

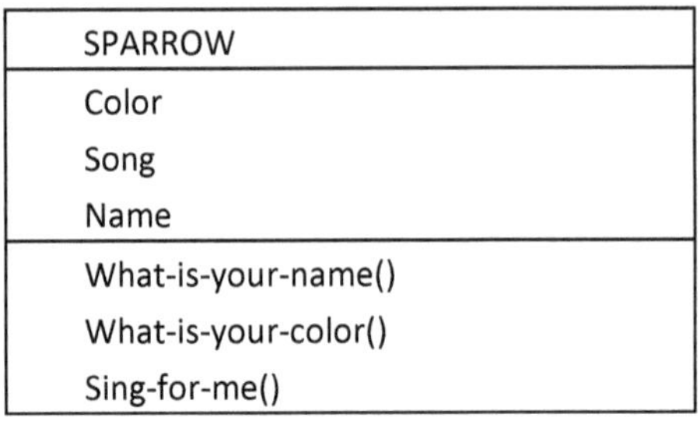

Fig. 6 Object instance

Table 2 Cognitive activities meanings and questions

Cognitive activity	Definition (www.dictionary.com)	Example of questions
Identify	To recognize or establish as being a particular person or thing	What do you see?
Perceive	To recognize, discern, envision, or understand	What is color of the object?
Associate	To connect or bring into relation, as thought, feeling, memory and many other	What category does this objet belong to?
Recognize	To identify as something or someone previously seen, known and many other	Which object is the smallest?
Locate	To assign or ascribe a particular location to (something)	Where can you put this object?

5 Experimental Evaluation

We conducted an experiment at a local primary school to assess the effect of our approach. The experiment was held in the morning and consisted of a one hour familiarization with the system session and two 45-min sessions, one for females and one for males. Participants were first grade divided into two groups consisting of 30 female student and 28 male students.

In the first hour of the experiment, the application was saved into the school's laptops. Moreover, the teachers arranged the tables and chairs and provided the students with headphones to give the chance for every student to play alone without interruption. In the experimental session, students were asked to open the application and start playing the game. They were informed that they will not receive any support from any one, including the teachers. The performance of each student was recorded via Steps Recorder. Once finished, the teachers collected their responses files. After files where collected the results were recorded based on the student's choice in the test part.

The results in Table 3 show the percentage of students who were able to recognize the different shape of a given letter. The leftmost instance of the letter is the

Table 3 Shape recognition results

Shape	ق	ـق	ـقـ	قـ
% Value	1	0.97	0.864	0.864

canonical shape that is introduced for learning the letter. Thus, all students were able to recognize it. The others shape were not introduced and the students were asked to identify them. The middle column presents a letter that is close to the original, thus the students did not have difficulties recognizing it. In the last two columns, even though the shapes presented a bigger challenge, 86% of the students were able to recognize them. We consider this process as a learning transfer activity, implying that the students are capable of using previous acquired knowledge in new learning situations. We attribute this gain to the multi-modality that GBL provides.

6 Conclusion

Current research confirms that game-based learning has potential benefits in the educational context. Being a novel approach, GBL presents several challenges, among them how to design a software model of GBL that is implementable. Our work addressed this challenge directly and provided a complete solution, demonstrating that given an abstract framework, we can capture it in software design and implementation. Our software model consists of a system architecture, a model-view-controller architecture, a structural design, and a behavioral design. Our implementation of the model uses agent technology, wherein agents represents smart object of the GBL world. In this context, agents model very nicely the behavior of learners, tutors, and other components in the GBL world. This technology supports the intelligent interactions between the player, non-player characters, and other game components, naturally lending itself to modeling of educational games. Hence, agent technology was used to derive our design and implementation of an educational game from the framework.

Acknowledgements We would like to thank Dr. Adel Serhani for his guidance and his comments. This work is partially supported by the UAE University NRF, grant # 31T042-UAEU/NRF (1) 23030.

References

1. Lenhart A et al (2008) Teens, video games, and civics. Pew Internet Am Life Project
2. Anderson J, Rainie L (2012) Gamification: experts expect game layers to expand in the future, with positive and negative results. Pew Internet Am Life Project
3. Salen K (2008) Toward an ecology of gaming. MA: The MIT Press, pp 1–20

4. Kirriemuir J, McFarlane C (2006) Literature review in games and learning. Future Lab Series Report 8
5. NSF Task Force on Cyberlearning (2008) Fostering learning in the networked world: the cyberlearning opportunity and challenge. National Science Foundation, Washington, D.C
6. Gwo-Jen H (2014) Definition, framework and research issues of smart learning environments—a context-aware ubiquitous learning perspective. Smart Learn Environ
7. Dunwell I, De Freitas S, Jarvis S (2011) Four-dimensional consideration of feedback in serious games. Digital Game Learn, pp 42–62
8. Van Staalduinen JP, de Freitas S (2011) A game-based learning framework: linking Game design and learning. Learning to play: exploring the future of education with video games, pp 53
9. Honey M, Hilton M (2011) Learning science through computer games and simulation. National Academies Press, Washington, D.C
10. Kincheloe J, Horn Jr. R (2007) The praeger handbook of education and psychology, Westport, CT: Praeger, vol. 1
11. Shaffer DW, Squire KR, Halverson R, Gee JP (2005) Video games and the future of learning. Phi Delta Kappan 87(2):11–104
12. Wouters P, van Nimwegen C, van Oostendorp H, Spek E (2013) A meta-analysis of the cognitive and motivational effects. J Educ Psychol 102(2):249
13. Sitzmann T (2011) A meta-analysis examination of the instructional effectiveness of computer-based simulation games. Pers Psychol
14. Connolly T, Boyle E, MacAurther E, Hainey T, Boyle J (2012) A systematic literature review of empirical evidence on computer games and serious games. Comput Edu
15. Ke F (2009) A qualitative meta-analysis of computer games as learning tools. IGT Global
16. Charsky D (2010) From edutainment to serious games: a change in the use of game characteristics. Game Cult
17. McClarty KL, Orr A, Frey PM, Dolan RP, Vassileva V, McVay A (2012) A literature review of gaming in education. Gaming Edu
18. Kiili K (2006) Evaluations of an experiential gaming model, human technology: an interdisciplinary. Humans in ICT Environments, pp 187–201
19. Amory A (2007) Game object model version II: a theoretical framework for educational game development. Edu Technol Res Dev, pp 51–77
20. http://www.yoyogames.com/studio

Representing Spatial Relationships Within Smart Cities Using Ontologies

Tristan W. Reed, David A. McMeekin and Femke Reitsma

1 Background

"Landscapes of knowledge" is the vision of a geosemantic web that supports the exploration of geographically referenced knowledge by its location, and conversely, the exploration of geographic landscapes through affiliated knowledge. Data that is geo-referenced facilitates the possibility that thee data can be loaded and displayed on a map so that whatever type of information the data represents can be viewed in relationship to its location.

Geographic information frameworks are an essential component of smart cities [1]. A wide variety of sensor data can be provided by Internet of Things (IoT) devices and it is important to be able relate the spatial context of these devices to that of the greater environment, alongside the significance of affiliated knowledge about the environment. By understanding this context, interpretation of the data the sensors produce is aided. Sensors can be geo-referenced to the physical phenomena that they are monitoring.

The vision of "Landscapes of knowledge" enables the exploration of such frameworks to better understand them. Adding semantic information to the geographic data presents the user with a visual and knowledge rich representation of the data [2].

This concept is shared by many, and much research has been undertaken to associate spatial data with ontologies [3–5]. An ontology is a formal representation of a concept, where it can be made up from definitions from the following different concepts: classes, relations and functions [6]. In associating spatial data and

T.W. Reed (✉) · D.A. McMeekin
Department of Spatial Sciences, Curtin University, Bentley, WA, Australia
e-mail: tristan.reed@curtin.edu.au

F. Reitsma
Department of Geography, University of Canterbury, Christchurch, New Zealand

L. Ismail and L. Zhang (eds.), *Information Innovation Technology in Smart Cities*,
https://doi.org/10.1007/978-981-10-1741-4_3

ontologies, the ability to discover and use geographic information in all its various forms and for all its varied applications is greatly enhanced. In semantically extending spatial data, the possibility to discover and use spatial data is increased as the data now richer in content hence the possibility that it will be discovered through search is higher as the search can be about the semantic information embedded within and/or the geographic information represented. Ontologies are extremely important in integrating data and knowledge [7].

Developing and implementing a geosemantic data model is of great value to independent applications for specific purposes, and more broadly to Spatial Data Infrastructures (SDIs). SDIs have been defined in many ways and generally relate to things such as technology standard agreements, policies, resources both computer based and human that facilitate gathering, storing, processing geospatial data [8, 9] or as one of the authors defined it: "an SDI is the infrastructure required to make geospatial data available for use by those who may or may not be experts in the geospatial domain" [10]. A spatial data infrastructure can form part of the geographic information framework of a smart city, allowing the provision of and exploration of relevant spatial data.

Semantic representation of a concept increases that concept's clarity, or "a little semantics goes a long way" [11]. What is not typically known about the concept is its location and its relationship to its surroundings. Hart and Dolbear [12] speak of the "unattributed belief that 80% of all information has a geographic component." Any data that has some kind of location reference in it, be it a city name, an address a set of coordinates can be considered location data. This information may not be central to the data but it certainly enriches that data.

Semantic annotations may assist to resolve some problems regarding a location and its relationship to its surroundings by linking an ontology concept to the feature that may be represented as an information object [13, 14]. What is known about the spatial characteristics of a feature is what needs to be linked to the semantics. Semantic annotations do not resolve this, as they cannot connect a feature to its semantics. It is a unidirectional link that needs to expand such that the flow between spatial location and semantic structure in both directions can occur.

The ability to browse spatial data and see the underlying semantics is needed, to know that a mountain as a geographic feature is also known as a peak or a munroe in other cultural contexts. This kind of knowledge about the mountain is typically hidden in a spatial database, which requires queries and symbology to represent. This symbology can typically only represent two fields of data at any one time effectively on a spatial representation such as a map. In the mountain example, these two fields may be elevation and substrate represented by colour or another symbol type. By representing the semantics of each feature as an array of concepts alongside the map view, individual features can be explored and associated features discovered through the conceptual ontology view, or vice versa discover features based on spatial proximity to the mountain. This is advantageous for usage within Smart Cities, as the concepts can be from both the natural and built environment.

This can be seen in Fig. 1, where an example ontology is shown relating elements of the built environment. In this ontology, both "Curtin University" and

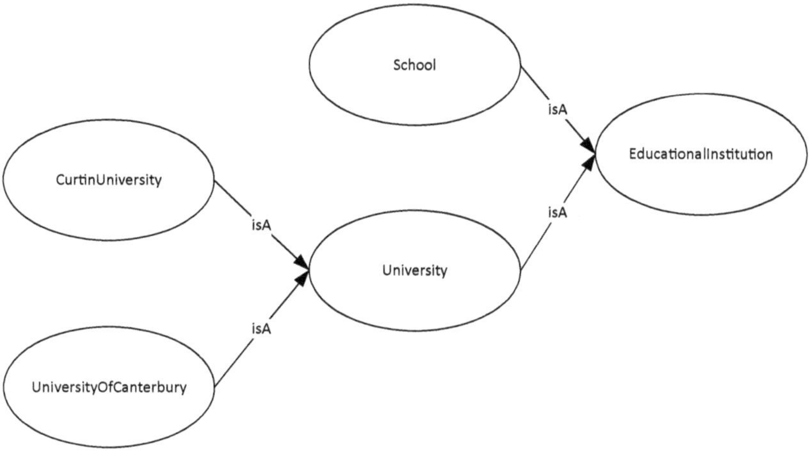

Fig. 1 A visual representation of the sample ontology

"University Of Canterbury" are instances of the class "University". The predicate between the subject "University" and the objects above is "isA", meaning that the semantic link is seen in that both of these objects are of the same type. As they correspond to real life locations, the ontology could contain geographic information about both locations embedded within in. It is in such a situation that a 'linked' view is advantageous.

Research in the field of geosemantics has explored the unidirectional link between spatial features and ontologies, where links from spatial data to ontologies have not yet been studied [15]. By enabling the user to explore spatial data through an exploration of its semantics, greater data discovery and use can be supported [16].

[17] presents an understanding of the role that semantics plays in integrating heterogeneous datasets where semantics are used to determine differences among concepts in geospatial datasets. Understanding the semantics of similarity for knowledge retrieval supports users in retrieving geographic information that is related to other bits of geographic information in some way. Measures of semantic similarity have been developed [18, 19], but fail to allow the user to browse geospatial information from both a semantic and spatial perspective.

Semantics in the geosciences has taken a back seat to research on Linked Data and removed the need for complex semantic descriptions in favour of links. Considering the potential for semantics to overhaul how geospatial information is discovered, the research challenges from [20] have highlighted the need to expand research on semantically linked geospatial data where the semantics can be equally explored alongside the geometric primitives.

LinkedGeoData [21, 22] supports the linking of data between projects where spatial datasets are converted to Resource Description Framework (RDF), and provide the potential for linking these datasets to ontologies. Geoknow is a recently established European Union research project [23], motivated by past work in the LinkedGeoData project, which aims to integrate multiple large-scale geographic linked data sets and make the results available to the public on the world wide web.

OpenStreetMap data available as an RDF knowledge base [24]. With such developments, a resource structure that highlights spatial data characteristics and the knowledge embedded within it provides great potential for enhancing knowledge discovery as well as spatial data reuse. Other linked data initiatives have been used or are planned to integrate cadastral datasets [25–27]. However, the overheads of converting spatial datasets to RDF or other semantic languages are significant as these languages are very verbose and spatial query and analysis runs across them are slow.

GeoRDF [28] is one format used to express linked geospatial data inside an RDF file. The format provides an RDF schema that allows primitives such as polygons and points to be represented as tags which are part of RDF instances located inside an RDF file [29]. GeoRDF has been used for the prototype application, which has been proposed by the World Wide Web Consortium for this purpose [30].

The following research explores the nature of the link between semantics and spatial features, representing that link as a bidirectional link that supports the application of spatial data through conceptual spaces and the exploration of knowledge through spatial features that are associated with that knowledge.

A description of our conceptual model is described that presents the fundamental specification of this bidirectional link at a data modelling level. Then, the design of the interface created in this research is presented with an example of how the semantics can be linked to geospatial features to more effectively support data discovery. Future extensions to the interface are also presented which further demonstrate use cases for bidirectional semantic links.

2 Methodology

A prototype web application called KnowledgeScape has been designed to demonstrate the ability to simultaneously explore spatial knowledge as an interlinked graph and map. The application allows the browsing of linked ontologies with embedded geographic data using a dual-view interface consisting of both a graph representation of the linked data instances in the RDF ontology, as well as a map representation showing the embedded GeoRDF primitives for each instance in the RDF ontology.

The application allows a user to select a node displayed on the graph corresponding to a physical object, and then view the embedded geographic information about the instance on the map.

To use the system, the user must supply the URL of an RDF ontology in the Web Ontology Language (OWL), which is entered into the web interface. The system then generates an interactive graph of the ontology, discriminating between

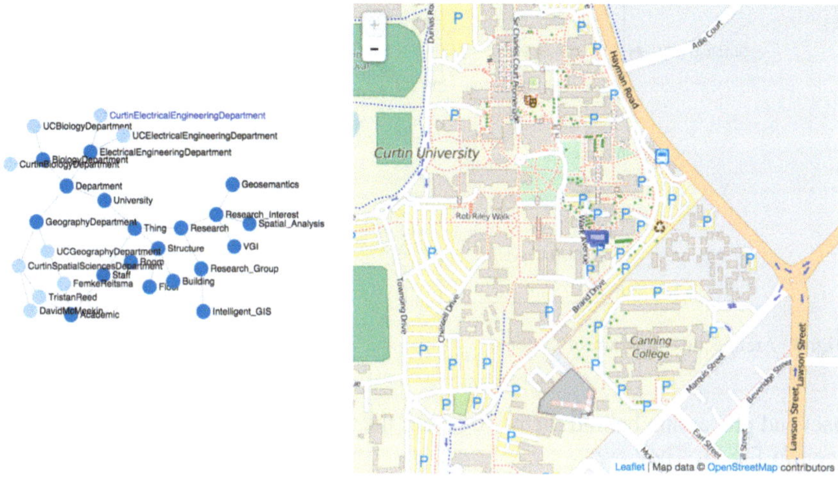

Fig. 2 Screenshot of KnowledgeScape user interface

nodes that are representative of classes and those representative of instances using different coloured node symbols (see Fig. 2). The links, drawn as lines between nodes are a visual representation of the predicates between subjects and objects.

When a user clicks on an instance, if there is embedded geographic information (namely, a GeoRDF tag describing either a single coordinate or group thereof) about that instance within the ontology, this is displayed on the map to the right hand side, with the selected node highlighted to show the connection.

The interface allows the user to compare differences between instances and discover geographic connections to semantic connections.

There are other formats of embedded geographic information, such as addresses used in the vCard RDF schema [31]. It is easily feasible for other types of embedded geographic information to be explored in the dual-view by modifying the parsing of the RDF and geocoding the addresses to coordinates. The modifications required are later described.

A range of server and client side technologies are used in the prototype. All server-side scripts are written in Python. The Django framework is used to allow the scripts to be turned into a web application, as well as the rdflib plugin to simplify some of the processing of the RDF/OWL ontologies.

JavaScript is used alongside HTML and CSS to create the user interface. The d3.js and Leaflet libraries are used for the graph and map visualization views respectively, with the jQuery library used to tie the interface together and to place Asynchronous JavaScript and XML (AJAX) calls to the backend to provide a rich user interface.

The design of the 'Graph to Map' direction of the system consists of both the backend server-side processing and frontend client interface components of the system. The system is presented as three Django views to the user with varying

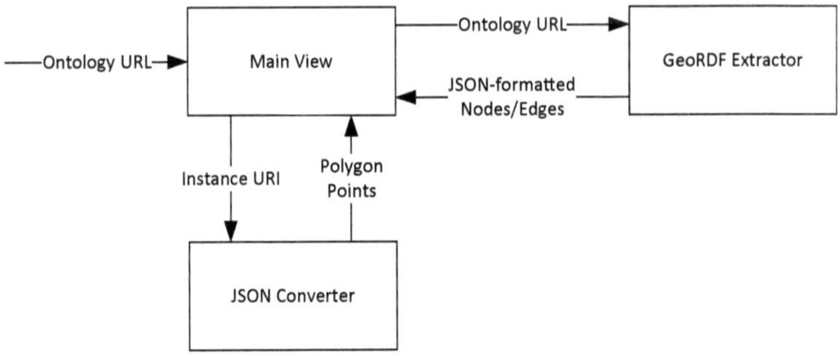

Fig. 3 Architecture diagram of system

backend processing behind them. Together, the three views form the architecture as seen in Fig. 3. They are:

- the 'Main View', which processes both the user's input as well as the interactions with the other views;
- the 'JSON Converter', which retrieves and processes the remote ontology;
- the 'GeoRDF Extractor', which extracts the GeoRDF tag from the selected instance and transforms it into a format understood by Leaflet.

The 'Main View' is the primary interface presented to the user when they first access the system. This view allows the user to specify a URL to a remote ontology. For the 'GeoRDF Extractor' to work, this ontology must contain embedded GeoRDF tags within instances, however the system will present a graph view of the ontology even if there are no embedded tags.

It is to be noted that although there are three views accessed by the user, only the 'Main View' is presented to the user as a traditional HTML page. The 'GeoRDF Extractor' and 'JSON Converter' both return data to be used by static JavaScript called within the 'Main View'. Therefore, another job of the 'Main View' is to proxy the GET and POST requests between the user and all of the views.

The 'Main View' also allows the user to specify a SPARQL query to be processed and displayed on the main page. This is not related to either of the map or graph views but rather is used as another discovery method for the user.

Once the user has specified the URL to an ontology, a POST request is made to the 'Main View'. Once the view has received a POST request, it then loads the JavaScript library files for both the graph and map views alongside a JavaScript configuration file for both views. If a SPARQL query has been specified by the user, this is processed by rdflib.

The configuration file then makes an AJAX request to the 'JSON Converter', supplying the ontology URL as a parameter. The result of this request is data in the format which the d3.js JavaScript library used to visualize the graph. This is required as d3.js cannot natively interpret OWL ontologies as input to generate a visual graph. Instead, a JSON dictionary is supplied which details information about each node (corresponding to a class or instance) and which nodes are linked to each other (corresponding to the predicates). The graph is then configured for display.

The ‘JSON Converter’ view first retrieves the requested OWL ontology and then processes it into the required JSON dictionary. The transformation consists of using the rdflib library to iterate over the entirety of the imported ontology. Iteration over the imported triples is then performed to reflect the axioms of the ontology in the JSON dictionary.

Firstly, all statements not related to instances and classes are ignored, as they are not relevant to the requirements of the visual graph. Secondly, the class and instance names are then transformed into a human-readable format to aid the user’s understanding when this information is displayed to the user.

The file is then iterated over twice to pull out all the classes and individuals respectively, which are then placed in the JSON dictionary which is supplied to the visualization library. The file is then iterated over for a final time to determine the links between the above objects.

As d3.js is expecting a JSON dictionary as input, this transformation enables the triples to be reduced from an RDF-based specification to a JSON list-based specification. The dictionary contains two lists: one which specifies the nodes and their properties and the other which specifies the edges (links between nodes) through listing the two nodes to be joined.

The nodes are also separated into two groups—one for classes and one for instances as this can aid in the user’s interpretation of the ontology. These two groups are indicated as part of the JSON dictionary—a different group number is given to instances compared to classes, which is then used to apply different styling visually to the node.

JavaScript onClick events are then attached to all nodes in the graph such that when they are clicked, the appropriate shape can be drawn on the Leaflet map. This is achieved by placing an AJAX call to the ‘GeoRDF Extractor’ which returns the shape data in the format stored within the GeoRDF tags, which is them processed by the JavaScript into a format suitable for Leaflet to display on a map.

To achieve this, the ‘GeoRDF Extractor’ is passed in the name of the selected instance as a parameter. It then looks up the geometric primitive stored in the GeoRDF tag of said instance and returns it into the script. Depending on whether the primitive is a line, a polygon or a “lat_long” point (currently the three types of shapes specified in the proposed ‘simple’ profile of GeoRDF), which can then be used by Leaflet to draw the primitive. This is achieved through running two SPARQL queries using the rdflib library: the first merely selects everything that is valid within the ontology but importantly transforms it into an iterable format. This is a query of the format:

```
SELECT ?s
WHERE {
         ?s ?p ?o .
}
```

This enables the subject to be extracted for the specified instance, which had previously been shortened for display from the full instance name. The second query then specifically grabs the shapes embedded within the GeoRDF tags associated with the user's selected instance. Leaflet requires an array of coordinates to draw a polygon, and as such a script must be developed to firstly grab the correct shape based on the user's selection of a particular instance and then convert the shape into the correct format that can be displayed. To achieve this, a SPARQL query is run for each shape type to select the relevant GeoRDF tags as follows:

```
SELECT ?g
WHERE {
        < subject > geo:polygon ?g
}
SELECT ?g
WHERE {
        < subject > geo:line ?g
}
SELECT ?g
WHERE {
        < subject > geo:lat_long ?g
}
```

The three separate queries allow the three different types of GeoRDF tags to be discriminated. It is expected that generally a subject will have only a single GeoRDF tag of the three, with the JavaScript processing operating under this assumption.

The contents of the tag are returned as a string, alongside an indicator as to which type of primitive the tag is. The contents are then converted in JavaScript to an array that is then used by Leaflet. Leaflet first clears the map of any previous shapes and then uses the array to draw a shape at the supplied coordinates. Depending on the type of primitive, a different type of shape is drawn in Leaflet.

To draw different shapes simply requires changing the name of the drawing function that is called between either "marker" for a point, "polygon" for a polygon or or "polyline" for a line.

The design of the reverse directional link—that is from 'Map to Graph' is very similar to the design of the 'Graph to Map' system. In this case, once the user has specified the URL to an OWL format ontology, the system displays all of the primitives embedded within the ontology that are located within the current viewport of the map. The user can then select any of the primitives on the map, which highlights the instance containing the embedded primitive on the visual graph view.

This method is still achieved using the existing three views with some modifications and additions. The 'Main View' achieves the same functionality in transferring the user's request to the other views, as well as providing the user interface.

However, the static JavaScript requires extra handlers to cater for the events generated by clicking on the primitives.

Fundamentally, the 'GeoRDF Converter' also functions as before but on top of this an extra function is required to pull out all the primitives within the viewport.

This function is called via an AJAX request to the page through a callback from Leaflet, such that when the viewport is changed the visible primitives are loaded with it. A GET parameter consisting of the viewport of the map is also passed in.

As functionality in rdflib is limited with regards to geographic operations, the entire RDF file is iterated over and all geographic primitives are extracted. These are then converted to geometric data types on the server, such that spatial operations can be run on the shapes to filter the shapes to see whether they are 'in' the viewport. As these shapes are associated with their instances, this information can be used to determine the reverse linkage, and is passed back with the geographic information. It is proposed for the above functionality to be implemented with the GeoDjango framework, which is an extension to the Django framework already used in the system.

Once these shapes are determined, they are then returned to the calling JavaScript which then uses the same Leaflet calls as before to draw the shapes on the map, as well as splitting out the instance name for use within onClick events used to highlight each instance on the graph. As Leaflet expects a list, JavaScript logic is used to make multiple draw calls and to decompose the dictionary into a format understood by Leaflet.

Similarly, the 'JSON Generator' has modifications such that the outputted JSON dictionary is filtered based on the user's input. This has the outcome of being able to highlight the related node to the geographic primitive. For each primitive drawn on the map, an onClick event is attached which passes the name of the instance corresponding to the primitive to the 'JSON Generator'.

The 'JSON Generator' then functions as before, except that the selected instance from the map is marked as a member of a third group inside the JSON dictionary that is generated for the nodes on the graph, which allows different styling when viewed on the graph. This information is then also used to 'fade out' the part of the graph that is not in view. Similarly, the current in-view polygons are assigned to a different group such that they can be styled differently. This information can either be stored using cookies and sessions or by being passed as additional GET variables in the AJAX request. Further investigation is required as to which is the preferred system for the average use case.

This is alongside being able to use the 'JSON Generator' in the same manner as before such that the first directional link is also able to be used.

3 Results

The prototype application seen in Fig. 2 has been evaluated with a sample ontology that describes the relationships between various components of Curtin University and the University of Canterbury as described in Sect. 1.

Further detail of the relationships can be seen in the diagram of Fig. 2 where a visualization of the ontology is shown as well as a more detailed sample of part of the ontology in Fig. 1.

The prototype application is able to show the embedded geographic information within the ontology on the map alongside the semantic relationship between ontology. This allows the user to explore the data by clicking on a node representing a physical object on the graph and then seeing how this is reflected on a map.

The prototype application demonstrates a unidirectional link from the graph instance to a primitive on a map. KnowledgeScape was developed as a web application primarily using Python and JavaScript, exploiting the Django, rdflib, d3.js and Leaflet libraries.

The design of the 'reverse link' is demonstrated to show how it is easy to extend the prototype application to form a fully bidirectional link for the visualization of geospatial linked data.

In this way, both the semantic and spatial context of the element can be explored, as is proposed for "Landscapes of knowledge". As described, the "Landscapes of knowledge" concept will allow greater understanding of the physical phenomena of Smart Cities.

The proposed 'complex' profile of GeoRDF that defines a more abstract way of defining all shapes as a group of points could also be implemented in the prototype. This would require modifying both the SPARQL query used to extract embedded geographic information alongside the JavaScript used to transform this information for display. For this same reason the "lat_long" tag is used from the "simple" profile rather than the more complex alternative of a "point" tag.

While the 'Map View Generator' view could be implemented using a GeoSPARQL query to select the relevant RDF tags that satisfy the bounding box that is equal to the map's viewport, the rdflib plugin used within KnowledgeScape currently does not support the standard. However, the design achieves similar functionality through a different method, as described.

As uniform resource addresses are to a degree ephemeral, the corresponding author may be contacted for the latest URL to a demonstration of the web-based system. The system is a representation of the main result of the research, namely allowing greater exploration of the relationships between semantic and spatial features of linked geographic data.

4 Conclusion

To achieve a vision of 'landscapes of knowledge', the geosemantic web, consisting of semantically linked geographic information, must grow alongside the introduction of spatial data infrastructures. Understanding is aided through the use of visualization tools to show the nature of the linked data.

While much work has been undertaken in generating geospatial linked data, tools to enable the semantic exploitation of the linked data are generally limited to basic unidirectional systems. Semantic data exploration allows greater data discovery for users which may yield greater knowledge and understanding around the area they are investigating. With improving the richness of the information embedded within the data there may also be an increase use of geospatial data leading to improved semantically enriched geospatial data which may lead to an increased use of the data and hence becoming self-perpetuating.

The KnowledgeScape interface described improves the ease of spatial data discovery through the use of a dual-view interface that allows the exploration of a bidirectional link between linked data shown in a graph format and as primitives on a map. This interface allows the user to see both the embedded GeoRDF from the current instance in a graph and a map simultaneously.

Future improvements to KnowledgeScape include the implementation of the reverse link detailed in Sect. 2. Currently, only the first directional link has been implemented in the prototype software.

Various performance and interface optimisations can be performed to improve the user experience. Namely, the POST request to the 'Main View' could be removed and replaced with client-side JavaScript logic. This would create a richer application experience as the page does not need to be refreshed.

It was suggested in Sect. 2 that the vCard RDF schema or other schema that embed geographic information as an address could also be used with KnowledgeScape. This would require the 'GeoRDF Extractor' to geocode the address after extracting the address from the RDF file using a SPARQL query. The result of the geocode request would then be returned to the JavaScript to be processed for display.

Finally, the graph does not display the content of the predicate between two objects but rather only that a link exists. By further customising the d3.js graph, text could be displayed on top of the link corresponding to the predicate. Currently, these values are thrown away by the 'JSON Converter'.

Acknowledgements The work has been supported by the Cooperative Research Centre for Spatial Information, whose activities are funded by the Business Cooperative Research Centres Program of the Australian Government. Support has also been received as an Australian Government Research Training Program Scholarship to one of the authors.

References

1. Su K, Li J, Fu H (2011) Smart city and the applications. In: 2011 international conference on electronics, communications and control (ICECC), pp. 1028–1031, IEEE
2. Chalmers M (1993) Using a landscape metaphor to represent a corpus of documents. In: Frank AU, Campari I (eds) Spatial information theory: a theoretical basis for GID (LNCS, vol. 716), Springer, Berlin, pp 377–390
3. Fonseca F, Egenhofer M, Agouris P, Camara G (2002) Using ontologies for integrated geographic information systems. Trans GIS 6(3):231–257
4. Fonseca FT, Egenhofer MJ (1999) Ontology-driven geographic information systems. In: 7th ACM symposium on advances in geographic information systems
5. Smith B, Mark DM (1998) Ontology and geographic kinds. In: Proceedings, international symposium on spatial data handling, Vancouver, Canada
6. Gruber T (1992) A translation approach to portable ontology specifications, Technical Report KSL 92–71, Knowledge Systems Laboratory, Stanford University
7. Lehmann J, Völker J (2014) An introduction to ontology learning, in perspectives on ontology learning. In: Lehmann J, Völker J (eds), IOS Press, Berlin, pp ix—xvi
8. Kuhn W (2005) Introduction to spatial data infrastructures. In: Presentation held on March 14 2005
9. What Is an SDI?—ArcNews Summer 2010 Issue. Retrieved 17 March 2016. http://www.esri.com/news/arcnews/summer10articles/what-is-sdi.html
10. McMeekin DA, West G (2012) Spatial data infrastructures and the semantic web of spatial things in Australia: Research Opportunities in SDI and the Semantic Web. In: Proceedings of IEEE 5th international conference on human system interactions (HSI2012), Perth, Australia
11. Hendler J (1998) A little semantics goes a long way. Web: http://www.cs.rpi.edu/hendler/LittleSemanticsWeb.html. Last accessed 15 Feb 2016
12. Hart G, Dolbear C (2007) What's so special about spatial? In: Scharl A, Tochtermann K (eds) The geospatial web. Springer, London
13. Reeve L, Han H (2005) Survey of semantic annotation platforms. In: Proceedings of the 2005 ACM symposium on applied computing, ACM, pp 1634–1638
14. Jones C, Rosin PL, Slade JD (2014) Semantic and geometric enrichment of 3D geo-spatial models with captioned photos and labelled illustrations. V&L Net 2014:62
15. Janowicz K, Scheider S, Pehle T, Hart G (2012) Geospatial semantics and linked spatiotemporal data past, present, and future. Semant Web 0 10(1). IOS Press
16. de Andrade FG, de Souza FG, Baptista C, Davis CA Jr (2014) Improving geographic information retrieval in spatial data infrastructures. GeoInformatica 18(4):793–818
17. Kuhn W (2005) Geospatial semantics: why, of what, and how?. J Data Semant III, pp 1–24
18. Janowicz K, Raubal M, Kuhn W (2011) The semantics of similarity in geographic information retrieval. J Spat Inf Sci 2:29–57
19. Sizov S (2010) GeoFolk: latent spatial semantics in Web 2.0 social media. In: Proceedings web search and data mining
20. Reitsma F, Laxton J, Ballard S, Kuhn W, Abdelmoty A (2009) Semantics, ontologies and eScience for the GeoSciences. Comput Geosci 35(4):706–709
21. Adding a Spatial Dimension to the Web of Data, LinkedGeoData.org, last accessed 16 Feb 2016
22. Stadler C, Lehmann J, Höffner K, Auer S (2012) Linkedgeodata: a core for a web of spatial open data. Semant Web 3(4):333–354
23. Geospatial Data and the Semantic Web, GeoKnow.eu, last accessed 22 July 2015
24. Garcia-Rojas A et al (2013) GeoKnow: leveraging geospatial data in the Web of data. In: Open Data on the Web (ODW13)
25. Saavedra J, Vilches-Blzquez LM, Boada A (2014) Cadastral data integration through Linked Data. In: Huerta J, Schade S, Granell C (eds) Connecting a digital Europe through location

and place. Proceedings of the AGILE'2014 international conference on geographic information science, Castelln, 3–6 June 2014. ISBN: 978-90-816960-4-3
26. Rojas LAR, Lovelle JMC, Bermúdez GMT, Marín CEM (2013) Open data as a key factor for developing expert systems: a perspective from Spain. IJIMAI 2(2):51–55
27. Khan Z, Anjum A, Kiani SL (2013) Cloud based big data analytics for smart future cities. In: Proceedings of the 2013 IEEE/ACM 6th international conference on utility and cloud computing, pp 381–386
28. Brodt A, Nicklas D, Mitschang B (2010) Deep integration of spatial query processing into native RDF triple stores. In: Proceedings of the 18th SIGSPATIAL international conference on advantages in geographic information systems
29. Chen H (2007) Geospatial semantic web. Image (Rochester, NY), pp 272–275
30. GeoRDF—W3C Wiki, w3.org. Last accessed 22 July 2015
31. vCard Ontology—for describing People and Organisations. Retrieved March 15, 2016, https://www.w3.org/TR/vcard-rdf/

An Improved Hammerstein Model for System Identification

Selcuk Mete, Hasan Zorlu and Saban Ozer

1 Introduction

System identification, the model of the system is achieved by utilizing data obtained from experimental or mathematical way [1–5]. System identification process can be used in the smart city concept. Since system identification can easily model practical applications such as peer to peer (P2P) file-sharing traffic, driver assistance system, road traffic state, ethernet-based traffic flows [6–9]. In recent years, various applications based on P2P file-sharing technology become prevailing. These applications provide convenience to the users; however, they also cause some problems such as noncopyright file-sharing and excessive network bandwidth occupation. In order to maintain a controllable network environment, network operators and network administrators begin to identify and control the P2P file-sharing traffic [6]. Road safety is one of the main objectives in designing driver assistance systems. On average, every 30 s, one person dies somewhere in the world due to a car crash. Among all fatal traffic accidents, 95% are caused by human errors. The obtained models can be used not only for the online identification of drunk drivers and, probably, stopping the car but also for designing proper controllers based on the configurations proposed in [7]. Road traffic congestion is a common problem all over the world. Many efforts have been done to reduce the impact of traffic congestion. One way is to use advanced Technologies such as sensor, communication and compute processing to traffic management field. These

S. Mete (✉) · H. Zorlu · S. Ozer
Department of Electrical and Electronic Engineering, Erciyes University,
Kayseri, Turkey
e-mail: selcuk.metes@gmail.com

H. Zorlu
e-mail: hzorlu@erciyes.edu.tr

S. Ozer
e-mail: sozer@erciyes.edu.tr

L. Ismail and L. Zhang (eds.), *Information Innovation Technology in Smart Cities*,
https://doi.org/10.1007/978-981-10-1741-4_4

technologies can provide traffic flow data to the traffic management center. In order to utilize such data in drawing decision-making foundations for traffic operator, data conversion into traffic state is desirable [8]. Many existing works in the field of internet traffic identification is available in. Most of the approaches use statistics-based methods to identify the wide variety of traffic found on the internet, due to the fact that methods based on port numbers (transport layer) are consider unreliable. But some approaches show a high degree of reliability when detecting flows for certain applications for instance identification of real-time Ethernet (RTE) traffic flows (TFs) [9].

System identification is proceeded through linear and nonlinear models as to the linearity of the system [1–5]. Linear system identification that the input and the output of the system stated with linear equations is mostly used because of its advanced theoretical background [3, 4]. However, many systems in real life have nonlinear behaviors. Linear methods can be inadequate in identification of such systems and nonlinear methods are used [1, 2, 5]. In nonlinear system identification, the input–output relation of the system is provided through nonlinear mathematical assertions as differential equations, exponential and logarithmic functions [10]. Autoregressive (AR), Moving Average (MA), Autoregressive Moving Average (ARMA) models are used for linear system identification in literature. Also Volterra, Bilinear and PAR (Polynomial Autoregressive) models are used for nonlinear system identification [1–5, 10–16]. Recently the block oriented models to cascade the linear and nonlinear system identifications as Hammerstein and Wiener models are also popular [17]. It's because these models are useful in simple effective control systems. Besides the usefulness in applications, these models are also preferred because of the effective predict of a wide nonlinear process [18, 19]. Hammerstein model is firstly suggested by Narendra and Gallman in 1966 and various models are tested to improve the model [20–22]. Generally, MPN (Memoryless Polynomial Nonlinear) model for nonlinear part and FIR (Finite Impulse Response) or IIR (Infinite Impulse Response) model for linear part are preferred in Hammerstein models in literature [23–29]. In this kind of cascade models, the polynomial representation has advantage of more flexibility and simpler use. Naturally, the nonlinearity can be approximated by a single polynomial. Also other benefit of these structures is to introduce less parameters to be estimated [30, 31]. To describe a polynomial nonlinear system with memory, the Volterra series expansion has been the most popular model in use for the last three decades [32–34]. The Volterra theory was first applied with nonlinear resistor to a White Gaussian signal. In modern DSP (Digital Signal Processing) fields, the truncated Volterra series model is widely used for nonlinear system representations. Also as the order of the polynomial increases, the number of Volterra parameters increases rapidly, and this makes the computational complexity extremely high. For simplicity, the truncated Volterra series is the most often preferred in literature [32–34]. The number of parameters of the Volterra model quickly increases with order of nonlinearity and memory length. As a consequence, large data sets are required in order to obtain an estimation of the model parameters with reasonable accuracy [28]. For these reasons, lots of block oriented applications Volterra model aren't preferred for the nonlinear part [30, 31, 35]. Hammerstein can easily model practical applications such as heat exchangers, electric drives, thermal microsystems, sticky control valves and magneto dampers [36].

In literature, classical algorithms, such as Recursive Least Squares (RLS) [26, 28], are used to optimize of Hammerstein models. These algorithms present better solutions when the model structure and some statistical data (model degree, input and noise distribution etc.) are known. Classical algorithms are mostly used because of their features such as lower hardware costs, convergent structure, and good error-analysis performance [10]. Evolution based on heuristic algorithms recently become more popular and are mostly used in system identification. These algorithms are developed especially to solve parameter optimization problem. Differential Evolution Algorithm (DEA), Clonal Selection Algorithm (CSA) and Genetic Algorithm (GA) are the examples of heuristic algorithms [29, 37–40].

The main motivation of this study is to suggest a simple and successful model structure. At this point authors designed an original and successful Hammerstein model by combining linear FIR model and nonlinear SOV Model. The structure of the proposed Hammerstein model is shown in Fig. 3. The proposed Hammerstein model is optimized with GA, CSA, DEA, and RLS. In simulations, different nonlinear and linear systems are identified by proposed Hammerstein model. Also, the performances are compared with different models in simulations.

The rest of this paper is organized as follows. Section 2 provides a summary of model structures. Types of algorithm are defined in Sect. 3. Simulation results demonstrating the validity of the analysis in the paper are provided in Sect. 4. Finally, Sect. 5 contains the concluding remarks.

2 Model Structures

2.1 MPN Model

MPN is a polynomial structure. In literature it is frequently used in the block oriented models [23–29]. A polynomial of a known p order in the input is expressed as follows:

$$y(n) = \sum_{l=1}^{p} c_l x^l(n) = c_1 x(n) + c_2 x^2(n) + \cdots + c_l x^l(n) \tag{1}$$

where c_l is the coefficient of the polynomial, l is an integer, and $l > 0$ [20]. $x(n)$ represents the model input.

2.2 SOV Model

SOV model is mostly preferred in identification of the nonlinear system [5, 15]

$$y(n) = \sum_{i=0}^{N} h_i x(n-i) + \sum_{i=0}^{N}\sum_{j=0}^{N} q_{i,j} x(n-i)x(n-j) \tag{2}$$

Here $y(n)$ represents the model output, $x(n)$ represents the model input, h_i represents linear and $q_{i,j}$ represents nonlinear parameters, "N" represents model length.

2.3 Hammerstein Model

Many systems can be represented by linear and nonlinear models [41]. Hammerstein model structure in Fig. 1 is formed by cascade of linear and nonlinear models. In Hammerstein model structure in Fig. 1, $x(n)$ is a nonlinear block input, $z(n)$ is a linear block input and $y(n)$ is a Hammerstein block output [21].

2.3.1 Hammerstein Model with MPN-FIR

In this structure in Fig. 2, MPN model is used as the nonlinear part and FIR model is used as the linear part. The nonlinear part is approximated by a polynomial function [25]. Let $x(n)$ and $y(n)$ be the input and the output data respectively of the nonlinear model. $z(n)$ is the unavailable internal data.

The Hammerstein model $H_H^{(p,m)}$ of order p and memory m can be described by the following equation:

$$y(n) = H_H^{(p,m)}[x(n)] \tag{3}$$

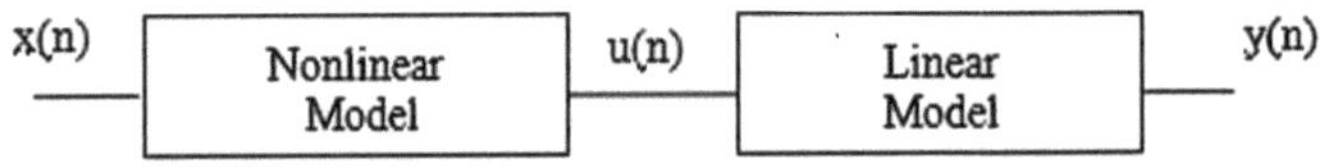

Fig. 1 Hammerstein model structure

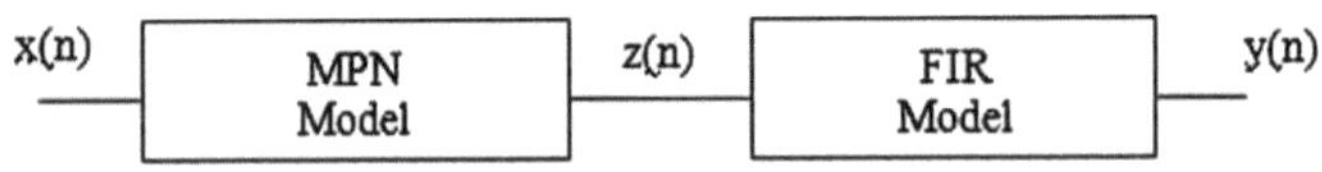

Fig. 2 Hammerstein model with MPN-FIR

Equation (3) could be expressed with an intermediate variable $z(n)$ as follows:

$$y(n) = \sum_{i=0}^{m} b_i z(n-i) \tag{4}$$

with $z(n) = \sum_{l=1}^{p} c_l x^l(n)$ the internal signal $z(n)$ cannot be measured, but it can be eliminated from the equation, by substituting its value in (3). We got [25]:

$$y(n) = \sum_{l=1}^{p} \sum_{i=0}^{m} c_l b_i x^l(n-i) \tag{5}$$

where b_i and c_l are the coefficients of the FIR and the MPN model respectively.

2.3.2 Hammerstein Model with SOV-FIR

In this structure, SOV model is used as nonlinear block and FIR model is used as linear block. Cascade structure is shown in Fig. 3 [42].

SOV model is defined as;

$$z(n) = \sum_{i=0}^{r} h_i x(n-i) + \sum_{i=0}^{r} \sum_{j=0}^{r} q_{i,j} x(n-i) x(n-j) \tag{6}$$

h_i represents linear and $q_{i,j}$ represents nonlinear parameters and linear FIR model output is defined as;

$$y(n) = \sum_{k=0}^{m} a_k z(n-k) \tag{7}$$

proposed Hammerstein output with the combine of these two defines is [42]:

$$y(n) = \sum_{i=0}^{r} \sum_{j=0}^{m} a_i h_j x(n-i-j) + \sum_{t=0}^{r} \sum_{z=0}^{m} \sum_{w=0}^{m} a_t q_{z,w} x(n-t-z) x(n-t-w) \tag{8}$$

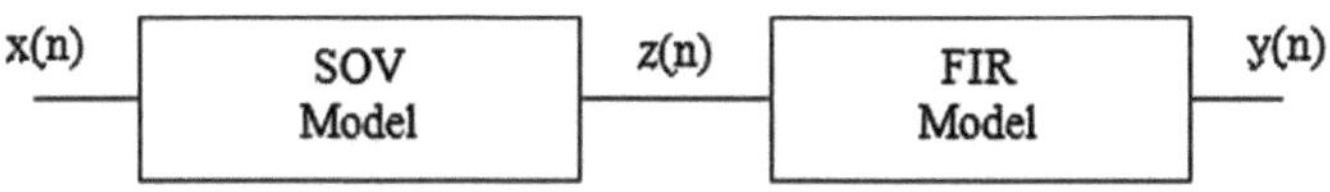

Fig. 3 Hammerstein model with SOV-FIR

3 Optimization Algorithms

In this study it was used classical RLS algorithm. This algorithm that is one of the adaptive algorithms presented to determine the model parameters to estimate chaotic time series, changes the model parameters to minimize the error in each iteration. Model parameters are calculated utilizing error and output values in each iteration by minimizing the difference between desired output and system output [43].

Also it was used heuristic CSA, GA and DEA algorithms in this study [44–51]. The other adaptive algorithms used in this study are working on basis of evolution principle [44]. Recently, evolutional calculation term is mostly used to present the techniques based on this principle. GA, evolutional programming, CSA, artificial immune algorithm and DEA are the examples of this class. The basic steps of an evolutional algorithm are as follows [10];

Create start population
Evaluate
Repeat the following steps until stop criteria provided

Step 1. New population created by selection of the individuals
Step 2. Change the new population
Step 3. Evaluate the new population

GA is first presented by Holland [45]. GA is one of the random search algorithms. Through natural selection, genetic development simulation is occurred. A basic GA selection operator consists of mutation operator, diagonally operator and selection operator. Because parallel structure of GA can efficiently search wide space and apply wide rules in operators. Moreover a GA has some disadvantages such as inadequate local search abilities and early convergence. In order to eliminate the disadvantages, DEA is developed by Price and Storn [46].

DEA is a simple but strong population based algorithm. DEA is developed for the solution of numeric optimization problems as a new algorithm that utilize the differences between the solutions [44, 46]. DEA is a population based algorithm and has advantages such as high local convergence speed, using least control parameters and the ability to find global minimum in a multi-mode search space. There are some studies in literature about applying DEA and GA to system problem [47–49].

Another development based evolutional optimization algorithm is CSA that is inspired by human immune system. This algorithm is used in solving various engineering problems because of its simple structure, applicability of all types of problems, being not parameter based and high convergence speed. CSA, basically is used to determine the basic features of immune reply to antigen alerts [50, 51]. According to this principle, cells that only recognize antigens are chosen to multiply. These cells are processed to affinity maturation process to multiply the simulation of selective antigens. In human body, when an antigen determined, this antigen is faced by antibodies produced by bone marrow. Antigens are tied to antibodies and warn the B cells to divide and turn into cells that excrete antibodies called plasma cells with the signals from T helper cells. Cell division process creates a clone, can create a cell or cell constellation from a single cell [10].

4 Simulation Results

In this study, identification structure is given in Fig. 4. The identification process is performed on three different systems that are all unknown; Linear ARMA, non-linear Hammerstein, Bilinear systems. In this process, model parameters are defined by minimizing the error (MSE) value between adapted algorithm and system output and model output with the help of a cost function. $y(n)$ is the system output, $y_m(n)$ is the model output and $e(n)$ is the error value. These systems have been tested on Gaussian distributed white noise (GDWN) input signal. GDWN input signal is given in Fig. 5. This signal is made with "*WGN*" command in Matlab platform. In simulation studies $x(n)$ input data is used for both system and model input.

Input sequence is GDWN of 250 data samples. Its variance is 0.9108. Unknown systems are identified with four different types of models. These models are described in (9), (10), (11) and (12). Hammerstein model with SOV-FIR in (9) is obtained from (8) with r = 1 and m = 1. Hammerstein model with MPN-FIR in (10) is obtained from (5) with p = 3 and m = 1. SOV model in (11) is obtained from (2) with N = 1. FIR model in (12) is obtained from (7) with m = 1. The memory length of the used FIR and Volterra model memories are chosen same as the FIR and Volterra in Hammerstein model block. In our study, it is aimed to have better solution in proposed Hammerstein structure than the sub models (FIR and SOV) that make up the block structure.

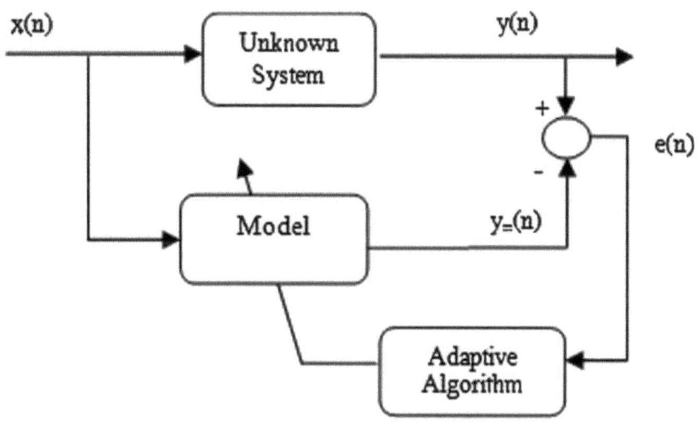

Fig. 4 General structure of system identification

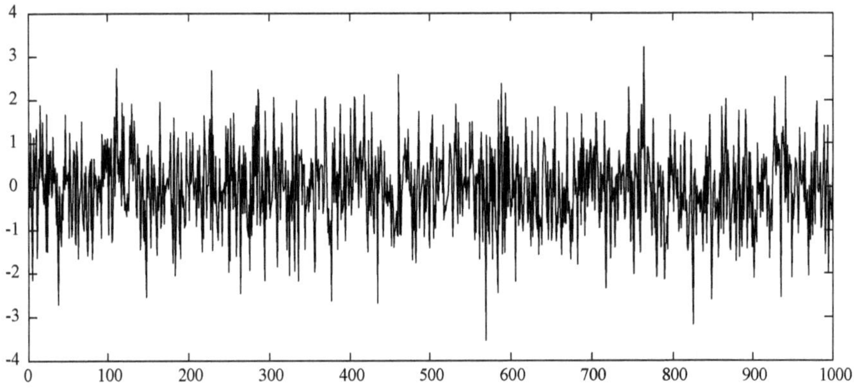

Fig. 5 GDWN input signal x(n)

$$\begin{aligned} Y_{m1}(n) = {} & a_0h_0x(n) + a_0h_1x(n-1) + a_0q_{0,0}x^2(n) + a_0q_{0,1}x(n)x(n-1) + a_0q_{1,0}x(n-1)x(n) \\ & + a_0q_{1,1}x^2(n-1) + a_1h_0x(n-1) + a_1h_1x(n-2) + a_1q_{0,0}x^2(n-1) \\ & + a_1q_{0,1}x(n-1)x(n-2) + a_1q_{1,0}x(n-2)x(n-1) + a_1q_{1,1}x^2(n-2) \end{aligned} \tag{9}$$

$$\begin{aligned} Y_{m2}(n) = {} & b_0c_1x(n) + b_0c_2x^2(n) + b_0c_3x^3(n) + b_1c_1x(n-1) \\ & + b_1c_2x^2(n-1) + b_1c_3x^3(n-1) \end{aligned} \tag{10}$$

$$\begin{aligned} Y_{m3}(n) = {} & h_0x(n) + h_1x(n-1) + q_{0,0}x^2(n) + q_{0,1}x(n)x(n-1) \\ & + q_{1,0}x(n-1)x(n) + q_{1,1}x^2(n-1) \end{aligned} \tag{11}$$

$$Y_{m4}(n) = a_0x(n) + a_1x(n-1) \tag{12}$$

Example-I (linear ARMA system), Example-II (nonlinear Hammerstein system) and Example-III (nonlinear Bilinear system) are chosen through various models published in journals and conferences. According to results of all simulations, the proposed Hammerstein model is more successful in terms of MSE and correlation value compared to other models. In Figs. 6, 7 and 8 and Tables 1, 2 and 3 the results are analysed. In these studies, all models are optimized till the error between the model output and system output is minimized by CSA, GA, DEA and RLS algorithm. The number of generation is selected in the CSA, GA, DEA training as 500.

Example-I In this example, considering the structure given in Fig. 4, unknown system, which is a Hammerstein system with SOV-FIR, is chosen as in (13).

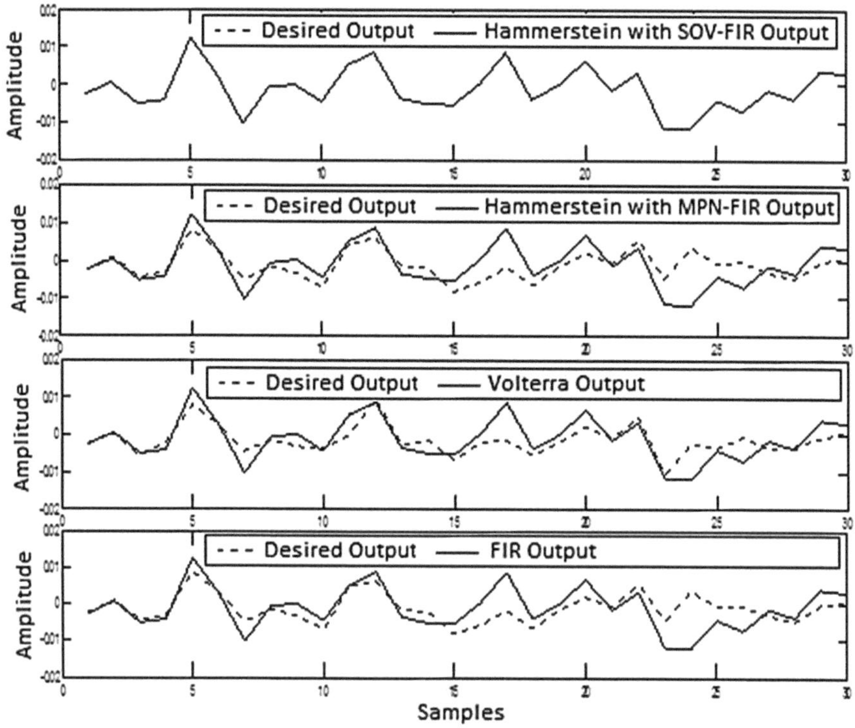

Fig. 6 Simulation results of all models with DEA

$$
\begin{aligned}
y(n) = {} & 0.0089\,[-0.4898x(n) + 0.3411x(n-1) - 0.0139x^2(n) \\
& + 0.1147x(n)x(n-1) + 0.1447x(n-1)x(n) + 0.0379x^2(n-1)] \\
& + 0.0013[-0.4898x(n-1) + 0.3411x(n-2) - 0.0139x^2(n-1) \\
& + 0.1447x(n-1)x(n-2) + 0.1447x(n-2)x(n-1) + 0.0379x^2(n-2)]
\end{aligned} \tag{13}
$$

It is identified with four different type models. All models are trained by CSA, GA, DEA, RLS algorithm and obtained MSE and Correlation values are given in Table 1. Also visual results are shown for 30 data points in Fig. 6.

Example-II In this example, considering the structure given in Fig. 4, unknown system, which is a Bilinear [42], is chosen as in (14). It is identified with different type models.

$$
\begin{aligned}
y(n) = {} & 0.25y(n-1) - 0.5y(n-1)x(n) + 0.05y(n-1)x(n-1) \\
& - 0.5x(n) + 0.5x(n-1)
\end{aligned} \tag{14}
$$

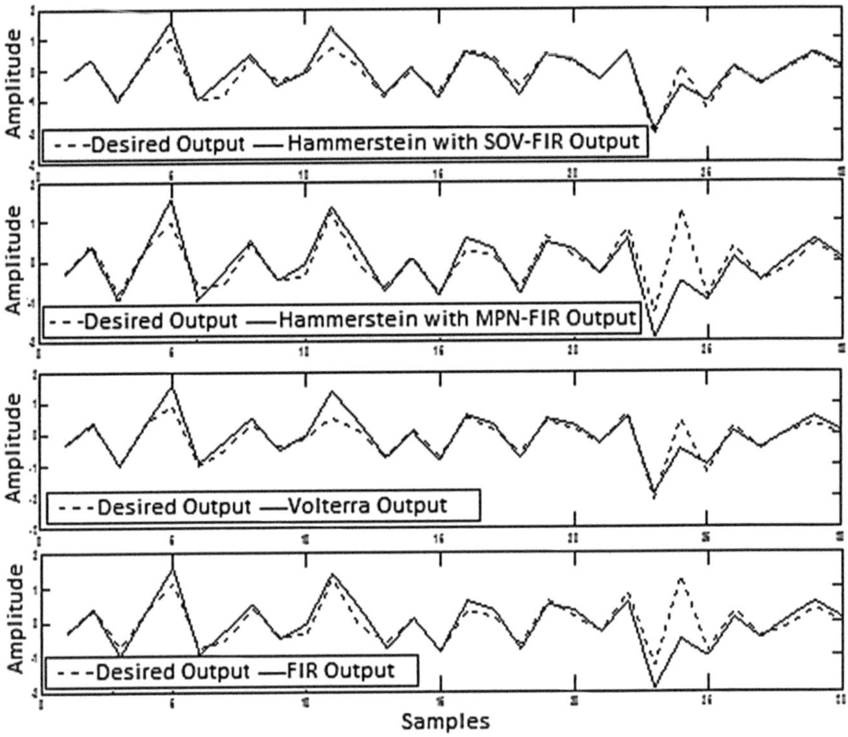

Fig. 7 Simulation results of all models with DEA

All models are trained by DEA, GA, CSA, RLS algorithm and obtained MSE and Correlation values are given in Table 2. Also visual results are shown for 30 data points in Fig. 7.

Example-III In this example, considering the structure given in Fig. 4, unknown system, which is an ARMA [12], is chosen as in (15). It is identified with different type models

$$\begin{aligned} y(n) = 0.7x(n) - 0.4x(n-1) - 0.1x(n-2) + 0.25y(n-1) \\ - 0.1y(n-2) + 0.4y(n-3) \end{aligned} \tag{15}$$

All models are trained by DEA, GA, CSA, RLS algorithm and obtained MSE and Correlation values are given in Table 3. Also visual results are shown for 30 data points in Fig. 8.

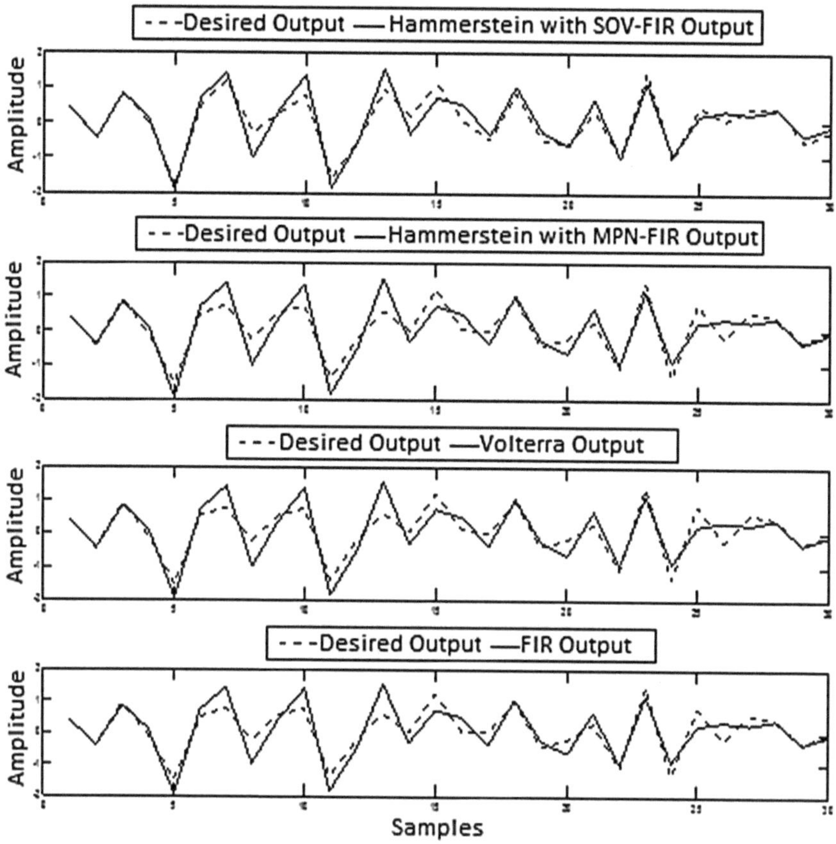

Fig. 8 Simulation results of all models with DEA

Table 1 MSE, correlation values for Example-I

		Model structure			
	Algorithm	Hammerstein with SOV-FIR	Hammerstein with MPN-FIR	Volterra	FIR
MSE	RLS	5.9836×10^{-10}	6.1226×10^{-6}	2.4982×10^{-7}	4.7425×10^{-6}
	CSA	2.8397×10^{-11}	1.6087×10^{-5}	1.1717×10^{-5}	1.6121×10^{-5}
	GA	5.4058×10^{-14}	1.6087×10^{-5}	1.1709×10^{-5}	1.6121×10^{-5}
	DEA	3.6754×10^{-22}	1.6087×10^{-5}	1.1709×10^{-5}	1.6121×10^{-5}
Correlation	RLS	0.9999	0.8798	0.9953	0.9068
	CSA	0.9999	0.7170	0.8036	0.7151
	GA	0.9999	0.7171	0.8033	0.7151
	DEA	0.9999	0.7170	0.8033	0.7150

Table 2 MSE, correlation values for Example-II

	Algorithm	Model structure Hammerstein with SOV-FIR	Hammerstein with MPN-FIR	Volterra	FIR
MSE	RLS	0.06169	0.15384	0.06615	0.13115
	CSA	0.05172	0.12724	0.06631	0.13115
	GA	0.05096	0.12710	0.06615	0.13115
	DEA	0.05096	0.12710	0.06614	0.13115
Correlation	RLS	0.9348	0.8257	0.9231	0.8425
	CSA	0.9403	0.8477	0.9232	0.8425
	GA	0.9414	0.8476	0.9231	0.8425
	DEA	0.9414	0.8476	0.9231	0.8425

Table 3 MSE, correlation values for Example-III

	Algorithm	Model structure Hammerstein with SOV-FIR	Hammerstein with MPN-FIR	Volterra	FIR
MSE	RLS	0.07612	0.11584	0.11313	0.11386
	CSA	0.06646	0.11369	0.11320	0.11386
	GA	0.06639	0.11356	0.11313	0.11386
	DEA	0.06637	0.11356	0.11313	0.11386
Correlation	RLS	0.9344	0.8927	0.8948	0.8940
	CSA	0.9396	0.8943	0.8948	0.8940
	GA	0.9397	0.8944	0.8948	0.8941
	DEA	0.9397	0.8944	0.8948	0.8940

5 Conclusion

System identification process can also be used in the smart city concept. Since system identification can easily model practical applications such as P2P file-sharing traffic, driver assistance system, road traffic state, ethernet-based traffic flows. This study aims to improve Hammerstein model for system identification area. Proposed Hammerstein model which is obtained by cascade form of the nonlinear SOV and linear FIR model is presented. System identification studies are carried out to determine the performance of the proposed model which is optimized by DEA, GA, CSA and RLS algorithm. So, different structure systems are identified with both proposed model and different type models. Proposed model has a complex structure as a disadvantage but has a successful identification tool as an advantage. According to the results, the systems can be identified with less error in proposed Hammerstein model with SOV-FIR compared to other model types although this model contains more parameters and is mathematically more complex. The performance comparison of algorithms has been realized, as well. As a result of this performance comparison, DEA has produced better results than others. Therefore, Hammerstein model may be preferred to model different type of system in smart cities.

Acknowledgements This work is supported by Research Fund of Erciyes University (Project Code: FDK-2014-5308).

References

1. Widrow B, Stearns D (1985) Adaptive signal processing. Prentice Hall, NJ
2. Honig HL, Messerschmitt DG (1984) Adaptive filters structures, algorithms and applications. Kluwer Academic Publishers, MA
3. Toderstrom S (1989) System identification. Prentice Hall, NJ
4. Ljung L, Soderstrom T (1983) Theory and practice of recursive identification. MIT Press, Cambridge
5. Isidori A (1985) Nonlinear control systems: An introduction. Lecture notes in control an information science. Springer, Berlin
6. Zhang R, Du Y, Zhang Y (2009) A BT traffic identification method based on peer-cache. In: Fourth international conference on internet computing for science and engineering, pp 320–323
7. Shirazi MM, Rad AB (2014) Detection of intoxicated drivers using online system identification of steering behavior. IEEE Trans Intell Transp Syst 15(4):1738–1747
8. Jiang GY, Wang JF, Zhang XD, Gang LH (2003) The study on the application of fuzzy clustering analysis in the dynamic identification of road traffic state. In: The IEEE 6th international conference on intelligent transportation systems-Shanghai, China, pp 1449–1452
9. Dominguez-Jaimes I, Wisniewski L, Trsek H (2010) Identification of traffic flows in ethernet-based industrial fieldbuses. In: IEEE conference on emerging technologies and factory automation (ETFA). doi:10.1109/ETFA.2010.5641008
10. Ozer S, Zorlu H (2012) Chaotic time series prediction using the nonlinear par systems. J Fac Eng Archit Gazi Univ 27:323–331
11. Fernández-Herrero A, Vaquer CC, Quirós FJC (2010) Adaptive identification of nonlinear MIMO systems based on Volterra models with additive coupling. In: IEEE sensor array and multichannel signal processing workshop (SAM), Jerusalem, pp 169–172
12. Chon KH, Cohen RJ (1997) Linear and nonlinear ARMA model parameter estimation using an artificial neural network. IEEE T Bio-Med Eng 44:168–174
13. Jingyu L, Pourbabak S (2004) A novel auto regression and fuzzy-neural combination method to identify cardiovascular dynamics. In: World Auto Congress, Seville, pp 31–38
14. Ozer S, Zorlu H (2011) Identification of bilinear systems using differential evolution algorithm. Sadhana-Acad P Eng S 36:281–292
15. Diniz PSR (2008) Adaptive filtering algorithms and practical implementations. Springer, USA
16. Zorlu H (2011) Identification of nonlinear systems with soft computing techniques. Dissertation, University of Erciyes
17. Hafsi S, Laabidi K, Lahmari MK (2012) Identification of wiener-Hammerstein model with multi segment piecewise-linear characteristic. In: IEEE Melecon, Tunisia, pp 5–10
18. Aguirre LA, Coelhoand MCS, Correa MV (2005) On the interpretation and practice of dynamical differences between Hammerstein and wiener models. IEE Proc-Control Theor Appl 152:349–356
19. Lee J, Cho W, Edgar TF (1997) Control system design based on a nonlinear first-order plus time delay model. J Process Control 7:65–73

20. Du Z, Wang X (2010) A novel identification method based on QDPSO for Hammerstein error-output system. In: Chinese control decision conference (CCDC), PRC, pp 3335–3339
21. Ozdinc TO, Hacioglu R (2007) Teleconferencing acoustic echo cancellation using adaptive Hammerstein block structure. In: IEEE signal processing and communications applications (SIU), Turkey, pp 1–4
22. Narendra KS, Galman PG (1966) An iterative method for the identification of nonlinear systems using a Hammerstein model. IEEE Trans Autom Control 11:546–550
23. Jeraj J, Mathews VJ (2006) Stochastic mean-square performance analysis of an adaptive Hammerstein filter. IEEE Trans Signal Process 54:2168–2177
24. Malik S, Enzner G (2011) Fourier expansion of Hammerstein models for nonlinear acoustic system identification. In: IEEE ICASSP, Prague, pp 85–88
25. Sbeity F, Girault JM, Ménigot S et al (2012) Sub and ultra harmonic extraction using several Hammerstein models. In: International conference on complex systems (ICCS), Morocco, pp 1–5
26. Yu L, Zhang J, Liao Y et al (2008) Parameter estimation error bounds for Hammerstein nonlinear finite impulsive response models. Appl Math Comput 202:472–480
27. Jeraj J, Mathews VJ, Dubow J (2006) A stable adaptive Hammerstein filter employing partial orthogonalization of the input signals. IEEE Trans Signal Process 54:1412–1420
28. Costa JP, Lagrange A, Arliaud A (2003) Acoustic echo cancellation using nonlinear cascade filter. In: International conference on acoustics, speech, and signal processing (ICASSP), Hong Kong, pp 389–392
29. Gotmare A, Patidar R, George NV (2015) Nonlinear system identification using a cuckoo search optimized adaptive Hammerstein model. Expert Syst Appl 42:2538–2546
30. Guo F (2004) A new identification method for wiener and Hammerstein systems. Dissertation, University of Karlsruhe
31. Kalafatis A, Arifin N, Wang L et al (1995) A new approach to the identification of pH processes based on the wiener model. Chem Eng Sci 50:3693–3701
32. Sappal AS (2011) To develop a linearization technique for mitigating the rf power amplifier's nonlinearity effects in a multi carrier W-CDMA base station. Dissertation, University of Punjabi
33. Chen Y, Zhang M, Wen XL (2010) Research of nonlinear dynamical system identification based on Volterra series model. In: ICIMA, China, pp 435–438
34. Mathews VJ, Sicuranza GL (2000) Polynomial signal processing. Wiley, NY
35. Hegde V, Radhakrsihnan C, Krusienski D et al (2002) Series cascade nonlinear adaptive filters. In: Midwest symposium on circuits and systems (MWSCAS), pp 219–222
36. Wang J, Zhang Q, Ljung L (2009) Revisiting the two-stage algorithm for Hammerstein system identification. In: Chinese control conference (CDC), Shanghai, pp 3620–3625
37. Akramizadeh A, Farjami A, Khaloozadeh H (2002) Nonlinear Hammerstein model identification using genetic algorithm. In: IEEE ICAIS, pp 351–356
38. Al-Duwaish HN (2000) A genetic approach to the identification of linear dynamical systems with static nonlinearities. Int J Syst Sci 31:307–313
39. Dewhirst OP, Simpson DM, Angarita N, et al (2010) Wiener-Hammerstein parameter estimation using differential evolution: application to limb reflex dynamics. In: International conference on bio-inspired systems and signal processing (BIOSIGNALS), Spain, pp 271–276
40. Nanda SJ, Panda G, Majhi B (2010) Improved identification of Hammerstein plants us ing new CPSO and IPSO algorithms. Expert Syst Appl 37:6818–6831
41. Haber R, Keviczky L (1999) Nonlinear system identification input-output modeling approach. Kluwer Academic, The Netherlands
42. Mete S, Ozer S, Zorlu H (2014) System identification using Hammerstein model. In: IEEE Signal processing and communications applications conference (SIU), Turkey, pp 1303–1306
43. Gauss KF (1963) Theory of the motion of heavenly bodies. Dover
44. Karaboğa D (2004) Yapay Zeka Optimizasyon Algoritmaları. Atlas Publishers, Istanbul
45. Holland JH (1975) Adaption in natural and artificial systems. MIT Press, Cambridge

46. Price K, Storn R (1997) Differential evolution: numerical optimization made easy. Dr. Dobb's J 78:18–24
47. Karaboga N (2005) Digital IIR filter design using differential evolution algorithm. EURASIP J Appl Signal Process 8:1269–1276
48. Chang WD (2006) Parameter identification of Rossler's chaotic system by an evolutionary algorithm. Chaos, Solitons Fractals 29:1047–1053
49. Chang WD (2007) Parameter identification of Chen and Lü systems: a differential evolution approach. Chaos, Solitons Fractals 32:1469–1476
50. De Castro LN, Von Zuben FJ (2001) Learning and optimization using clonal selection principle. IEEE Evol Comput (Special Issue on Artifi Immune Syst) 6:239–251
51. Aslantas V, Ozer S, Ozturk S (2007) A novel clonal selection algorithm based fragile watermarking method. LNCS 4628:358–369

An Image Based Automatic 2D:4D Digit Ratio Measurement Procedure for Smart City Health and Business Applications

Frode Eika Sandnes and Levent Neyse

1 Introduction

Digit ratio measurements are used in several avenues of research within healthcare, medicine and psychology [1–3]. The digit ratio is defined as the ratio of the index finger length (D2) divided by the ring ringer length (D4) and the ratio is also often referred to as the 2D:4D ratio. The 2D:4D ratio can be used as a crude indication of exposure to prenatal sex hormones [4]. The lengths are typically measured from the tip of the fingers to the basal crease of the finger where the fingers join the palm.

Still, such measurements are acquired manually from photocopies of the hands, or flatbed scans. However, recently a few experimental approaches have been proposed for the automatic measurement of the 2D:4D-ratio [5–8]. The first ever-reported attempt [5] was designed for one hand measurement of the hand using a mobile handset camera. The idea was to use the built in camera flash to make it easy to separate the hand from the background since the eliminated hand is much brighter than the background. Based on successful binarization of the hand images an outline was extracted and converted to angular coordinates. The derivatives of the angular representations were used as basis for determining the feature points of interest, that is, the fingertips and finger cervices. This approach was designed for one hand, as the other is used to handle the mobile device. Moreover, it was based on the assumption that the fingers are sufficiently spread out and that the hand constitute a plane perpendicular to the viewing angle of the camera.

F.E. Sandnes (✉)
Oslo and Akershus University College of Applied Sciences, Oslo, Norway
e-mail: frode-eika.sadnes@hioa.no

F.E. Sandnes
Westerdals Oslo School of Art, Communication and Technology, Oslo, Norway

L. Neyse
Kiel Institute for the World Economy, Kiel, Germany

L. Ismail and L. Zhang (eds.), *Information Innovation Technology in Smart Cities*,
https://doi.org/10.1007/978-981-10-1741-4_5

A two-hand approach for flatbed scans intended for finger ratio research was proposed [6] and later with an improved binarizer [7]. This binarization approach was algorithmically complex and vulnerable to certain background configurations. A different initiative to automate digit ratio measurements used so called colour structure codes to separate the hand from the background has been proposed [8], however, the reported results are limited. Although several advanced binarization algorithms are described in the literature [9, 10] it seems necessary with domain specific algorithms. The vast literature on skin detection is testament to this [11, 12].

Other research that share commonalities with automatic finger ratio extraction has been conducted into gesture recognition [13, 14]. However, the emphasis of gesture recognition is to classify hand postures while the objective of finger ratio algorithms is to make accurate and precise finger length measurements.

This paper describes a method that builds on the methods reported in [6, 7] were binarized images are scanned from one side to the other to construct the finger outlines for the two hands. The two hand scans are assumed to constrain the hand orientation and position. These constraints simplify the recognition process.

However, it is difficult to fully spread the fingers of two large hands on the glass plates of small A4 scanners. Moreover, when the hand is pressed against the scanner glass the fingers flatten and touch each other. This study therefore proposes a robust method for measuring digit ratios automatically to overcome many of the problems that present with manual measurements [15] such as non-standard and divergent measurement procedures and the presence of human bias when measuring ambiguous points of interest. Although it extends the methods in [6, 7] the method proposed herein is simplified and more robust as it relies on a standard clustering technique to obtain binarizing thresholds dynamically instead of finding these through manual experimentation. It is thus able to adapt to a wider range of lighting conditions, backgrounds and skin colours than the previous algorithms [6, 7]. Moreover, this study also provides results with the algorithm on a larger set of hand scans.

The proposed algorithm could for instance be used for the unassisted capture of the digit ratio and hand measurements of individuals in various smart city public locations, shops or homes, using standard imaging technologies such as camera enabled self-service kiosks and flatbed scanners. Possible applications include providing customers with tailor made products and services while achieving increased privacy and increased measurement accuracy. Sensitive information such as fingertip patterns [16] can be immediately discarded and thus not transmitted or stored electronically.

2 Method

The 2D:4D finger ratio measurements are achieved by first binarizing the scanned image of the hands. Then the outline of the hands is generated based on the binarized image. Next, the curve of the hand image is analysed to determine the measurement points that serve as the basis for the 2D:4D measurement.

2.1 Hand Background Separation

A binarization operator separates the hand from the background. The separation procedure utilizes the characteristic difference that a hand is saturated to some degree and the background is completely unsaturated. However, the background may vary in brightness from black, via shades of grey to white due to shadows and characteristics of the scanner hardware. The saturation level is therefore used to classify each pixel in the image as hand or background. A simple measure of saturation can be calculated by projecting the pixel in RGB space onto the plane defined by the normal going along the diagonal of the colour cube from black to white, namely

$$x = r - b \tag{1}$$

$$y = g - b \tag{2}$$

The saturation s is the distance from the projected point to the origin or the colour cube

$$s = \sqrt{x^2 + y^2} \tag{3}$$

This is used to define a saturation function $d(r, g, b)$

$$d(r, g, b) = \sqrt{(r - b)^2 + (g - b)^2} \tag{4}$$

The saturation is calculated for a subset of regularly sampled pixels, and these saturation values are clustered as either hand or background using the K-means++ clustering algorithm [17]. The K-means++ algorithm is an extension of the well-known K-means algorithm improved with a randomized seeding technique.

The binarizing threshold is computed as the midpoint between the maximum value of the less saturated background cluster and the minimum value of the more saturated hand cluster, that is

$$T_1 = \frac{1}{2}[\max(background) + \min(hand)] \tag{5}$$

Each pixel is then binarized according to the threshold T_1, that is

$$binary(x, y) = \begin{cases} 1, & T < d(image(x, y)) \\ 0, & otherwise \end{cases} \tag{6}$$

Here, $image(x, y)$ is the image pixel at pixel position x, y, $binary(x, y)$ is the binary pixel at position x, y and $d(p)$ is the saturation function.

Next, a second pass is performed to emphasize the divide between fingers. There may not be any background pixels between these fingers and the darkness values are therefore used instead. Only the pixels classified as hand pixels during the first pass are processed to compute a second threshold T_2. This threshold is found by clustering a regularly sampled set of hand pixels into hand and background according to pixel intensity using the K-means++ algorithm. The following intensity function is used

$$I(r,g,b) = \frac{1}{3}(r+g+b) \tag{7}$$

All the hand pixels found through the first pass is re-classified as background pixels if their intensity is below the threshold T_2. This second pass ensures that the dark cracks between fingers sticking close together are classified as background even though their pixels are saturated.

2.2 *Landscape Portrait Adjustments*

The finger-ratio detection algorithm assumes that the two-hand finger-scans are oriented such that all the fingers point rightwards. To ensure that images satisfy these assumptions a hand orientation step and possible rotation step are employed.

First, a check is performed to determine if the image is in portrait orientation. If the image is in landscape orientation it is rotated 90° to ensure that it is in portrait orientation. For an image to be in portrait orientation the height must be larger than the width.

2.3 *Micro Image Rotation*

Next, the tilt of the two hands is determined. The tilt detection is performed in several steps. First, the centroid of the hands is computed by finding the midpoint of all the hand pixels. That is:

$$[x_c, y_c] = \left[\frac{1}{N}\sum_{i=1}^{N} x_i, \frac{1}{N}\sum_{i=1}^{N} y_i\right] \tag{8}$$

Here x_i, y_i are all the pixels labelled to be part of the hands and N is the number of hand pixels. This centroid is assumed to lie between the two hands. Therefore, the hand image is divided into two halves separated by the vertical line y_c. The centroid computation is thus repeated for the upper half and the lower half, respectively, giving the two new centres $[x_{top}, y_{\mathrm{top}}]$ and $[x_{bottom}, y_{bottom}]$. The angle of tilt A is then

$$A = 180 - a\tan2\left(x_{bottom} - x_{top}, y_{top} - y_{bottom}\right) \quad (9)$$

To reduce the computational load it is only necessary to consider a subset of the image pixels in order to get a sufficiently accurate result. A total of 100 × 100 regularly spaced pixels were sampled in the above computation. Moreover, the comparatively expensive image rotation step is only performed if the tilt is larger than 5° as small angles have little effect on subsequent processing.

2.4 Macro Image Rotation

The final rotation detection step is to determine whether the fingers are pointing left or right. If the fingers are pointing left the image must be rotated by 180° in order for the finger to point right.

Fingers are detected by scanning the image vertically. The finger side will lead to more transitions between background and hand pixels compared to the palm side (see Fig. 1).

The finger direction detection involves dividing the image into two halves along the vertical line x_c, that is, the vertical midpoint of the hand. For each side the binarized image is scanned vertically from top to bottom from one side to the other. If two consecutive pixels are different the difference is counted. After this step, there will be a sum of difference for each respective image half, namely left and right.

If the sum of differences on the right side is larger than the sum of differences on the left side, it is an indication that the finger are pointing rightwards. However, if the sum of differences for the left side is larger than that of the right side, the image needs to be rotated 180°, that is

$$left > right \quad (10)$$

To speed up computation only a subset of 100 × 100 pixels was considered.

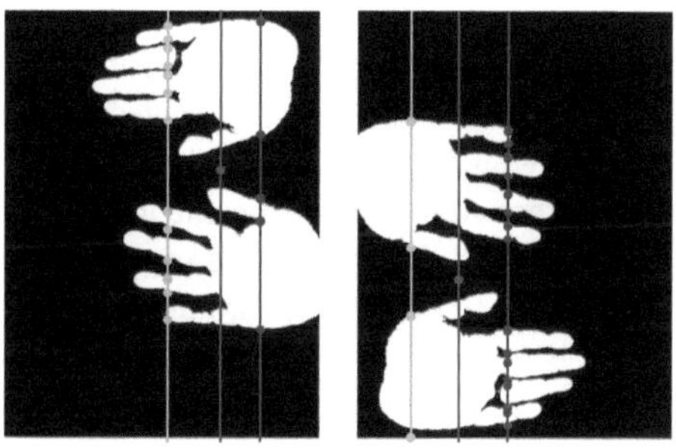

Fig. 1 Determining the finger pointing direction

2.5 Noise Removal

To further eliminate noise in the binarized image and enhance the hand contour a median filter is used. Experimentation reveals that a median filter with a size similar to 1% of the image gives good results. For example a 31 × 31 median filter was used for images with a resolution of 2548 × 3508 pixels. To achieve computational efficiency an adaptation of a generalised median filter was implemented. It comprises a sliding window that is moved across all the images of the image. For each pixel assessed the majority of pixels determine the final pixel value. For example, if the 961 pixels of a 31 × 31 window contains 481 or more white pixels the current pixel is set to white, otherwise it is set to black.

2.6 Hand Outline Extraction

The outline of the hand is obtained by scanning the image from right to left with one vertical line at a time from top to bottom (starting at $y = 0$). That is, the image is scanned in the direction from the end of the image towards the fingertips. Once a pixel change is detected a potential finger candidate is found and the start point y_{start} is recorded. That is if

$$binary(x, y_i) \neq binary(x, y_{i-1}) \tag{11}$$

The end of the finger candidate is detected once the pixel changes back to the background colour which point y_{end} is recorded.

A check is performed to see if the finger candidate segment $[y_{start}, y_{end}]$ is the result of noise or not by determining if the length of the segment is above a threshold W. The segment start and end-points that pass this test are stored in a list M_x. The threshold W was set to

$$W = \frac{y_{\max}}{100} \tag{12}$$

Initially, the number of segments resulting from the vertical scan is zero. Once the tip of the first finger is detected the number of segments increases to 2. As we scan downwards the x-axis the number of segments will increase to 16, which means that all the fingers beside the thumb have been detected. Next, the number of segments will decrease as the scan reaches the crevices of the fingers, and of course increase once the thumbs are detected. In other words, an interesting finger feature point is encountered once there is a change in the number of segments resulting from a scan. The second part of the hand outline extraction involves combining the traces stored in M into two continuous hand outline curves. These continuous curves are found by connecting neighbouring points in M.

2.7 Finger Vector Detection

The finger vectors are computed using fine-tuned fingertip points *D* and the crease *R* of each finger. Fine-tuning is needed since the vertical scan line may have a different angle to the fingertip tangent. First, lines passing through the middle of each finger are found by using the first quartile point and the third quartile point on the curve going from the crevice point and fingertip point defining the curve segment as basis for the midline. The first quartile point is found by taking the ¼ point along the outline trace between the two neighbouring points of interest (see Fig. 2). Similarly, the third quartile points of the finger are found by taking the 3/4 point between the estimated fingertip point and the two neighbouring root points.

The finger midline is defined by the two midpoints between the two points on the ¼ along the sides of the finger, and between the two points ¾ along the sides of the finger (see Fig. 2). The crevice and fingertip points are then updated according to where midlines cross the hand outline curve.

The creases *R* of the ring fingers are simply found as the midpoint between the two neighbouring crevice points at the left and right side of the finger. The crease point for the index finger has only one reliable crevice point. The estimate of the

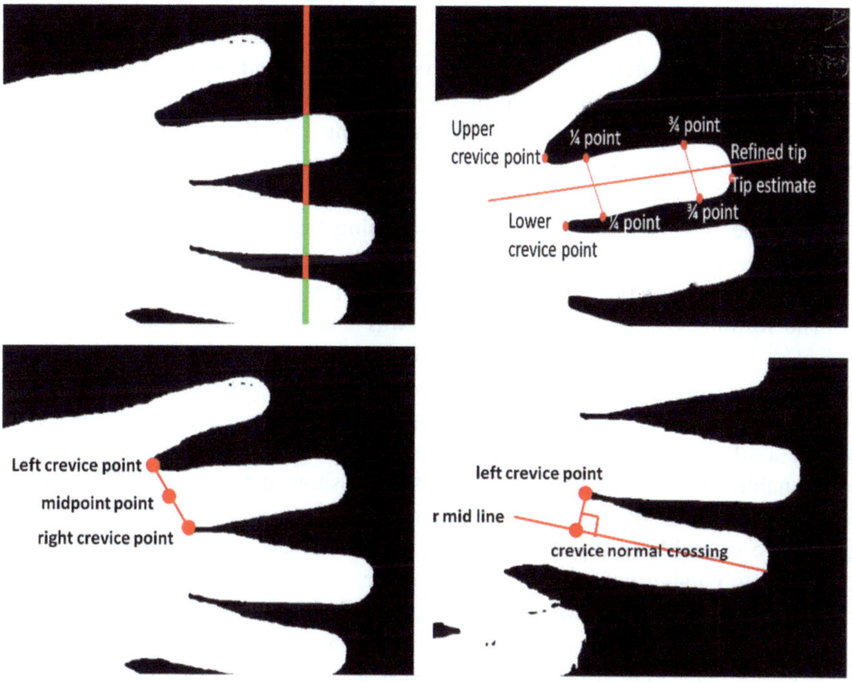

Fig. 2 Scanning for the hand outlines (*top left*), fine-tuning the fingertip point (*top right*), determining the finger root points of the ring finger using the midpoint (*bottom left*) and detecting the index finger root using the normal (*bottom right*)

crease is therefore defined as the point where the normal of the midline intersects the crevice (see Fig. 2).

2.8 Digit Ratio Computation

Once reliable finger crease points R and fingertip points D have been determined, the finger ratio for finger i and j is simply the ratio of the two finger lengths defined by the vector going from the finger crease R to the fingertip D:

$$DR(i,j) = \frac{|R_i - D_i|}{|R_j - D_j|} \tag{22}$$

Note that the digit ratio is computed for the left and the right hands, respectively.

3 Experimental Evaluation

The method proposed herein was tested with 284 high quality hand images acquired using flatbed scanners. The scans were acquired by different researchers with different equipment, different setups and a diverse range of individuals. Common to all scans were that both hands of the individual were pressed palm-down against the scanner glass. The index and ring finger lengths for all the hand images were manually measured using the standardized procedures outlined in [14] as reference.

The algorithm was implemented in Java using a custom-made image analysis library. The Apache Commons machine learning library was used for clustering. The results were run on a high-spec Windows laptop. The unoptimized code took about three hours to process the images, that is, less than one minute per image.

The result revealed that the algorithm was unable to process 44 of the scans (15.5%). A manual inspection of the problematic images revealed that the main reasons were inability to separate the hands from the background or inability to acquire all the points of interest. For example, some of the images had backgrounds with a similar colour to the skin colour of the hands, which the binarizing algorithm was unable to handle. Some of the scans also showed hands where the fingers where facing each other instead of pointing in one direction. Clearly, the scanline procedure is unable to capture the points of interest successfully if the fingers are not close to parallel with respect to each other.

The algorithm was able to acquire finger length measurements for the remaining 84.5% of the scans. A filtering step was performed to discard measurements that were out of range. The manual measurements revealed that the finger length varied from 5.92 to 9.07 cm. Therefore, results outside the more generous range of 5.5 and 9.5 cm were discarded. In total, 119 images (42.3%) had to be discarded due to their out of range values since these would give unreliable digit rates. The remaining 121 scans (42.6%) were therefore used in the subsequent analysis.

Figure 3 shows the mean finger lengths acquired using manual measurements and the automatic procedure. The figure shows the two sets of measurements are similar, but the mean does not show discrepancies that exist for individual measurements. Moreover, t-tests revealed significant differences between the manually and automatically acquired digit ratios for the left (t(119) = 1.98, $p < 0.001$) and the right hands (t(119) = 1.98, $p < 0.001$).

To more closely explore the discrepancies between the manual and the automatic procedures the histograms of the differences between the digit ratios for the left and the right hands are plotted in Fig. 4. The histograms reveal large differences between the manually acquired digit ratios and those acquired automatically. The width of the distribution reveals that variations occur with more than 0.1, which is above the desired level for the method to be considered accurate. Although the histograms centre around zero, it is a bias towards the positive side for both hands. This suggests that there is a systematic error with the automatic procedure.

Manual inspection of the individual results reveals several possible explanations for these results (see Figs. 5 and 6). First, not all the points of interest are identified correctly, especially for hands that are at an angle in relation to the scanning

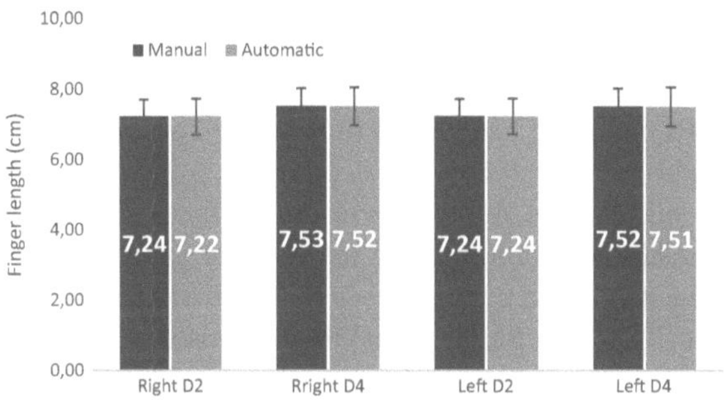

Fig. 3 Finger length measurements. *Error bars* shows standard deviation

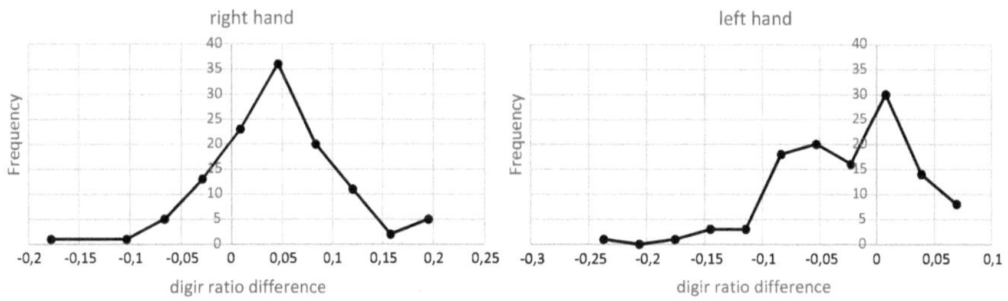

Fig. 4 Histograms of the differences between the manually and automatically measured digit ratios for the *right* and the *left hand*

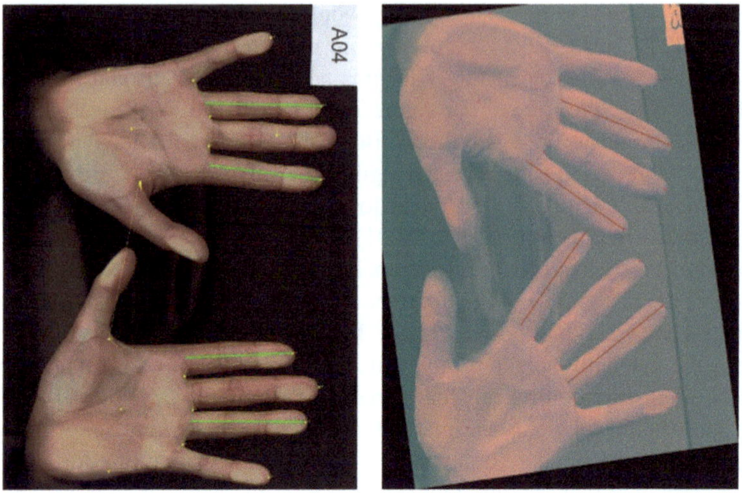

Fig. 5 Successful measurements uncorrected (*left*), rotation correction (*right*)

direction. Moreover, there seems to be frequent discrepancies between the actual finger crease at the root of the finger and the crease estimate acquired using the midpoint or normal (see Fig. 6 top left). The quality of the scanned images also affects the results. In particular if there are visual artefacts caused by poor scanners (see Fig. 6 top right), skin coloured background (see Fig. 6 bottom left) and incorrect skin detection if the hand is not fully pressed against the scanner glass (see Fig. 6 bottom right).

These results suggest that future efforts should focus on further improving the robustness of the skin detection, making the point of interest detection invariant to the direction of the hands and consider each hand individually. Moreover, it may be necessary to optically search for the actual finger creases at the root of the fingers in order to achieve a sufficient accuracy, as estimations based on the hand outline appears to be unreliable.

4 Conclusions

Automatic detection of digit ratios has potential for several smart city applications. An automatic procedure for measuring digit ratios was therefore presented. The method was tested on a set of 284 images and the results compared with measurements acquired manually. Although the results are encouraging, the accuracy and robustness is still not yet sufficient for professional applications, confirming that the automatic detection of digit ratios is a harder problem than it seems. Future work will therefore focus on improving the algorithm for separation of hands from the background, handle each hand separately and develop an angle invariant hand

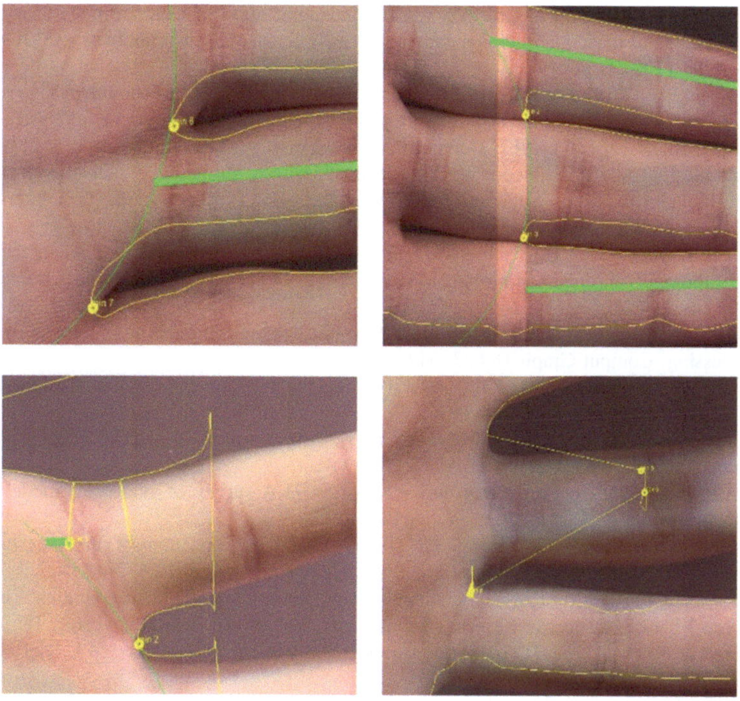

Fig. 6 Unsuccessful measurements. Inaccuracy of crease point based on outline (*top left*), scanner noise (*top right*), skin-background similarity (*bottom left*) and finger above the glass (*bottom right*). The *yellow curves* shows the binarization boundaries

outline analyser. Most importantly, an improved automatic procedure needs to perform an optical local search for the actual finger crease point where the finger joins the palm of the hand. The automatic detection of finger creases is a challenging problem due to the large diversity in the hand colour, texture and shape.

References

1. Putza DA, Gaulinb SJC, Sporterc RJ et al (2004) Sex hormones and finger length what does 2D:4D indicate? Evol Hum Behav 25:182–199
2. Voracek M, Pietschnig J, Nader IW et al (2011) Digit ratio (2D:4D) and sex-role orientation: further evidence and meta-analysis. Personality Individ Differ 51:417–422
3. McIntyre MH, Barrett ES, McDermott R et al (2007) Finger length ratio (2D:4D) and sex differences in aggression during a simulated war game. Personality Individ Differ 42:755–764
4. Manning JT, Scutt D, Wilson J et al (1998) The ratio of 2nd to 4th digit length: a predictor of sperm numbers and concentrations of testosterone, luteinizing hormone and oestrogen. Hum Reprod 13:3000–3004

5. Sandnes FE (2014) Measuring 2D: 4D finger length ratios with Smartphone Cameras. In: Proceedings of IEEE international conference on systems, man and cybernetics (SMC), IEEE, pp 1697–1701
6. Sandnes FE (2015) An automatic two-hand 2D:4D finger-ratio measurement algorithm for flatbed scanned images. In: Proceedings of IEEE international conference on systems, man and cybernetics (SMC), IEEE Computer Society Press, pp 1203–1208
7. Sandnes FE (2015) A Two-stage binarizing algorithm for automatic 2D:4D finger ratio measurement of hands with non-separated fingers. In: Proceedings of 11th international conference on innovations in information technology (IIT'15), IEEE, pp 178–183
8. Koch R, Haßlmeyer E, Tantinger D et al (2015) Development and implementation of algorithms for automatic and robust measurement of the 2D: 4D digit ratio using image data. Curr Dir Biomed Eng 1:220–223
9. Fukumoto M, Suenaga Y, Mase K (1994) Finger-pointer: pointing interface by image processing. Comput Graph 18:633–642
10. Sauvola J, Pietikäinen M (2000) Adaptive document image binarization. Pattern Recogn 33:225–236
11. Vezhnevets V, Sazonov V, Andreeva A (2003) A survey on pixel-based skin color detection techniques. Proc Graphicon 3:85–92
12. Kakumanu P, Makrogiannis S, Bourbakis N (2007) A survey of skin-color modeling and detection methods. Pattern Recogn 40:1106–1122
13. Pavlovic VI, Sharma R, Huang TS (1997) Visual interpretation of hand gestures for human-computer interaction: a review. IEEE Trans Pattern Anal 19:677–695
14. Freeman WT, Roth M (1994) Orientation histograms for hand gesture recognition. Technical report. Mitsubishi Electric Research Laboratories, Cambridge Research Center, TR-94–03a
15. Neyse L, Brañas-Garza P (2014) Digit ratio measurement guide. No. 1914. Kiel Working Paper
16. Coetzee L, Botha EC (1993) Fingerprint recognition in low quality images. Pattern Recogn 26:1441–1460
17. Arthur D, Vassilvitskii S (2007) k-means++: the advantages of careful seeding. In: Proceedings of the eighteenth annual ACM-SIAM symposium on Discrete algorithms (SODA'07). Society for Industrial and Applied Mathematics, Philadelphia, PA, USA, pp 1027–1035

Image Segmentation as an Important Step in Image-Based Digital Technologies in Smart Cities: A New Nature-Based Approach

Seyed Jalaleddin Mousavirad and Hossein Ebrahimpour-Komleh

1 Introduction

Smart city is an urban space that integrates digital technologies to increase the quality of life. To this purpose, data are gathered from citizens and devices using sensors of monitoring systems. Images as one of the principal data types can be seen in different technologies in smart cities such as intelligent transport systems, tourism applications, augmented reality, wearable devices, indoor and outdoor video surveillance, and real-time science understanding. Therefore, it is necessary to provide efficient algorithms to process of images.

One of the first steps in image processing algorithms is image segmentation. Image segmentation is the process of separating an image into meaningful objects. It is considered as an important preprocessing step for image analysis [1]. Many image segmentation methods have been presented in recent years such as normalized cut [2, 3], region growing and merging [4], and fuzzy c-mean [5].

Image thresholding is a popular method for image segmentation. In this method, an image is separated into different regions using on (for bi-level thresholding) or more (for multilevel thresholding) threshold values. Image thresholding is widely applied to many image processing applications such as optical character recognition [6], video change detection [7], rice identification [8], and medical image segmentation [9].

In general, thresholding can be divided into two approaches namely parametric and non-parametric. The first approaches contain the approaches that try to estimate the parameters of a known distribution. Kittler and Illingwor [10] presented a

S.J. Mousavirad (✉) · H. Ebrahimpour-Komleh
Department of Computer Engineering, Faculty of Computer
and Electrical Engineering, University of Kashan, Kashan, Iran
e-mail: jalalmoosavirad@gmail.com

H. Ebrahimpour-Komleh
e-mail: ebrahimpour@kashanu.ac.ir

L. Ismail and L. Zhang (eds.), *Information Innovation Technology in Smart Cities*,
https://doi.org/10.1007/978-981-10-1741-4_6

thresholding approach based on mixture of normal distribution and minimizes the classification error probability. In another work, Wang et al. [11] integrated histogram with the Parzen window technique to estimate the spatial probability distribution of gray-level image values. Dirami et al. [12] used Heaviside functions to approximate the gray level histogram and then a new multilevel thresholding is applied.

In the non-parametric approaches, the threshold values are determined based on a given criterion. Otsu [13] selected the optimal threshold values using maximizing the between-class variance. This method is very time-consuming. To overcome this problem, Liao et al. [14] proposed a fast Otsu's method that uses a look-up-table. In another work, Kapur et al. [15] proposed a multilevel thresholding technique using the entropy of the histogram.

One of the problems of traditional methods is the long computation time when the number of thresholds rises. To eliminate such problems, metaheuristic techniques such as particle swarm optimization [16, 17], differential evolution [18], bacterial foraging algorithm [19], bat algorithm [20], artificial bee colony [21, 22], and differential evolution [3, 23] are widely applied for multilevel image thresholding.

There are two metaheuristic algorithms that are inspired by the lifestyle of cuckoo: cuckoo search (CS) [24] and cuckoo optimization algorithm (COA) [25]. There are some works about image thresholding using CS [26, 27] but so far, COA has not been applied to image thresholding problem. Cuckoo optimization algorithm (COA) is a new metaheuristic that is inspired by the lifestyle of a species of bird called cuckoo. Specific egg laying and breeding of the cuckoo are the basis of this algorithm. This algorithm has presented a good performance for various optimization algorithms such as fuzzy controller [28], scheduling [29], feature selection [30], data clustering [31], and neural network training [32].

In this paper, cuckoo optimization algorithm is proposed for multilevel image thresholding using the entropy criterion. The rest of this paper is organized as follows: Sect. 2 introduces the cuckoo optimization algorithm. Section 3 presents the proposed multilevel image thresholding. The experimental results are evaluated in Sect. 4. Finally, some conclusions are made in Sect. 5.

2 Cuckoo Optimization Algorithm

Cuckoo optimization algorithm is a new metaheuristic which is inspired by a bird named cuckoo. This algorithm is based unusual egg laying and breeding of cuckoos.

In this algorithm, each solution is called a habitat. Cuckoos are divided into two groups: mature cuckoos and eggs. Similar to other nature-based metaheuristics, it starts with an initial population of cuckoos.

These cuckoos lay eggs in other bird's nests. Some of the eggs that are more similar to the host bird's eggs have the opportunity to mature. Cuckoos with less similarity are detected by host birds and are deteriorated. Grown eggs represent the suitability of the area. The more eggs survive in an area; the more profit is assembled in that area.

Cuckoos search the best area to lay eggs. Grown eggs represent the suitability of the area. After eggs grow and change to a mature cuckoo, they construct some communities. Cuckoos in other communities immigrate toward the best community. They will live somewhere near the best habitat.

An egg laying radii is assigned to each cuckoo. It is related to the number of eggs each cuckoo and the cuckoos' distance to the best habitat. Cuckoos start to lay eggs in some random nests inside her egg-laying radius. Cuckoos start to lay eggs in some random nests.

This iterative process continues until the best position is achieved. Finally, most of cuckoo population are assembled near the best habitat. Figure 1 indicates the flowchart of COA.

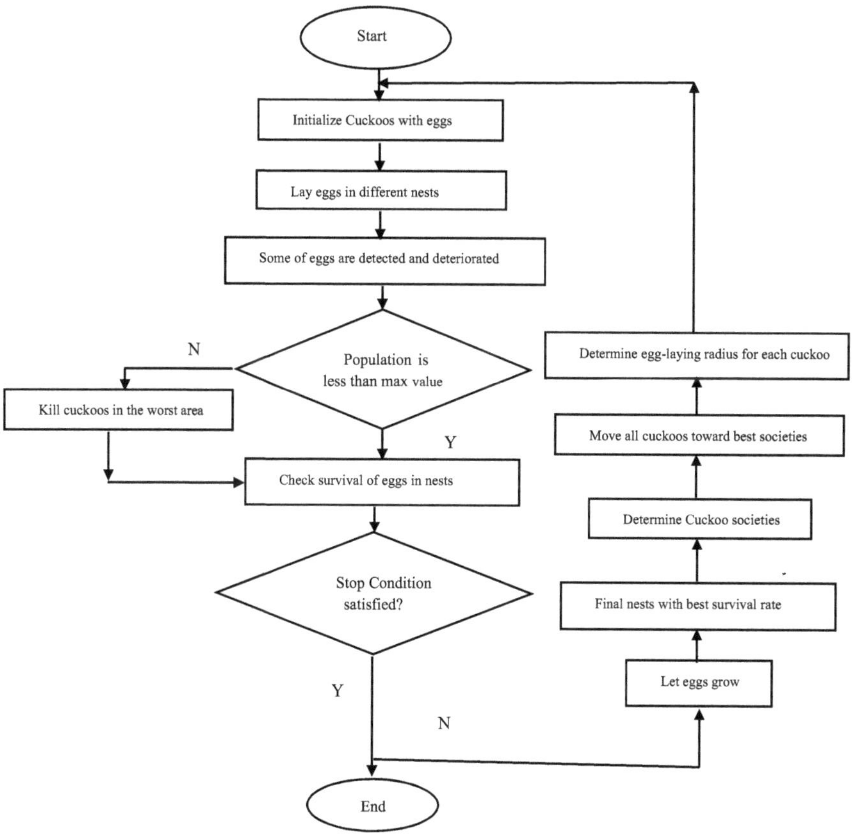

Fig. 1 The flowchart of the COA

3 The Proposed Approach

3.1 Entropy Based Criterion

This criterion tries to maximize the entropy of the segmented image histogram. Assume that there be L gray levels in image I. Image I contains N pixels. Let $h(i)$ denotes the number of pixels with gray level I, and $p(i) = h(i)/N$ is the probability of occurrence of gray level i in the image I.

The objective function f to select D thresholds $[t_1, t_2, \ldots, t_D]$ is defined as follows.

$$
\begin{aligned}
& f([t_1, t_2, \ldots, t_D] = H_0 + H_1 + \cdots + H_D \\
& \omega_0 = \sum_{i=0}^{t_1-1} p_i, \quad H_0 = -\sum_{i=0}^{t_1-1} \frac{p_i}{\omega_0} \ln \frac{p_i}{\omega_0} \\
& \omega_1 = \sum_{i=t_1}^{t_2-1} p_i, \quad H_1 = -\sum_{i=t_1}^{t_2-1} \frac{p_i}{\omega_1} \ln \frac{p_i}{\omega_1} \\
& \omega_2 = \sum_{i=t_2+1}^{t_3} p_i, \quad H_2 = -\sum_{i=t_2+1}^{t_3} \frac{p_i}{\omega_2} \ln \frac{p_i}{\omega_2} \\
& \omega_D = \sum_{i=t_D+1}^{L} p_i, \quad H_D = -\sum_{i=t_D+1}^{L} \frac{p_i}{\omega_D} \ln \frac{p_i}{\omega_D}
\end{aligned}
\tag{1}
$$

3.2 Maximum Entropy Based Cuckoo Optimization Thresholding (MECOAT)

In this paper, a new maximum entropy based optimization is proposed. The details of MECOAT are described as follows.

Step 1. Initialize the parameters of COA

In the first step, parameters of COA are initialized such as The number of habitat ($Num_{Habitat}$) lower (LB) and higher (HB) band of parameters, maximum iterations of the algorithm (Max_{Iter}). Denote the current iteration with $Iter$. Initialize the starting iteration Iter = 1.

Step 2. Definition of ELR

Cuckoos lay eggs within a maximum distance of their habitat. This parameter is called "Egg laying radius(ELR)". It is defined as follows:

$$\mathrm{ELR} = \alpha \times \frac{\text{Number of Current Cuckoo's eggs}}{\text{Total Number of Eggs}} \times (HB - LB), \tag{2}$$

where α is a user-defined constant.

Step 3. Generating of initial population

In the nature-based optimization, each solution is represented by an array. This array is called chromosome, and particle in the genetic algorithm and particle swarm optimization, respectively. In COA, this array is called "habitat". In the multi-level thresholding problem, a habitat is a $1 \times Num\ Of\ Threshold$ array which *Num Of Threshold* is the number of threshold values. This array is defined as follows:

$$habitat = [x_1, x_2, \ldots, x_{Num\ Of\ Threshold}]$$

Figure 2 shows a habitat for a typical histogram. This habitat represents a thresholding with three threshold values. This algorithm starts with $NUM_{Habitat}$ initial habitats randomly.

Step 4. Calculate the profit value of each cuckoo according to the entropy-based criterion

Step 5. Laying eggs in different nests

Each cuckoo starts laying eggs randomly in some other host birds' nests in the range of ELR. After egg laying process, eggs with less similarity to the host birds'

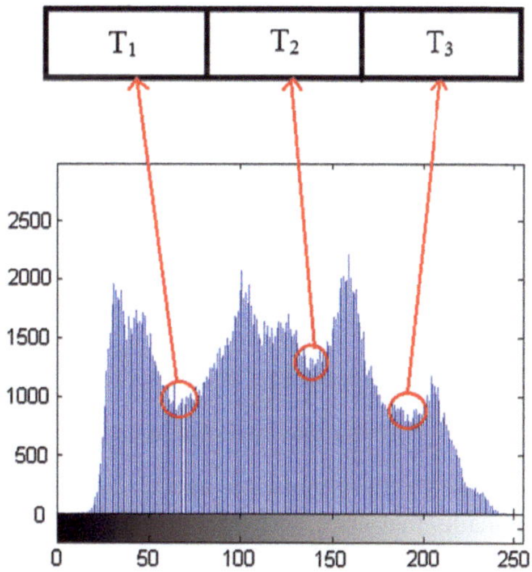

Fig. 2 An example of a habitat for a typical histogram

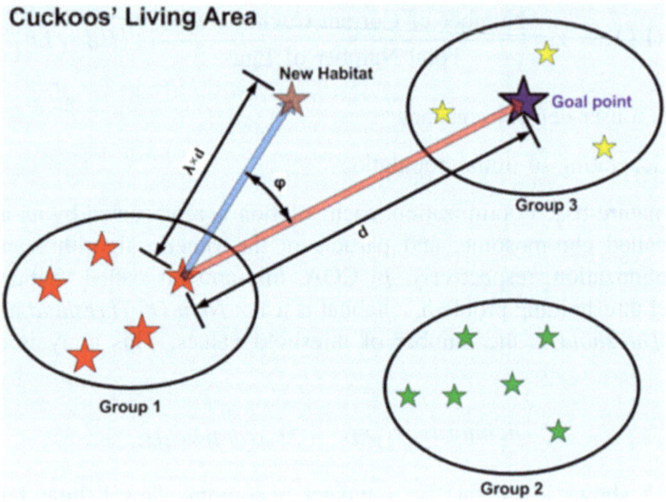

Fig. 3 Cuckoos' movement toward the best habitat [25]

eggs will be detected and deteriorated. For this purpose, P% all eggs with less profit will be deteriorated after egg laying.

Step 6. Immigration of cuckoos

When young cuckoos grow up, they construct communities. In the time of the egg laying, they immigrate toward better community with more similarity of eggs to host birds. A k-means clustering algorithm is applied to construct communities (K is 3–5 seems to be sufficient). Then, mean profit value of each community is computed, and the maximum profit specifies the goal point.

Cuckoos' movement toward the best habitat depicts in Fig. 3. In Fig. 3, the distance between cuckoo and the goal point is shown by d. Each cuckoo only moves λ% of all distance toward goal point and also has a deviation of ϕ radians. The value of ϕ and λ is generated by uniform distribution

$$\begin{aligned} &\lambda \sim U(0,1) \\ &\varphi \sim U(-\omega, +\omega) \end{aligned} \tag{3}$$

where ω is a parameter that constrains the deviation from the goal point.

Step 7. Eliminating cuckoos in the worst habitat

Due to the population balance of cuckoos in nature, N_{max} number of cuckoos remains that have better profit value, and other cuckoos disappear.

4 Experimental Results

The proposed MECOAT algorithm is performed in MATLAB 2014a on a desktop computer with 3.19 GHz CPU and 12 GB RAM. Five well-known images are used for evaluating the proposed algorithm. These images are the popular test images, which are also applied in [19, 33]. The test images and the corresponding histograms are shown in Fig. 4. The parameters of the MECOAT algorithm are shown in Table 1.

4.1 *Solution Quality*

The results of MECOAT algorithm for image thresholding are presented in Table 2. As can be seen, computational time slightly increases when the number of thresholds rises. Table 3 compromise the fitness values of MECOAT algorithm with three other algorithms. According to this table, MECOAT algorithm provides better results than other algorithms in most cases.

4.2 *Peak Signal to Noise Ratio (PSNR)*

Peak signal to noise ratio (PSNR) is a popular performance indicator that is used to compare different multilevel thresholding techniques in the literature [34, 35]. This value is expressed in decibels (DB). The higher value of PSNR indicates the better quality of the thresholded images. The PSNR is defined as follows:

$$PSNR = 20\log_{10}\left(\frac{255}{RMSE}\right) \tag{4}$$

where RMSE is the root mean square error, and defined as

$$RMSE = \sqrt{\frac{\sum_{i=1}^{M}\sum_{j=1}^{N}\left(I(i,j) - \tilde{I}(i,j)\right)^2}{MN}} \tag{5}$$

where I and $\tilde{I}$ are original and segmented images, and M and N are the dimension of images.

Table 4 shows the PSNR values for various optimization algorithms. It can be seen that for almost all the images, MECOAT algorithm provides higher PSNR compared with other methods. In addition, as the number of thresholds increases, The PSNR raises.

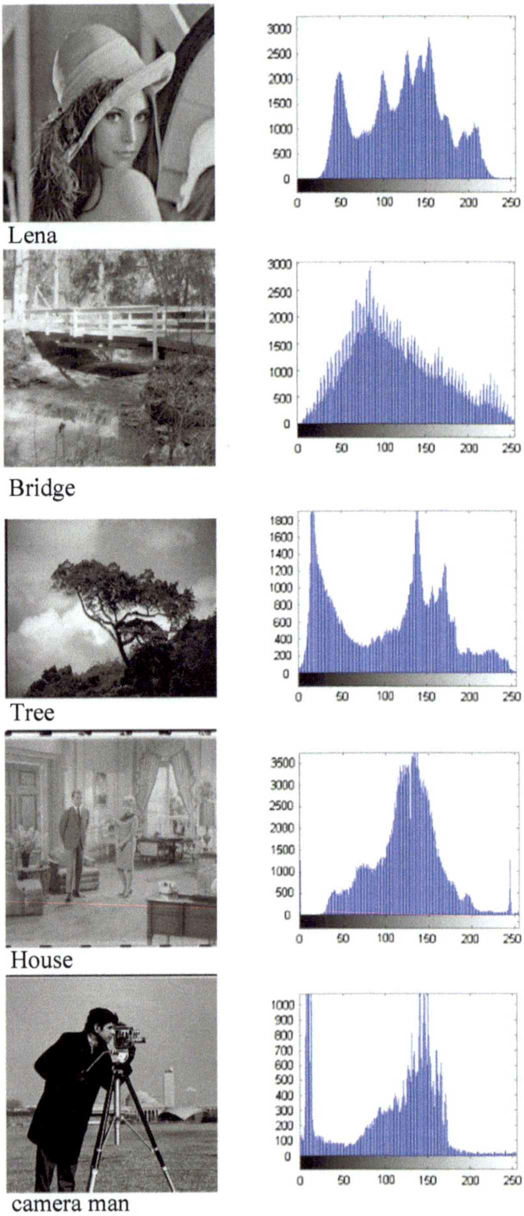

Fig. 4 Test images and their histograms

Table 1 The parameters used in the MECOAT algorithm

Parameters	Value
Number of cuckoos	50
Maximum iteration	200
Number of clusters	3

Table 2 Threshold values, computational times, and fitness for test images by using the MECOAT algorithm

Test images	No. of thresholds	Optimal threshold value	Time	Fitness
Lena	2	79,146	3.6473	15.7071
	3	79,146,228	3.8090	16.0781
	4	65,109,158,228	3.9437	18.9027
	5	58,99,135,170,228	4.1224	19.9439
Bridge	2	98,172	3.4765	13.3356
	3	64,125,189	3.6038	16.8686
	4	56,104,152,200	3.7757	18.7346
	5	49,89,129,171,209	3.9640	20.2005
Tree	2	105,185	3.2363	14.5571
	3	59,122,185	3.4169	17.3668
	4	49,91,132,186	3.5608	19.1296
	5	49,90,131,182,216	3.7186	19.7464
House	2	101,171	3.4620	14.48
	3	66,111,173	3.6273	17.859
	4	65,108,155,191	3.7917	19.0005
	5	63,99,132,166,200	3.9723	20.7955
Cameraman	2	111,175	3.2039	14.5609
	3	77,124,175	3.3278	18.8738
	4	29,76,126,175	3.5022	19.7591
	5	28,69,105,143,182	3.6410	21.3072

4.3 Stability Analysis

Generally speaking, nature-based optimization algorithms are stochastic and the results are not the same in each run. Therefore, it is necessary to compare the stability of the MECOAT algorithm.

To study the stability of the nature-based optimization algorithm, the following formula is used:

$$STD = \sqrt{\sum_{i=1}^{k} \frac{(\sigma_i - \mu)^2}{K}} \quad (6)$$

where STD is the standard deviation, σ_i is the fitness value obtained by the ith run, K is the number of runs, and μ represents the mean value of σ. The lower value for the STD represents the more stability of the algorithm.

The standard deviation value for 10 runs are shown in Table 5. From this table, it can be seen that MECOAT is more stable than other algorithms in most experimental conducted.

Table 3 Comparison of fitness values for various optimization algorithm

Test images	No. of thresholds	Fitness values			
		MECOAT	PSO	GA	BAT
Lena	2	**15.7071**	15.6949	15.6572	**15.7071**
	3	**16.0781**	15.5226	15.5681	15.7045
	4	**18.9027**	18.3083	17.9748	18.4994
	5	**19.9439**	19.7668	19.434	19.8346
Bridge	2	**13.3356**	13.3108	13.2386	13.2869
	3	**16.8686**	16.7977	16.5507	16.7150
	4	18.7346	18.6867	18.5632	**18.7698**
	5	**20.2005**	20.044	19.9691	20.0093
Tree	2	**14.5571**	14.55	14.505	14.5571
	3	**17.3668**	17.3535	17.1177	17.3426
	4	**19.1296**	19.0998	18.7378	18.5615
	5	19.7464	19.7547	19.97	**19.8097**
House	2	**14.4800**	14.4722	14.4117	**14.4800**
	3	**17.8590**	17.8579	16.9874	17.3252
	4	19.0005	18.915	18.6533	**19.0147**
	5	**20.7955**	20.7071	20.4138	20.7597
Cameraman	2	14.5609	14.5418	14.4542	**14.5677**
	3	**18.8738**	18.7703	17.6822	18.3829
	4	19.7591	19.4248	19.5716	**19.7793**
	5	**21.3072**	20.3239	20.6797	20.8291

The best results for each row were boldfaced

4.4 Statistical Analysis

In this section, we apply statistical analysis on the proposed approach. Statistical analysis is an impotent process because COA has a stochastic nature. Statistical analysis methods are divided into two categories: parametric and non-parametric. Parametric analysis methods have an assumption that instances derive from a probability distribution whereas non-parametric methods have no assumption on the instances. Therefore, Parametric methods have some parameters which it increases with the amount of instances but non-parametric methods don't have any parameters.

In this paper, some non-parametric statistical analysis methods are applied to consider the efficiency of proposed algorithm. Two hypotheses are defined in non-parametric methods. The first one is the null hypothesis (H_0) that shows "no difference" whereas the second one, the alternative hypothesis (H_1), indicates a different (here, a significant different among algorithms). A level of significance α shows the probability of rejecting H_0 while it is true. *P*-value is shown the validity of a hypothesis. There is strong evidence against H_0 when a *P*-value is less than α so H_0 will be rejected. Conversely, H_1 will be rejected when *P*-value is greater that α.

Table 4 Comparison of PSNR values for various optimization algorithms

Test images	No. of thresholds	PSNR			
		MECOAT	PSO	GA	BAT
Lena	2	**12.3545**	12.3543	12.3265	**12.3545**
	3	**15.6856**	15.6672	15.4634	15.6379
	4	**18.6485**	18.6269	18.4221	18.5295
	5	**21.3096**	21.2314	21.0109	21.2145
Bridge	2	**12.8954**	12.8953	12.8911	12.8953
	3	**16.1603**	16.1544	16.1126	16.1554
	4	**19.1470**	19.1373	19.0877	19.1330
	5	**21.9029**	21.8682	21.8031	21.8295
Tree	2	**12.8902**	12.8900	12.8775	**12.8902**
	3	16.2397	16.2337	16.1497	**16.2401**
	4	**19.1435**	19.1176	19.0356	19.0889
	5	**21.9332**	21.882	21.7963	21.9052
House	2	**12.3703**	12.3700	12.3594	**12.3703**
	3	**15.3678**	15.3644	15.3181	15.3543
	4	**18.2675**	18.2535	18.1128	18.2620
	5	**20.9195**	20.8564	20.7872	20.9070
Cameraman	2	**12.2465**	12.2461	12.1984	12.2465
	3	**15.2279**	15.2190	15.1594	15.2170
	4	18.2887	18.2640	18.0614	**18.2930**
	5	**20.9405**	20.8751	20.8077	20.9344

The best results for each row were boldfaced

Non-parametric statistical methods can be divided into two classes: pairwise comparisons and multiple comparisons. Pairwise comparisons are used between two algorithms while multiple comparisons perform a comparison among more than two algorithms.

In this paper, Wilcoxon signed-rank test as a pairwise comparison and Friedman test as one multiple comparisons are applied. Wilcoxon signed-rank test is a pairwise comparison that shows if there is a difference among algorithms. in the following, the details of Wilcoxon signed-rank test are explained.

1. Calculate the difference between the values of two algorithms (d_i) on the *i*th out of n problems.
2. Rank the absolute value of *d*.
3. Calculate the test statistic W as follows:

$$W = |R^{+} - R^{-}| \tag{7}$$

where R^+ is the sum of ranks which the first algorithm outperforms the second one and R^- is the sum of ranks which the second algorithm outperforms the first one. Moreover, Ranks of $d_i = 0$ equally distributed among sums. R^+ and R^- can be calculated as follows:

Table 5 Comparison of standard deviation values for various optimization algorithms

Test images	No. of thresholds	Standard deviation			
		MECOAT	PSO	GA	BAT
Lena	2	**0**	0.0002	0.0411	0
	3	**0.0019**	0.0148	0.1175	0.1168
	4	**0.0063**	0.0297	0.1208	0.2707
	5	**0.0214**	0.0578	0.1997	0.2424
Bridge	2	**0**	0.0001	0.0033	0.0001
	3	**0.0012**	0.003	0.0253	0.0057
	4	**0.0026**	0.008	0.0485	0.0153
	5	**0.0026**	0.0166	0.0447	0.0644
Tree	2	**0**	0.0002	0.0111	**0**
	3	0.0004	0.0044	0.0826	**0**
	4	**0.0074**	0.0198	0.073	0.0442
	5	**0.0033**	0.0398	0.0883	0.0315
House	2	**0**	0.0006	0.0118	**0**
	3	**0.0009**	0.0031	0.0268	0.0221
	4	**0.0045**	0.0085	0.0828	0.0078
	5	0.0095	0.0322	0.0684	**0**
Cameraman	2	0.0001	0.0013	0.0475	**0**
	3	**0.0019**	0.0086	0.0677	0.0164
	4	0.0041	0.0232	0.1047	**0.0004**
	5	**0.0126**	0.0465	0.0633	0.0429

The best results for each row were boldfaced

$$\begin{aligned} R^{+} &= \sum_{d_i > 0} rank(d_i) + \frac{1}{2}\sum_{d_i = 0} rank(d_i) \\ R^{-} &= \sum_{d_i < 0} rank(d_i) + \frac{1}{2}\sum_{d_i = 0} rank(d_i) \end{aligned} \tag{8}$$

4. Calculate T = min (R^+, R^-).
5. The null hypothesis will be rejected if T is less than value of the distribution of Wilcoxon.
   More details can be found in [36]. Table 6 shows the results of Wilcoxon statistical test for the fitness values. According to the results, MECOAT shows a high improvement compared to PSO, GA, and BAT algorithms with the level of significance $\alpha = 0.05$.

Table 6 *P*-value for the Wilcoxon signed rank test

Comparison	*P*-value
MECOAT versus PSO	1.6268E−04
MECOAT versus GA	2.9316E−04
MECOAT versus BAT	0.0065

Table 7 Friedman ranks and the corresponding *P*-value

Test images	No. of thresholds	Ranks			
		MECOAT	PSO	GA	BAT
Lena	2	1.5	3	4	1.5
	3	1	4	3	2
	4	1	3	4	2
	5	1	3	4	2
Bridge	2	1	2	4	3
	3	1	2	4	3
	4	2	3	4	1
	5	1	2	4	3
Tree	2	1.5	3	4	1.5
	3	1	2	4	3
	4	1	2	4	3
	5	4	3	1	2
House	2	1.5	3	4	1.5
	3	1	2	4	3
	4	2	3	4	1
	5	1	3	4	2
Cameraman	2	2	3	4	1
	3	1	2	4	3
	4	2	4	3	1
	5	1	4	3	2
Average ranks		1.425	2.8	3.7	2.075
P-value		3.2207E−07			

Another statistical test applied to this paper is Friedman test. It is a multiple comparison test. In this algorithm, first, each algorithm will be ranked separately. The best algorithm has the rank 1, the second one has the rank 2, etc. then. The average rank should be obtained for each algorithm. Table 7 shows the rank obtained for each algorithm and their average. According to this table, MECOAT presents the best rank among other compared algorithms. In addition, *P*-value shows the existence of significant differences among the algorithm considered.

5 Conclusion

In this paper, we have proposed a method, called MECOAT algorithm, for multilevel image thresholding based on entropy criterion. This algorithm is based unusual egg laying and breeding of cuckoos. The performance of MECOAT is evaluated using fitness function, PSNR, and standard deviation. The MECOAT algorithm is applied to several images. The MECOAT algorithm presents competitive performance compared with other algorithms.

References

1. Sanei SHR, Fertig RS (2015) Uncorrelated volume element for stochastic modeling of microstructures based on local fiber volume fraction variation. Compos Sci Technol 117:191–198
2. Shi J, Malik J (2000) Normalized cuts and image segmentation. IEEE Trans Pattern Anal Mach Intell 22(8):888–905
3. Ayala HVH, dos Santos FM, Mariani VC, dos Santos Coelho L (2015) Image thresholding segmentation based on a novel beta differential evolution approach. Expert Syst Appl 42 (4):2136–2142
4. Tremeau A, Borel N (1997) A region growing and merging algorithm to color segmentation. Pattern Recogn 30(7):1191–1203
5. Gong M, Liang Y, Shi J, Ma W, Ma J (2013) Fuzzy c-means clustering with local information and kernel metric for image segmentation. IEEE Trans Image Process 22(2):573–584
6. White JM, Rohrer GD (1983) Image thresholding for optical character recognition and other applications requiring character image extraction. IBM J Res Dev 27(4):400–411
7. Su C, Amer A (2006) A real-time adaptive thresholding for video change detection. In: 2006 IEEE International conference on image processing, 2006. IEEE, pp 157–160
8. Mousavirad SJ, Tab FA, Mollazade K (2012) Design of an expert system for rice kernel identification using optimal morphological features and back propagation neural network. Int J Appl Inf Syst 3(2):33–37
9. Manikandan S, Ramar K, Iruthayarajan MW, Srinivasagan K (2014) Multilevel thresholding for segmentation of medical brain images using real coded genetic algorithm. Measurement 47:558–568
10. Kittler J, Illingworth J (1986) Minimum error thresholding. Pattern Recogn 19(1):41–47
11. Wang S, Chung F-l, Xiong F (2008) A novel image thresholding method based on Parzen window estimate. Pattern Recogn 41(1):117–129
12. Dirami A, Hammouche K, Diaf M, Siarry P (2013) Fast multilevel thresholding for image segmentation through a multiphase level set method. Sig Process 93(1):139–153
13. Otsu N (1975) A threshold selection method from gray-level histograms. Automatica 11(285–296):23–27
14. Liao P-S, Chen T-S, Chung P-C (2001) A fast algorithm for multilevel thresholding. J Inf Sci Eng 17(5):713–727
15. Kapur JN, Sahoo PK, Wong AK (1985) A new method for gray-level picture thresholding using the entropy of the histogram. Comput Vis Graphics Image Process 29(3):273–285
16. Liu Y, Mu C, Kou W, Liu J (2014) Modified particle swarm optimization-based multilevel thresholding for image segmentation. Soft Comput 19(5):1311–1327
17. Liu Y, Mu C, Kou W, Liu J (2015) Modified particle swarm optimization-based multilevel thresholding for image segmentation. Soft Comput 19(5):1311–1327
18. Mlakar U, Potočnik B, Brest J (2016) A hybrid differential evolution for optimal multilevel image thresholding. Expert Syst Appl
19. Sathya P, Kayalvizhi R (2011) Modified bacterial foraging algorithm based multilevel thresholding for image segmentation. Eng Appl Artif Intell 24(4):595–615
20. Alihodzic A, Tuba M (2014) Improved bat algorithm applied to multilevel image thresholding. Sci World J

21. Cuevas E, Sención F, Zaldivar D, Pérez-Cisneros M, Sossa H (2012) A multi-threshold segmentation approach based on artificial bee colony optimization. Appl Intell 37(3):321–336
22. Sağ T, Çunkaş M (2015) Color image segmentation based on multiobjective artificial bee colony optimization. Appl Soft Comput 34:389–401
23. Cuevas E, Zaldivar D, Pérez-Cisneros M (2010) A novel multi-threshold segmentation approach based on differential evolution optimization. Expert Syst Appl 37(7):5265–5271
24. Yang X-S, Deb S (2009) Cuckoo search via Lévy flights. In: World congress on nature & biologically inspired computing, 2009. NaBIC 2009. IEEE, pp 210–214
25. Rajabioun R (2011) Cuckoo optimization algorithm. Appl Soft Comput 11(8):5508–5518
26. Bhandari AK, Singh VK, Kumar A, Singh GK (2014) Cuckoo search algorithm and wind driven optimization based study of satellite image segmentation for multilevel thresholding using Kapur's entropy. Expert Syst Appl 41(7):3538–3560
27. Agrawal S, Panda R, Bhuyan S, Panigrahi B (2013) Tsallis entropy based optimal multilevel thresholding using cuckoo search algorithm. Swarm Evol Comput 11:16–30
28. Balochian S, Ebrahimi E (2013) Parameter optimization via cuckoo optimization algorithm of fuzzy controller for liquid level control. J Eng
29. Azizipanah-Abarghooee R, Niknam T, Zare M, Gharibzadeh M (2014) Multi-objective short-term scheduling of thermoelectric power systems using a novel multi-objective θ-improved cuckoo optimisation algorithm. IET Gener Transm Distrib 8(5):873–894
30. Mousavirad SJ, Ebrahimpour-Komleh H (2014) Wrapper feature selection using discrete cuckoo optimization algorithm. Int J Mechatron Electr Comput Eng 4(11):709–721
31. Ameryan M, Totonchi MRA, Mahdavi SJS (2014) Clustering based on cuckoo optimization algorithm. In: 2014 Iranian conference on intelligent systems (ICIS). IEEE, pp 1–6
32. Ebrahimpour-Komleh H, Mousavirad SJ (2013) Cuckoo optimization algorithm for feedforward neural network training. Paper presented at the 21th Iranian conference on electrical engineering (ICEE2013), Ferdowsi University of Mashhad
33. Gao H, Xu W, Sun J, Tang Y (2010) Multilevel thresholding for image segmentation through an improved quantum-behaved particle swarm algorithm. IEEE Trans Instrum Meas 59 (4):934–946
34. Horng M-H (2011) Multilevel thresholding selection based on the artificial bee colony algorithm for image segmentation. Expert Syst Appl 38(11):13785–13791
35. Fan C, Ouyang H, Zhang Y, Xiao L (2014) Optimal multilevel thresholding using molecular kinetic theory optimization algorithm. Appl Math Comput 239:391–408
36. Derrac J, García S, Molina D, Herrera F (2011) A practical tutorial on the use of nonparametric statistical tests as a methodology for comparing evolutionary and swarm intelligence algorithms. Swarm Evol Comput 1(1):3–18

Part II
New Generation of Big Data Processing

Information Management in Collaborative Smart Environments

Shravan Sridhar Chitlur and Achim P. Karduck

1 Introduction

The envisioned Senseable Smart Cities will consist of systems and citizens involving large-scale Cyber-Physical Environments with supporting analytics infrastructures to enable predictive insightful actions, complementing this can be the nature of collaboration enabled by key IT systems form the building blocks. Within these Senseable Smart cities, data driven decision making and collaboration between the systems and citizens should be promoted to optimize resource utilization and to improve the quality of life, e.g. with respect to mobility, resource and infrastructure sharing etc. Existing challenges in mobility, security and safety, logistics, health and lifestyle support, user and citizen involvement, or energy prosumer models will have new avatars in the coming future for the envisioned Smart Cities. Sustainable smart environments enabled by collaborative data analytic infrastructures provide scalable performance and high availability to deliver insightful actions in order to achieve defined goals [1]. The future smart environments certainly constitute the most complex infrastructures, with respect to operation and evolution. Thus, to provide operational efficiency and to reduce the risk of failure by a sustainable decision support system will play a crucial role.

Mobile technologies are endorsing the usage of sensor-based digital assets in daily activities and thus accumulating large amounts of data which we can use to assess citizens behavior and preferences. Analysis of this data promises to provide us with actionable insights for sustainable decision making in smart environments [2].

S.S. Chitlur (✉) · A.P. Karduck
Furtwangen University, Furtwangen im Schwarzwald, Germany
e-mail: Shravan44@gmail.com

A.P. Karduck
e-mail: Karduck@hs-furtwangen.de

L. Ismail and L. Zhang (eds.), *Information Innovation Technology in Smart Cities*,
https://doi.org/10.1007/978-981-10-1741-4_7

The digital assets for Senseable Cities are beginning to be designed to promote collaboration among themselves and also with the users/citizens. Communication, coordination, and collaboration between activities of the digital assets and humans within the resulting Cyber-Physical Environments are more and more sensors based e.g. app-based personalized service delivery. Enabling users with data driven decision making for personalization and localization can enhance user experience and reliability.

Replumbing our existing technologies of infrastructures to prepare for and enable the future smart environments in lines with ubiquitous computing and digitization can efficiently and effectively optimize the performance leading to operational swiftness. Smart Environments might first be witnessed on reliable operational bases like enterprises and organizations. Similar to Senseable Smart Cities, they constitute complex, large-scale Cyber-Physical Environments, and we believe that substantial insights can be gained from the implementation and validation of our strategies in Senseable Enterprise/Organization, and in turn this can propel the adoption of data management strategies in our envisioned Cities. In this context, our research aims at providing an investigation into the components of latest data analysis infrastructures for insightful actions. Figure 1 provides the context of the research outlining the various factors leading to an improved and optimized smart environment. Several features of Operational excellence and Innovation driven by big data analysis are showcased to promote sustainability.

Much of the collaborated data obtained from present day smart environments are in unstructured form and is characterized by heterogeneity, complexity, scale,

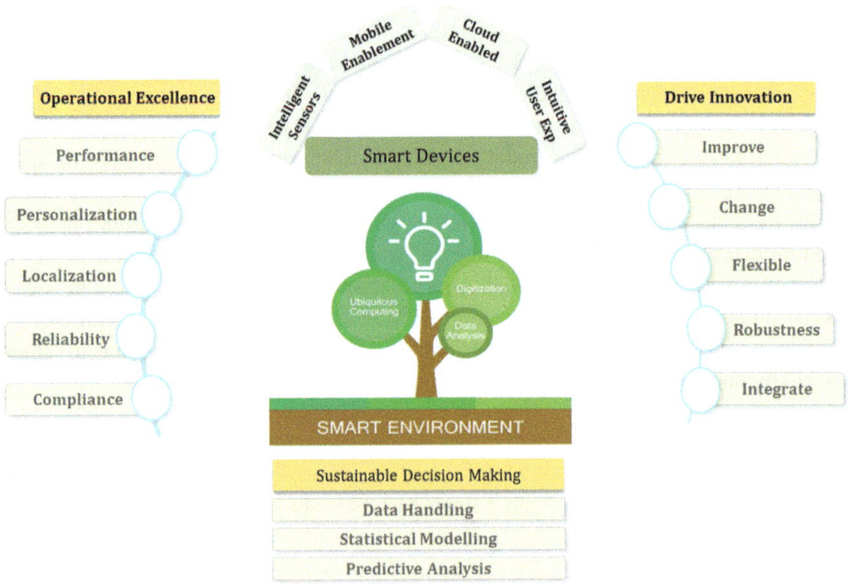

Fig. 1 Data driven decision making in SE—overview

timeliness and authenticity. Traditional relational databases are less efficient and sometimes incompetent to meet the growing demands of big data analytics. Hence, leading us towards a wider adoption of big data analytics comprising of distributed computing and data storage, distribution systems like Hadoop, high performance in memory databases like SAP HANA, and predictive analysis tools like R are investigated here. The ambition of the research is to test and propose a combination of these path-breaking technologies to support data analysis for insightful actions in smart environments. This can be adopted in organizations, but as well for the future smart habitats to make user experience more intuitive and enriching [3].

2 System Architecture for Insightful Decision Making in Smart Environments

The scope of this paper includes technology and infrastructure recommendations for collaborative data analytics in the envisioned smart environments, recommendations are inspired from the momentum in enterprises and organizations. For smart cities to leverage from the current wave of digitization and ubiquitous computing, analysis of data from transactional systems can be a key factor in achieving its long term objectives such as agility, insightful actions and operational efficiency. Transactional systems governing several business processes are collecting huge amounts of data. Performing predictive analysis on this data with the help of statistical modelling can offer organizations a distinct competitive edge. The latest technological advancements in this arena offer faster data analytical processes, reducing batch like analytical processes to almost real time. This relates to the ability of making better decisions and enabling meaningful actions at the right time. It signals the dawn of a new era in which systems and infrastructures begin to complement human life at natural speed, whether in business, private affairs, or as citizens [4].

It will be shown subsequently, as to how the recommended components Hadoop, Hana and Liferay can provide the collaborative big data infrastructures leveraging the predictive analytics capabilities of R in order to assist real time decision making. The results thus obtained can also be presented using dashboarding tools like Tableau and promises to derive meaningful insights in an intuitive way, e.g. in a business security and safety context. Several tool combinations can be integrated to obtain similar kind of results in today's big data scenario, but the selected technical stack can provide an optimal solution as outlined in the coming sections. Figure 2 shows the Testbed components within our proposed architecture.

Hadoop is open source framework for processing, storing and analyzing massive amounts of distributed and unstructured data. It is designed to handle Petabytes and Exabyte's of data distributed over multiple nodes in parallel [5].

Data storage capabilities of Hadoop are impeccable and utilities built around the framework to ease out the process of data administration and handling are developing at a very rapid pace to cater the growing requirements. Hadoop can handle large

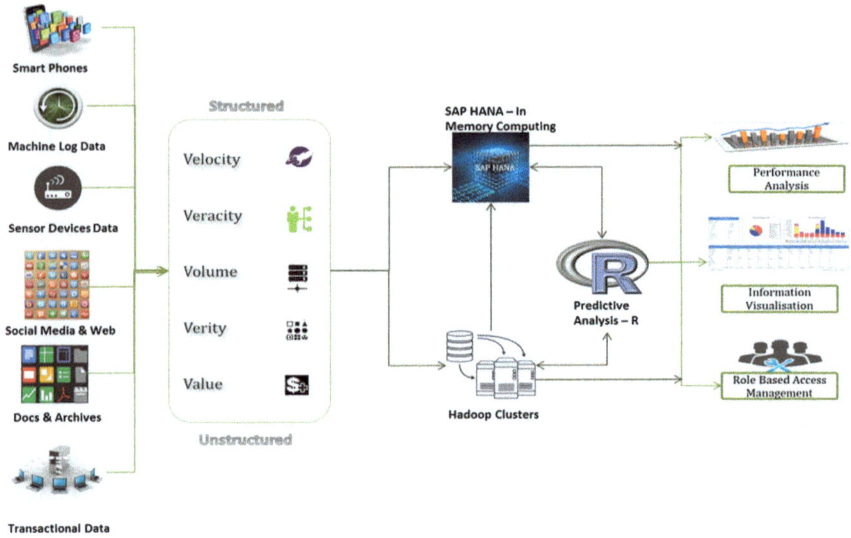

Fig. 2 Proposed system architecture

amounts of unstructured data coming from applications, sensors, social media mobile devices, logs etc. Hadoop follows the principles of distributed storage and computing, hence can be scaled up from one system to several thousands of machines.

The Hadoop ecosystem consists of many independent modules like Hadoop Distributed File System (HDFS), MapReduce, and HBase. Instead of dealing with large amounts of data in one single machine, Hadoop breaks data into multiple parts which can be processed and analyzed at the same time across different machines. Hadoop is scalable, cost effective, flexible and fault tolerant. Hadoop's No Structured Query Language (SQL) technology can handle unstructured files with relative ease. Table 1 describes the components of the Hadoop architecture of Fig. 3, as adopted here.

Table 1 Components—Hadoop architecture

Component name	Description
Master Name Node	Gives instruction to Data Node (DN) to perform input/output
Slave Name Node	Assists Name Node (NN) in monitoring HDFS cluster
Slaves (DN)	Read/writes HDFS blocks in local hard drive of slave machine and communicates with NN and DN
Master Job Tracker	Interface with client applications and monitors execution plans
Slave Task Tracker	Responsible for executing individual tasks given by Job Tracker

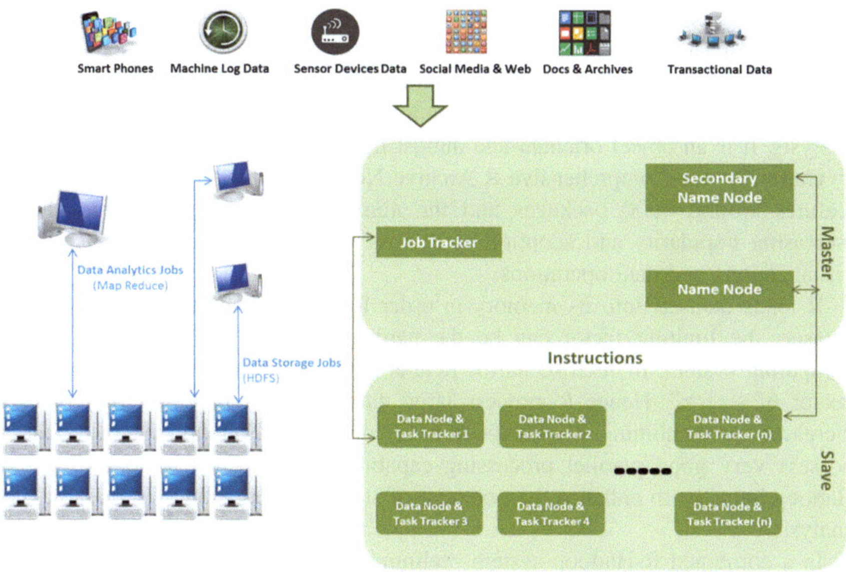

Fig. 3 Hadoop architecture

Configuring data jobs (data replications, flow etc.) in Hadoop environment is the primary step in setting up data storage and analytics. Data from several sources are induced into the Hadoop environment and this initiates a data storage job, these jobs are segregated by Job Tracker onto respective task trackers. Name Node identifies data nodes to store packets of data and corresponding replications in other data nodes as backups. Metadata of the data stored in Hadoop clusters will be present in name nodes and a backup of the same is maintained in Secondary Name Nodes.

3 Statistical Predicative Modelling Using R-Hadoop

Predictive modelling is one of the most common data mining techniques. As the name implies, it is the process of taking historical data (from past), identifying hidden patterns in the data using statistical models and then use these models to make predictions about what may happen in the future. The biggest challenge in predictive analysis is to validate the authenticity of the models being built. The methodology to achieve the same can be to prepare the data, perform exploratory data analyses, build a first model and iteratively build concurrent models. Using this champion challenger model R has a proven record of statistical modelling in various domains.

R is an integrated suite of software facilities for data manipulation, calculation and graphical display. It is an open source technology and hence offers some inherent advantages like lowered Total cost of Ownership (TCO) and easy wide-spread adoption.

R offers excellent data handling and storage facilities along with a suite of operators for calculations on arrays. It also consists of a large coherent and integrated collection of intermediate tools for data analysis and graphical displays. R being a programming language has limited user friendly interface for data analysis. It is an object oriented and almost non-declarative language.

Currently, the Comprehensive R Archive Network (CRAN) package repository features around 6000 packages and the number is growing. This unveils the increasing popularity and adoption of R. It is widely used for machine learning, visualizations and data operations.

R loads datasets into its memory in order to analyze the data, hence for large datasets the limiting factor can be the hardware capabilities of the underlying computing system. In such scenarios R fails with exceptions like "cannot allocate vector of size x". Hence to process large datasets the capabilities of R can be increased by combining it with distributed computing Hadoop systems. Hadoop possess very good parallel processing capabilities and hence R combined with Hadoop helps us to enhance the potential of statistical modelling for massive data analysis.

In a combined R-Hadoop system preliminary data analysis functions like data loading, exploration, analysis and visualizations are handled by R. Data storage/retrieval and distributed computing is handled by Hadoop. It's an established fact that advanced machine learning algorithms works better with large data sets hence using R with Hadoop is recommended [6].

In [7] we have shown how R on Hadoop and Hana can be enabled. Further, we explain in the paper how R used in conjunction with Hadoop can help to overcome several of R's predominant drawbacks such as limited data size and performance. With respect to SAP HANA, R code is embedded in SAP HANA SQL code in the form of a RLANG procedure. To achieve this, the calculation engine of the SAP HANA database was extended. The calculation engine supports data flow graphs (calculation models) describing logical database execution plans. Database execution plans are inherently parallel and therefore, multiple R processes can be triggered to run in parallel [8]. R, HANA and Hadoop are conceived and built to serve different and specific purposes. In [7] we combine the best use cases of the three platforms in order to help smart environments to perform better. It became evident that the combination of R, HANA and Hadoop can play a crucial role in nurturing knowledge discovery and insightful processing.

4 Strategic Decision Making and Data Analytics

We are aware, that adoption of change in Smart Environments requires strategic decision making, commonly defined by Key Performance Indicators (KPIs). Within a focused case based setting, in this case a University context, we have deployed our R-Hana-Hadoop infrastructure in the context of a KPI-defined process. We are convinced, that the approach can as well be adopted for large-scale environments, to support consensus among the stakeholders. Our approach here is that strategic decision making should be aligned with the data analytic practices in any organization, and thus as well for the envisioned Smart Environments. Our process of deriving actionable insights from business transactional data started with defining the high level business process taxonomy for the organization. For each of the identified processes, key performance indicators (KPI's) were derived. These KPI's are instrumental subsequently in determining the process maturity and performance. Near to long term business strategic objectives of the organization were identified and there after the derived KPI's were prioritized and aligned to the respective objectives of the organization as shown in the Fig. 4. The process outlined above formed the basis for the Extraction, Transformation and Loading (ETL) of the data into their respective data marts. Data latency levels for the KPI's were also agreed upon for delta data uploads into the data marts [9].

Subsequent to identifying the KPI's, we performed source systems analysis to understand the feasibility for fulfilling the data requirements to measure KPI's. There after a prioritized KPI implementation plan considering technical readiness (data availability) and impacted business objectives was derived to facilitate a phase wise implementation of suggested analytical solution. Data volumes and data transformations levels were determined and the technical stack shown in the Fig. 5 was used to derive predictive insights. Historical transactional data is staged in Hadoop through an open source ETL tool Talend and from there we can have the hot data (most accessed and highly important data in organizations) placed in Hana for near real-time in-memory processing. This can be directly fed into R system for deriving predictive insights on the data. These insights were fed into the dashboarding tool Jasper Reports for improved visualization and customization. Characteristics of the dashboards based on different user roles were determined to provide a role based visualization.

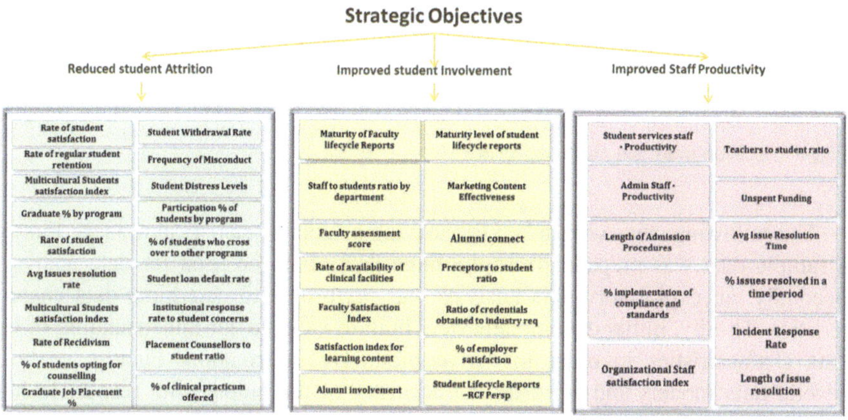

Fig. 4 KPI's by strategic objectives

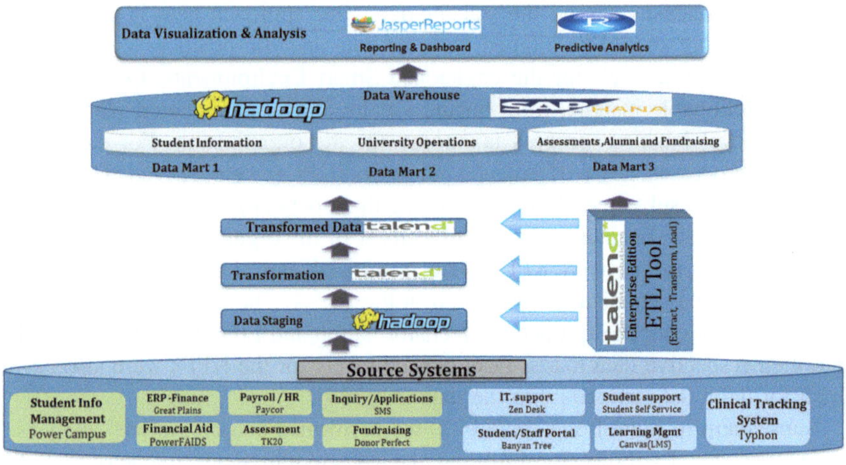

Fig. 5 Utilized technical stack

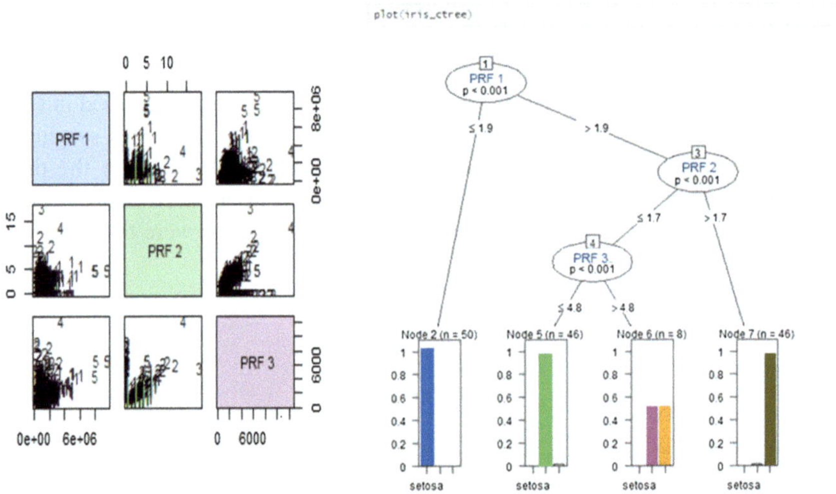

Fig. 6 Decision trees and clustering in R

The following steps were taken for arriving at suitable statistical models. Firstly, a sample data was prepared from the huge data set to be used for data analysis. Secondly, an exploratory data analysis was performed to determine data characteristics. Thirdly, initial data models were built to validate our understanding of the data characteristics in order to derive predictive information. Finally we built several data models and compare the accuracy of the output with respect to historical data and conclude on an agreed upon model for further analysis and development (Fig. 6).

5 Collaboration in Smart Environments

Collaboration in Smart Environments relates to communication, coordination, and cooperation between actors in Cyber-Physical Environments, such as in the future Smart Cities. With mobile and sensor based technologies, a fundamental improvement is observed in the resource efficiency e.g. logistics in mobility. More streamlined and personalised engagement channels for citizens in the form of personalised social media portals, collaboration tools etc. are arising, including standards for their integration. Real time/near real time collaboration within Cyber-Physical Environments is a key determining factors for the success of any smart project initiative.

Stratification of processes and information at various levels become quintessential to achieve desired outcomes, new Digital Ecosystems should enable data driven, insightful decision making along the process orchestration, knowledge management, identity infrastructures etc. these components together form building blocks for the complex Senseable Smart Environments and in-turn Smart Cities. It should also be noted, that these environments should allow citizens to define and operate their own communities by grouping citizens of similar roles and responsibilities forming basically user groups or teams with target workflow or knowledge support. Further, collaboration environments should allow these groups to be aggregated at much higher levels, thus enabling cross group information access and sharing. Portals enabling collaboration should also provide us with another angle of social collaboration, that cuts across both group and system wide collaboration. Like minded citizens from different formal groups and forums should be able to form their own informal groups to share best practices, e.g. for social engagement in their fields of interest. Collaboration in Smart environments can be enriched based on information management and insightful data driven decision making, which are outlined in the research presented in the former section (Fig. 7).

To promote collaboration in smart environments, we have decided to leverage Liferay portals. It is an open source web platform that contains many default portlets to facilitate the need of collaboration in various forms like wiki, message

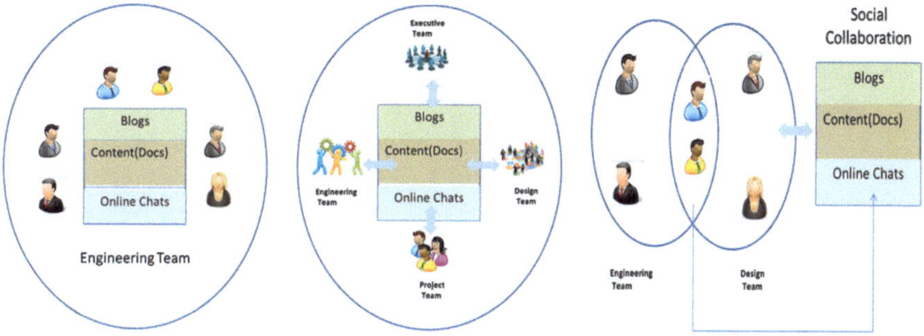

Fig. 7 Lifecycle—system of engagement

boards, blogs, forums etc. Some of these portlets are plug and play and come by default in the standard package, some of the more specific ones are also available from their market place along with abilities to have custom developed portlets, thus helping the cause of easy implementation and adoption and optimised development time.

Liferay portals are designed and built to suit the needs of role based content delivery and user management, thus reducing the need for customization to a great extent. Liferay even supports features like Single Sign on (SSO), custom fields, workflow and rules engine, user personalisation, multi-language support, contextual search and many more. It is one of the few packages available which can provide such collaboration building blocks out of the box.

We believe that the adoption of Liferay portals to promote collaboration in smart environments can drive digital transformation. In addition, one can leverage the content management system (CMS) capabilities of Liferay and digitize documents wherever required.

Also upon aggregating various systems of engagement and transactional system of records, we can derive actionable predictive insights by using the previously presented combination of R, HANA and Hadoop. Figure 8 provides us with an overview of how Liferay can be integrated into existing infrastructures. In a nutshell, various portlets of Liferay provide user friendly applications with both presentation and business logic capabilities. To integrate them with other applications, Liferay works with Representational State Transfer (REST) Application Programing Interface (API), Web Services JSON, XML etc. We can use Enterprise Serial Bus (ESB) to connect with BI reporting tools, BPM engines and so on. To promote ease of usage, one can incorporate the advantages of Single Sign on (SSO), Lightweight Directory Access Protocol (LDAP), Apache Solr Search etc.

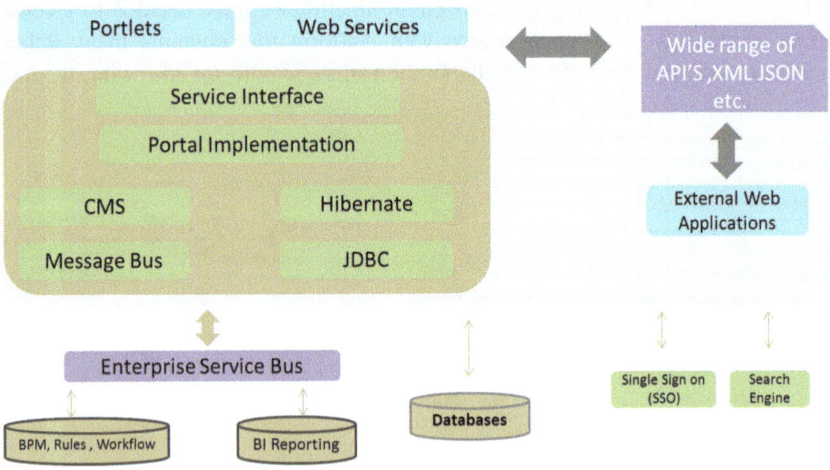

Fig. 8 Liferay integration architecture

6 Conclusion and Outlook

Our work has outlined, how the envisioned Senseable Smart Cities depend on large-scale analytic infrastructures to enable insightful actions, and how this can be complemented to promote collaboration in these Cyber-Physical Environments. It has been shown, how the adoption of predictive analysis driven decision making supported by statistical data modelling can strongly benefit from a combined deployment of R, HANA and Hadoop. In terms of operational efficiency, scalability and reliability, the outlined architecture contributes to a robust and scalable data analysis infrastructure for insightful actions. The scope of our work presented includes technology and infrastructure recommendations for collaborative data analytics in the envisioned smart environments, derived from the momentum and best practices in enterprises and organizations for achieving long term objectives such as agility, insightful actions and operational efficiency, incorporating similar infrastructure in Senseable City Environments makes it robust, provide the large-scale performance required, and are evolvable over time. We can benefit from combined real-time big data analytics and real time collaboration to build sophisticated systems that can provide us with the leading predictive indicators on defined KPI's. We have shown, that this enables us with actionable insights on events/trends that were almost impossible to predict until recently.

Complex systems are never built from scratch, but rather utilize the product innovation ecosystem available. Numerous technologies and methodologies have emerged recently to fulfil the growing demands of data gathering and storage, insightful analysis, and collaboration between the digital assets and among citizens. It is a big challenge for architects of smart environments to design a collaborative system integrated with traditional transactional systems and then to manage the data to offer an ever improving and agile Digital Ecosystem. The work presented sheds some light on determining the right set of technologies depending upon the context of usage.

The entire concept outlined in this paper is to synthesize vast amounts of collaborated transactional data and to derive meaningful actionable insights to ensure an improved user experience with a sustainable growth in smart environments, such as the envisioned Senseable Smart Cities. Future work relates to the validation of the operational performance and scalability of the proposed components for large-scale environments, including the complementing collaboration support. Further, the strategic decision making in the context of data science opens an important area for future research.

References

1. Ratti C (2013) Smart city, smart citizens. Egea Publ, s.l.
2. Lee D, Felix JRA, He S, Offenhuber D, Ratti C (2015) CityEye: real-time visual dashboard for managing urban services and citizen feedback loops. s.n, Cambridge, USA

3. Poslad S (2009) Ubiquitous computing—smart devices, environments and interactions. Wiley, West Sussex
4. Barlow M (2013) Real time big data analytics—emerging architecture. s.n, Boston
5. Kelly J (2014) Big data: Hadoop, business analytics and beyond. [Online] Available at: http://wikibon.org/wiki/v/Big_Data:_Hadoop,_Business_Analytics_and_Beyond. Accessed 7 Mar 2016
6. Revolution Analytics (2011) Advanced 'big data' analytics with R and Hadoop. [Online] Available at: www.RevolutionAnalytics.com. Accessed 11 Feb 2016
7. Chitlur SS, Karduck AP (2015) Data driven decision making for sustainable smart environments. In: Innovations in information technology (IIT), 2015 11th international conference on IIT, Dubai, UAE
8. SAP AG (2014) SAP HANA R integration guide. SAP, s.l.
9. A community white paper—Jagadish HV, U. o. M. (2012) Challenges and opportunities with big data. Community White Paper, United States
10. Senseable City Laboratory. [Online] Available at: http://senseable.mit.edu/. Accessed 10 Mar 2016
11. Anon. (n.d.) What is SAP HANA platform. [Online] Available at: http://www.freehanatutorials.com/2013/01/what-is-sap-hana-platform.html. Accessed 10 Mar 2016
12. Apache (2013) HDFS architecture guide. [Online] Available at: http://hadoop.apache.org/docs/r1.2.1/hdfs_design.html#Introduction. Accessed 10 Mar 2016
13. Ellis B (2014) Real-time analytics—techniques to analyze and visualize streaming data. Wiley, Indianapolis
14. Jain K (2014) SAS vs. R (vs. Python). [Online] Available at: http://www.analyticsvidhya.com/blog/2014/03/sas-vs-vs-python-tool-learn/. Accessed 1 Mar 2016
15. Janert PK (2011) Data analysis with open source tools. O'Reilly, Sebastopol
16. Prajapati V (2013) Hadoop, big data analytics with R and Hadoop. Packt Publishing Ltd., Birmingham

Business Behavior Predictions Using Location Based Social Networks in Smart Cities

Ola AlSonosy, Sherine Rady, Nagwa Badr and Mohammed Hashem

1 Introduction

Analyzing business behavior is a very challenging process. To understand business behavior, lots of field research have to be performed. To attain a field research about business behavior in a city, Extensive of accumulated experiences have to be acquired from consumers and business owners who live in that city. This kind of data about business places can be provided by Location Based Social Networks (LBSNs) instead of performing exhaustive field research [1]. This is because LBSNs can be considered as a connection between real life movements of people around places and the social media transactions. People physical turnouts for places in real life can be projected through the users' online checkins, because an online checkin for a venue involves the physical location of that venue on earth. Consequently, understanding business behavior in a city can be deduced from a global view of the users' checkins to venues in LBSNs.

There have been vast research essences for LBSNs Data. Mining LBSNs for friendship, as well as venue and trajectory recommendations are commonly prevailing areas of research that provided multiple services for LBSNs users. Also, analyzing the tenor of LBSNs data for community detection has been presented in various research too for human behavior exploration.

Up till now, it is unclear how certain business is predicted to be popular and how can an investor understand business terrains in a city in order to decide which business are expected to run well and where. LBSNs provide global information about business behavior. This kind of information can be exploited by business decision makers for predictions in specific terrains. Understanding and answering

O. AlSonosy (✉) · S. Rady · N. Badr · M. Hashem
Faculty of Computer and Information Science, Information Systems Department, Ain Shams University, Cairo, Egypt
e-mail: ola.a.alsonosy@cis.asu.edu.eg

L. Ismail and L. Zhang (eds.), *Information Innovation Technology in Smart Cities*, https://doi.org/10.1007/978-981-10-1741-4_8

these questions can be beneficial in urban planning, recommending hotspot business openings terrains and directs advertising campaigns in LBSNs.

In this research work, an urban analysis exploiting data collected from Foursquare about business turnouts is done for business prediction purposes. The research observes Foursquare venues' behavior from the investor's perspective. The venues observations were analyzed to help in predicting business turnouts for new business openings or other related business needs in certain geographical terrains. The research suggests a spatial interpolation technique, which is commonly used for predicting spatially correlated data in ecological fields, to predict business turnouts. Additionally, a similarity embedded spatial interpolation technique is introduced to suit LBSNs data. The similarity embedded spatial interpolation considers multiple characteristics amongst neighbor venues that can affect the turnout prediction accuracy in addition to their spatial closeness. The proposed similarity embedded spatial interpolation is compared to classical spatial interpolations commonly used in predictions of spatially correlated data values. The designed model shows a significant alleviation of prediction error than the classical spatial interpolation techniques.

The rest of the document; is organized as follows; in Sect. 2 an overview of the previous research performed over LBSNs is presented. Section 3 reviews the data observations about business behavior over Texas, which will be used in the case study performed later. Section 4 views the suggested spatial interpolation techniques used for predicting business turnouts in LBSNs along with an introduction of the proposed similarity embedded spatial interpolation prediction technique. Section 5 assesses the proposed prediction techniques through an experimental case study predicting business turnouts in Texas using the suggested techniques and presents comparative results. Finally, Sect. 6 concludes the paper and suggests future extension of this research.

2 Related Work

Location added service social networks and LBSNs have attracted many researchers in the last decade. Research in the area of LBSNs can be categorized into four domains which are shown in Fig. 1:

1. Location aware recommendation systems
2. Location aware community detection
3. Human mobility analysis research based on LBSNs
4. Location aware business enhancement systems

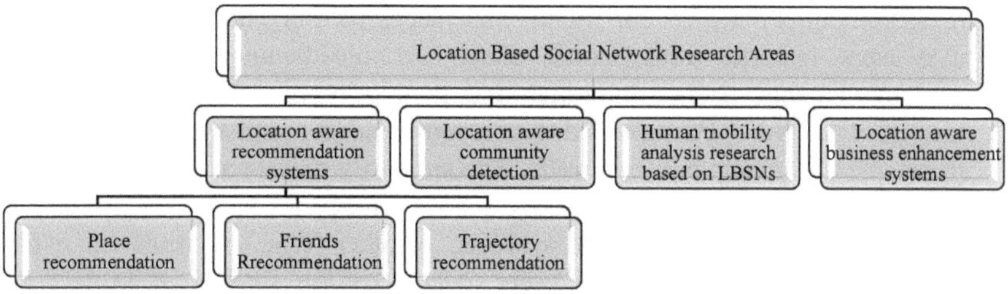

Fig. 1 LBSNs research domains

2.1 *Location Aware Recommendation Systems*

Some of the research focused on improving location aware social networks for recommendation systems production purposes. Mainly, these recommendations systems aim to predict information in order to help the individual LBSN's users in their daily life decision making. This is done through analyzing their checkins history in the LBSN using collaborative filtering methods based on similarities between users or venues. The evolved systems include venues recommenders based on users' checkins, friends recommeder systems based on their checkins and trajectory recommender systems.

- **Place Recommendation systems** The research in this area uses individual transactions made by the LBSNs' users in order to recommend venues or places that might be of users' interests. In [2, 3] the author exploits checkins of users in Foursquare and MovieLens in an item based collaborative filtering in order to recommend venues for users. The papers suggest a travel penalty parameter to involve the location feature in the item based collaborative filtering technique used. Also a user portioning method is applied to support system scalability. In [4] a location-based and preference-aware recommender system based on user based collaborative filtering is proposed. The system offers a particular user a set of venues within a geospatial range with the consideration of user preferences that are automatically learned from his/her location history and social opinions, which are mined from the location histories of the *local experts*. The research in [5] fuses user preferences, social influences and geographical influences of points of interests in a user-based collaborative filtering to enhance venues recommendations to users. In [6, 7], the authors exploit GPS data along with LBSN's transactions in learning from user histories for place recommendations.
- **Friends Recommendation systems** Exploiting LBSNs' data for building friends recommendation systemsare implemented in several research. In [8], link predcition technique is developed from Gowalla for friends recommendations. The system takes into consideration place features, social features and other global features based on similarity measurements between users who do not share any friends. In [9] a hybrid content-based and social-based collaborative filtering technique is developed for building a hierarchial similarity

measurement framework for friends and places recommendation. The systems uses users' histories from Geolife taking into considerations the sequences of user movements, users similarities and popularity of venues. Also in [10], Geolife data with GPS provided data recommended friendships based on similarity measurements between users. The similarity measurements are inducced from venues semantics concluded by venues' categories visited by users.

- **Trajectory Recommendation systems** Another type of research, which exploits LBSNs' data to mine users trajectories are studied. A project like GeoLife, which is introduced and studied in [11–13], is a social networking service incorporating users and locations from user generated GPS data, to visualize and understand user trajectories. Geolife searching for trajectories using user movements learned by users' online checkins between locations, spatio-temporal users' queries and GPS provided logs. Geolife provided data were used along with GPS traces to formulate a tree based hierarchical graph for travel recommendations in [14]. In [15, 16] data from Foursquare checkins along with GPS provided taxi trajectories are used in providing trajectories to construct an online trip planning system that builds a routable graph from these trajectories. In [17], LBSNs data were exploited to make a compound module of regular movements, irregular movements and novelty seeking for mobility prediction purposes.

2.2 Location Aware Community Detection

Users are studied in LBSNs for Location aware community detection. In [18], a study of the social and spatial properties of communities in Gowalla and Twitter was presented. The authors found that in Twitter popular users hold communities together, while in Gowalla community members tend to visit the same places. Also the study in [19] involves the use of social theories and community detection algorithms over location information provided by LBSNs and geographic features to capture how communities form, and to define accurate models of community evolution. In [20], the researcher obeserves that the social network can be formulated based on places as the focis of the network by investigating the relationship between the types of places where people meet and the likelihood that those people are friends. In [1], large-scale data from Foursquare is analyzed across three cities (London, New York, and Paris) in order to produce an inter-urban analysis. The research in [21] provides an attempt to clustering social activities using LBSN's data gathered from publicly available foursquare related tweets in the region of Colonge in Germany. Clusters are formulated by modeling the difference of the overall temporal distribution of check-ins. Furthermore, the paper presented a technique of multidimensional scaling to compute a classification of all clusters and visualizes the results. Also the project Livehoods presented in [22] arranges community clusters of 18 million Foursquare checkins based on an affinity matrix constructed based on both spatial and social affinities. A map based tool Hoodsquare is presented in [23], in which neighborhoods are constructed based on geographical features along with users' visits deduced from twitter. In [24], the authors propose to model human activities and geographical areas by means spectral clustering technique that uses feature vectors crawled from Foursquare users in New York and London.

2.3 *Human Mobility Analysis Based on LBSNs Data*

Some other research have analyzed human mobility patterns based on their activities of checkins in LBSNs. In [25], a case study analyzed the spatiotemporal dependent user behavior from Foursquare in Lisbon Metro in order to find potential correlation in user behavior patterns in a working week in two different time instances. In [26], a model of human mobility dynamics is developed. The model combines the periodic day-to-day movement patterns provided by a cell phone data with the social movement effects coming from the friendship network provided by LBSN's data. Furthermore, An identification of sequential activity transitions in weekday and weekends are explored from foursquare provided data in [27]. More global data analysis studies have been also conducted. In [28], users' mobility patterns are analyzed in 34 cities around the world using Foursquare data. The analysis was conducted to study the effect of places distributions across different urban environments on the human mobility patterns. Also in [29], the author studied 22 million checkins across 220,000 users and reported a quantitative assessment of human mobility patterns by analyzing the spatial, temporal, social, and textual aspects associated with footprints from LBSN. In [30] a study in China exploring footprints from the Chinese twitter Wibo were made.

2.4 *Location Aware Business Enhancement Systems*

The research in business enhancement area according to LBSN's data is not very popular. In [31], PLUTUS systems is introduced to suggest a best set of customers for venues owners to propose more offers for them. The system uses an item based collaborative filtering for recommendation. Also in [32], users' movements histories were gathered from Foursquare checkins to choose from numbers of suggested locations where to locate a new branch of chain stores. The system compared different machine learning techniques taking into consideration users' transitions sequences, geographical features and competitive features of the suggested locations. A case study targeting popular chain stores was applied in New York City, e.g. Starbucks, Mcdonalds and Dunckin Donuts.

This work extends those research directions by suggesting the exploitation of LBSNs data for predicting business behavior in certain terrains, which can serve business owners in their decision makings. The prediction uses spatial machine learning methods. A Spatial Interpolation method is applied and experimented for prediction in LBSNs. Additionally, a similarity embedded spatial interpolation method is proposed for more accurate predictions. The next sections introduce and discuss those spatial prediction techniques.

3 Data Observations

One of the most popular and widely used LBSN is Foursquare [33]. Foursquare was launched in March 2009. By August 2011, it had about 1 million checkins and this number is accumulated to 7 billion checkins by March 2015. This vast amount of data transactions was a motivation of this research for performing urban analysis to infer business behavior through Foursquare users' online checkins.

Foursquare is a local search and discovery service mobile application which helps its users to find places of interest. Its database contains information about locations (venues) and people (users). Venues in Foursquare are categorized into 10 different main categories, which are further divided into about 400 subcategories in a three level hierarchal tree [34]. When a user visits a venue, he/she goes through Foursquare and *checks in*to that venue. A user can *like* or *dislike* a venue that he/she had checked in. Also a user may leave a *tip* about the venue he/she had visited for other users to see.

The venues' density has been used in some research for interpreting business terrain popularity [23, 29, 32]. In this work, business terrain popularity is interpreted by the terrain's venues' turnouts. The turnout of a certain venue is inferred through the number of checkins for that venue over a specific time period. In the presented case study here, observations of the number of checkins in Texas venues over a year of study (between March 2014 and March 2015) are attained to infer the business turnouts, the detailed extraction module will be explained later in Sect. 5.1.

When plotting Texas venues heat map, it has been identified that venues tend to be dense in certain areas rather than others, this can be seen in Fig. 2a. In Fig. 2b, the checkins observed in the year of study for Texas venues are observed. When comparing the two figures, it is observed that densely cluttered terrains do not often imply bigger values of checkins. Therefore, the checkins is used as a measurement of usage and popularity for venues rather than venues densities.

To confirm the previous hypothesis, observations about the number of venues in each main category in Texas have been counted in Table 1. Also, a summation of business turnouts for all venues belonging to each main category was conducted. We concluded that the number of venues a category may not always express its business popularity. For example, shops and services have the largest number of venues in Texas and yet not the most popular category, while travel and transport are the most popular, because of their having the largest number users' checkins, while shop and service comes as the third popular category.

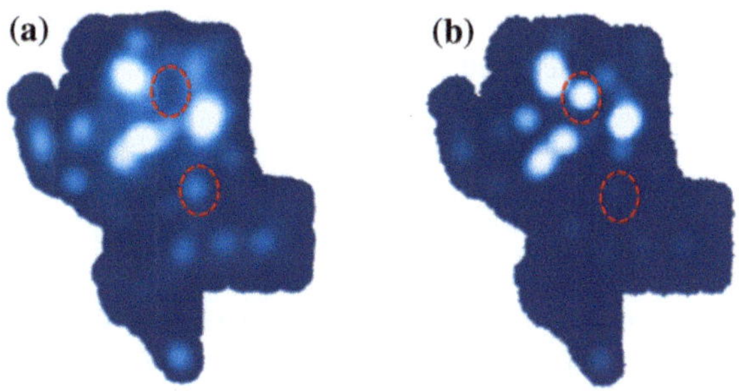

Fig. 2 **a** Venues densities heat map. **b** Venues checkins heat map

Table 1 Category—number of venues versus category—number of turnouts

Category	Number of venues	Category	Number of turnouts
Shop and service	35,594	Travel and transport	5,659,092
Food	33,750	Food	1,456,103
Outdoors and recreation	29,680	Shop and service	933,421
Professional and other places	24,857	Outdoors and recreation	617,089
Travel and transport	18,467	Professional and other places	570,288
Nightlife spot	14,215	Nightlife spot	552,965
Arts and entertainment	11,730	Arts and entertainment	512,925
College and university	9702	Residence	238,613
Residence	4233	College and university	149,868
Event	685	Event	45,880

4 Business Turnouts Prediction Techniques

Spatial data mining discovers patterns of a certain spatially correlated variable in data sets. These spatial patterns can be used for prediction purposes using spatial machine learning techniques. Spatial machine learning techniques formulate a mathematical model that presents spatial patterns and applies it later to predict unknown values of the spatially correlated variable in study.

From the data collected about venues from Foursquare over the year of study, there are two types of features observed, spatially correlated features and non-spatially correlated ones. Business turnouts are one of the spatially correlated features, and can be predicted based on their location among other neighboring observations for the same feature using spatial machine learning.

The spatial machine learning method suggested to be applied in this study is the spatial interpolation. Furthermore, a similarity learning embedded spatial interpolation technique is introduced. The proposed technique introduces other venues' features, which are non-spatially correlated, in the interpolation process, which has an impact of obtaining better prediction performance.

4.1 Spatial Interpolation

Interpolation is a numerical analysis method that is used in predicting new data points within the range of a discrete set of known data points. The general formulation of the interpolation problem can be defined as follows:

The data set of size N provides discrete observations that are acquired for the variable under study y_i, where $i = 1,\ldots,$ N. From the data observations, the solution model tries to find a mathematical function $F(y_i)$ that fits through all the values of y_i. After that the mathematical function F can be applied to predict an unknown value of the studied variable $\hat{y}$.

Spatial Interpolation techniques were developed for the cases in which the variable under study y, is assumed to be spatially correlated. They predict the unknown value of the variable y from observations of the same variable in positions located in the neighborhood of the unknown variable $\hat{y}$ location [35]. The techniques were usually applied to predict values for environmental observations in missing spots that are not sampled by a non-fully covered network of sensors; e.g. temperature, wind direction, etc. [36, 37]. The general formulation of the spatial interpolation problem can be defined as follows:

The data set provides N observations or measurements that are acquired for the variable under study y_i at discrete locations l_i, where $i = 1,\ldots,$ N. Instead of finding a mathematical function $F(y_i)$ that models the relationship of the variable values y_i, as in classical interpolation, the model is explicitly built over the assumption based on Tobler's first law hypothesis of geography. The hypothesis states that, "*Everything is related to everything else, but near things are more related than distant things*" [38]. Therefore, in predicting the unknown variable $\hat{y}$ at location $\hat{l}$, spatial interpolation uses the set of values of the same variable y_i at discrete locations l_i for all locations in K, where K is a set of locations $\{l_1, l_2, \ldots l_k\}$ representing the spatial neighborhood of the unknown variable's location $\hat{l}$.

Accordingly, the application of the Spatial Interpolation technique for business turnouts' prediction is presented in the following steps:

1. **Neighborhood identification** This involves selecting the neighborhood K for the target venue location $\hat{l}$ of the unknown business turnout $\hat{y}$. The selection criteria is based on the distance between the target location $\hat{l}$ and other locations l_i of different venues v_i in the dataset.
2. **Sampling phase** This involves collecting the business turnouts y_i at locations l_i belonging to the neighborhood K.
3. **Prediction phase** This involves the estimation for the unknown business turnouts $\hat{y}$ by applying a model to the values collected from the sampling phase.

Spatial interpolation is considered as a machine learning process, because the choice of the model or the mathematical function used in prediction is explicitly embedded in the system. In this study we will test K nearest neighbors Spatial Interpolation KNN [35] and Inverse Distance Weight Interpolation IDW [39]. In the KNN Spatial Interpolation, predicting the unknown business turnout $\hat{y}$ is simply by calculating the mean value of the known business turnouts y_i for venues in the neighborhood K defined by:

$$\hat{y} = \sum_{i=1}^{k} y_i \tag{1}$$

where k is the number of venues formulating the neighborhood of the unknown variable.

In IDW, Spatial Interpolation distances between the target venue location $\hat{l}$ and locations of other venues observed by LBSN l_i weigh their influence in the calculation of the unknown business turnout. Estimating the unknown business turnout $\hat{y}$ in IDW Interpolation is expressed as:

$$\hat{y} = \frac{\sum_{i=1}^{k} \frac{1}{d_i} y_i}{\sum_{i=1}^{k} \frac{1}{d_i}} \tag{2}$$

where d_i is the Euclidean distance from location of observed venues y_i to the unknown target venue location $\hat{l}$ and k is the number of venues formulating the neighborhood of the unknown variable.

The Spatial Interpolation introduced here are believed to be more practical choice for predicting spatially correlated features in LBSNs than the classical spatial regression machine learning techniques, for the two following reasons:

- The LBSNs large data size makes the learning process in classical spatial regression machine learning techniques very complex and time consuming, while the spatial interpolation uses only the neighborhood observations as a training set for learning.
- The formulation of LBSNs data is user dependent, this means that the data is dynamic, thus cannot provide a solid base for the learning phase in spatial regression. Therefore, the rapid predictions that might rise from the LBSNs users' sides need rapid learning phases with recently refreshed data, which are provided in spatial interpolation techniques.

4.2 Similarity Embedded Spatial Interpolation

Similarity Embedded Spatial Interpolation is based on a modification over the hypothesis of Tobler's first law of geography. The proposed modified hypothesis is "*Everything is related to everything else, but near things* ***that are more similar to each other*** *are more related than distant things.*" According to this hypothesis, other features, rather than the spatially correlated feature, can be exploited to contribute to more accurate predictions for the spatially correlated feature.

The Similarity Embedded Spatial Interpolation is attained through the following phases:

1. **Similarity based Neighborhood selection** that involves selecting the neighborhood venues that will be used for prediction. The selection criterion is based on a *closeness weight* between the target venue and other venues observed by the LBSN's data. The *closeness weight* is calculated based on both geographical distance and other observed non-spatially correlated features similarities between the target venue and other observed venues in the network.

The closeness weight $C(v_i, v_j)$ between two venues v_i and v_j, can be calculated through:

$$C(v_i, v_j) = \frac{sim(v_i, v_j)}{d(v_i, v_j)} \tag{3}$$

where $d(v_i, v_j)$ is the distance between the two venues v_i and v_j, and $sim(v_i, v_j)$ is the similarity measurement between the two venues v_i and v_j.

The distance $d(v_i, v_j)$ between two venues v_i and v_j is the Euclidean distance, which calculated by:

$$d(v_i, v_j) = \sqrt{\left|lat_i - lat_j\right|^2 + \left|long_i - long_j\right|^2} \tag{4}$$

where the locations of venue v_i and venue v_j are provided by the LBSN, through their latitudes and longitudes $(lat_i, long_i)$ and $(lat_j, long_j)$, respectively.

The similarity measurement between venues is calculated based on other features observed about venues and not spatially correlated. The similarity $sim(v_i, v_j)$ between two venues v_i and v_j is calculated by the *simple matching coefficient* similarity measurement:

$$sim(v_i, v_j) = \frac{\left|\sum_{x=1}^{FN} \left[f_x(v_i) = f_x(v_j)\right] : \forall f_x \in F\right|}{NF} \tag{5}$$

where $F = \{f_1, f_2, \ldots, f_{NF}\}$ is the set of features used in the similarity assessment, $f_x(v)$ is the feature x value in venue v, NF is the total number of features used in the similarity assessment.

2. **Sampling phase** After selecting neighborhood K venues based on the *closeness weight* between the target venue and other venues observed in the LBSN. Sampling involves collecting business turnouts y_i for all venues belonging to the neighborhood K.
3. **Prediction phase** The two spatial interpolation techniques KNN [35] and IDW [39] are applied, as described in the previous section, using business turnouts already sampled for venues in the similarity based neighborhood.

5 Experimental Case Study

An experimental case study is implemented to assess the prediction performance of spatial interpolation techniques and similarity embedded spatial interpolation for application over LBSNs data. The study aims to predict business turnouts for venues in Texas in the United States of America using data extracted from Foursquare.

5.1 Data Extraction

A data extraction module has been designed to load data populated by Foursquare about Texas venues in the United States of America. The Texas venues' data have been extracted using a specially designed crawler implemented using python programming language version 2.7. The crawler extracted data from Foursquare API for developers without breaching privacy restrictions imposed by the API's administration.

Each venue in Foursquare is defined by number of features. Only three of them are compulsory definition features; the venue's *name*, *location* (latitude and longitude) and its *category*. There are other numerous features that a venue may have, but not compulsory for a venue definition, such as *country*, *city*, *state* and *subcategory*.

To have an overall view of the business terrain in Texas, we could not crawl venues that have Texas as its defined state, because state is not a compulsory feature definition as illustrated previously. Therefore, we had to go through the following phases, shown in Fig. 3, to extract as much venues as possible registered in Texas in Foursquare:

1. **Location Based Scanning** First venues that have its location coordinates within Texas bounding Box of coordinates were extracted. This was implemented through an exhaustive scanning process. The scanning was implemented by crawling 10 km × 10 km sequential bounding boxes searching areas limited within the geographical Texas bounding box. The output of this phase is a set venue records in the form of (*venue-id, latitude, longitude*).

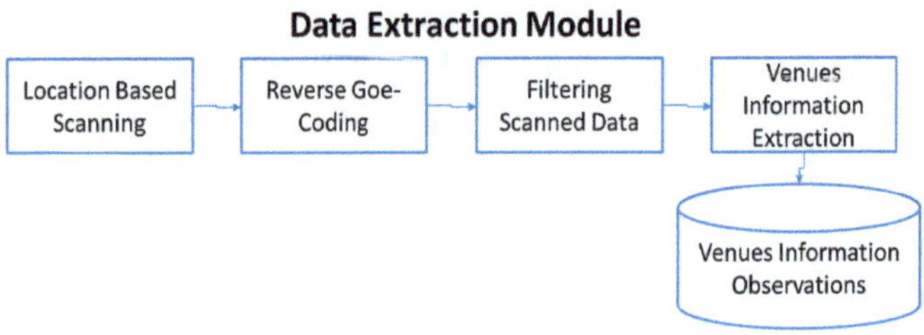

Fig. 3 Venues extraction module

2. **Reverse Geo-Coding:** A Reverse Geo-coding is performed over the venues extracted from the previous step using MapQuest API [40]. The output of this process is an address record for each of the extracted venues.
3. **Filtering Scanned Data** The venues are filtered to include only venues belonging to Texas. This phase is performed because some observations are carried out of venues that do not belong to Texas. The reason of this is that Texas bounding box may include some other venues outside Texas at the bounding box margins and corners.
4. **Venues information extraction** Other features about the filtered venues is extracted, e.g. *category*, *total number of checkins*, *subcategory*. Only categorized venues are included in this study. Also, venues that have subcategory as "home" is excluded from this experiment because it is believed that it will have a negative effect on the research results, as homes are not business venues and have the greatest number of checkins naturally. This results in a total of 184,879 categorized non-home venues in Texas.

Business turnouts is inferred by performing Venues Information Extraction phase in two different time instances (March 2014 and March 2015). A venue turnout, in this observation period, is calculated by subtracting the *total number of checkins* for that venue in the former observation from the *total number of checkins* in the later observation. A cleanup process is implemented to filter inactive venues and include only active venues for the study. A venue is considered to be active venue if it has more than 10 checkins in the observation year of study.

The result of the data extraction module is 41,954 total net categorized non-home active venues scattered over Texas region. 90% of these venues are randomly chosen as the training set with known business turnouts, while the 10% of venues are considered with unknown business turnouts to be used as a test set.

5.2 Data Prediction

To predict business turnouts using spatial interpolation, a spatial correlation of business turnouts has to be proven first. Moran's I test is used for this purpose. Moran's I test is a test used to estimate the spatial correlation among observations of variable y in the data set [41]. The result of the Moran's test is a value p calculated by:

$$p = \frac{N}{\sum_i \sum_j w_{ij}} \frac{\sum_i \sum_j w_{ij}(y_i - \bar{y})(y_j - \bar{y})}{\sum_i (y_i - \bar{y})^2} \tag{6}$$

where N is the number of observations for the variable y indexed by i and j; $\bar{y}$ is the mean of y; and w_{ij} is an element of a spatial weights matrix projecting the neighborhood relationships between observations based on the Euclidean distance between each pair of observations i and j.

The value of p ranges between −1 and 1, where −1 indicates a weak spatial correlation among the tested variable and 1 indicates a strong spatial correlation of the variable.

The Moran's I test was implemented over the calculated values for business turnouts over the year of study resulting in a p-value = 0.4385, which indicates a strongly spatial correlation among the business turnouts in our study.

The prediction accuracy is measured by calculating the root mean square error (RMSE) between the predicted value and the ground truth of each venue in the test set, which is given by:

$$\text{RMSE} = \sqrt{\frac{\sum_{i=1}^{n} (\hat{y}_i - y_i)^2}{n}} \tag{7}$$

where $\hat{y}_i$ is the predicted business turnout of venue v_i, y is the ground truth of business usage for the same venue, and n is the number of venues in the test set.

To access the effectiveness and accuracy of the proposed technique, a comparison between the spatial Auto Regression model, SAR [42], and the Spatial Interpolation as spatial machine learning has been conducted. For Spatial Interpolation, the KNN, IDW and Similarity Embedded Spatial Interpolation are implemented as parts of the predicting model. In the Similarity Embedded Spatial Interpolation, *category* and *subcategory* are the non-spatially correlated features that are used to calculate the similarity between venues in the similarity based neighborhood selection step. The prediction accuracy was calculated for the aforementioned techniques using the effect of a neighborhood surrounding each observation. The neighborhood is determined according to the k nearest neighbors with a value of k = 10. Table 2 summarizes the results of the comparison.

The results indicates a reduction for the RMSE in the KNN spatial interpolation prediction over the SAR model prediction with an average value of 78% while the

Table 2 A comparison between SAR, KKN Spatial Interpolation, IDW Spatial Interpolation, Similarity Embedded—KNN Spatial Interpolation, Similarity Embedded—IDW Spatial Interpolation with $k = 10$

	Average % RMSE reduction for $k = 10$
KNN/SAR	78.75
IDW/SAR	83.75
Similarity embedded—KNN/SAR	87.50
Similarity embedded—IDW/SAR	87.50

IDW spatial interpolation reduces the RMSE with an average value of 83.75% than SAR. Both similarity embedded spatial interpolation with KNN prediction and with IDW prediction have nearly the same results in reducing the RMSE with about 78.5% than SAR. The results obtained obviously show that using spatial interpolation techniques outperforms SAR. Also taking into consideration similarities for venues in the proposed similarity embedded approach has a significant effect in reducing RMSE and increasing prediction accuracy.

Figure 4 studies the different Spatial Interpolation techniques performances thorough another experiment with investigation for the parameter k. The figure shows a better performance of the IDW spatial interpolation than the KNN spatial interpolation for the different values of k. The reason behind this is that the IDW spatial interpolation involves distances related weights in the prediction process; therefore, it is more realistic in the expression of the spatial closeness effect to the prediction process. Moreover, the involvement of the features (category and subcategory) in the similarity embedded spatial interpolation clearly reduces the prediction RMSE.

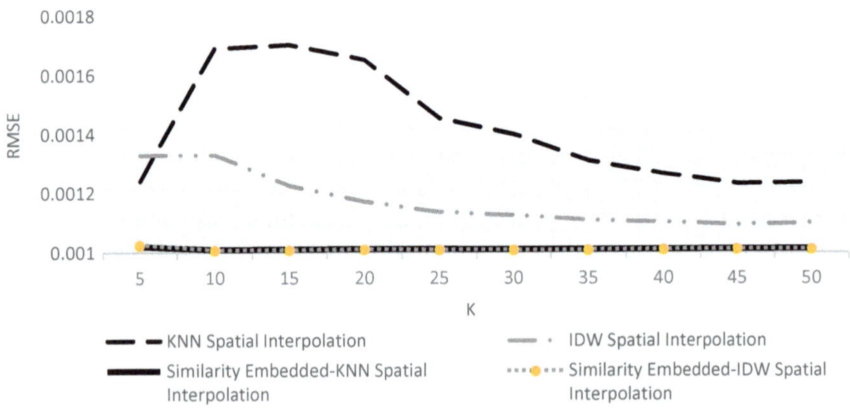

Fig. 4 A comparison between KKN Spatial Interpolation, IDW Spatial Interpolation, Similarity Embedded—KNN Spatial Interpolation, Similarity Embedded—IDW Spatial Interpolation for $k = [5:50]$

Table 3 A comparison between KKN Spatial Interpolation, IDW Spatial Interpolation, Similarity Embedded—KNN Spatial Interpolation, Similarity Embedded—IDW Spatial Interpolation

	Average % reduction in RMSE
IDW/KNN	16.40
Similarity embedded—KNN/KNN	27.45
Similarity embedded—IDW/IDW	13.19
Similarity embedded-IDW/Similarity embedded-KNN	0.01

Table 3 summarizes the results of the average reduction in the RMSE between the spatial interpolation pairs shown in the table for experiments implemented over different neighborhood sizes ranging between $k = 5$ and $k = 50$.

The results shows that the IDW spatial interpolation reduces the RMSE with an average of 16.4% more than the KNN spatial interpolation, while the similarity embedded KNN spatial interpolation results in a reduction of about 27.5% more than the KNN spatial interpolation. Furthermore, the similarity embedded IDW spatial interpolation reduces the average RMSE with about 13% more than the IDW spatial interpolation. Also the similarity embedded spatial interpolation using IDW and KNN in their prediction phase almost have the same RMSE.

The results obtained from the previous experiments show that applying spatial interpolation as machine learning techniques is a better choice when predicting spatially correlated variables in LBSNs than classical SAR machine learning model. The IDW spatial interpolation showed a better projection of the effect of spatial closeness over the prediction accuracy than KNN spatial interpolation. Additionally, considering more features provided by LBSNs in formulating the neighborhood to be included in the prediction process in the proposed similarity based spatial interpolation results in better prediction accuracies than IDW and KNN spatial interpolations.

6 Conclusion

Understanding business behavior in cities is useful for business owners to help them in decisions about new openings or other related business needs. This paper suggested the exploitation for LBSNs data for predicting business behaviors in future smart cities. The research proposes an application of spatial machine learning techniques to predict business turnouts in certain geographic locations. Spatial Interpolation techniques are suggested to be applied for learning and prediction purposes instead of regular spatial regression techniques. *K* nearest neighbors and Inverse Distance Weighted Spatial Interpolations have been proposed to enhance predicting the business turnouts in LBSNs. A Similarity Embedded Spatial Interpolation technique is furthermore proposed, which involves the use of features provided by LBSNs in the interpolation process to more enhance the accuracy of the prediction results.

An experimental case study was held inspecting venues behavior in Texas to test the proposed prediction methods and techniques against business turnouts. Data was extracted from Foursquare LBSN for the test. The results of the experiments has shown better prediction accuracy of the KNN and IDW spatial interpolation techniques application than the Spatial Auto Regression technique with about 79 and 84% reduction in RMSE respectively. Also the proposed Similarity embedded spatial interpolation prediction has shown outperformance over SAR regression model with about 88% RMSE reduction.

An additional comparison between the classical KNN and IDW spatial interpolation techniques and the proposed similarity embedded spatial interpolation was implemented. The techniques were tested over different neighborhood sizes ranging from 5 to 50 neighbors. The experiment showed a better prediction performance for the similarity embedded spatial interpolation than the KNN spatial interpolation with about 27%. Also similarity embedded spatial interpolation results in about 13% more prediction accuracy than the IDW spatial interpolation.

For more future insights about business behavior, the temporal factor for the business turnouts can be a useful extension for this research to achieve more enhancements in the prediction accuracy.

References

1. Bawa-Cavia A (2011) Sensing the urban: using location-based social network data in urban analysis. In: Workshop on pervasive and urban applications. San Francisco, CA, pp 1–7
2. Sarwat M, Levandoski JJ, Eldawy A, Mokbel MF (2014) LARS*: an efficient and scalable location-aware recommender system. IEEE Trans Knowl Data Eng 26(6):1384–1399 (archive)
3. Levandoski JJ, Sarwat M, Eldawy A, Mokbel MF (2012) LARS: a location-aware recommender system. In: 2012 IEEE 28th international conference on data engineering, vol 1. pp 450–461
4. Bao J, Zheng Y, Mokbel MF (2012) Location-based and preference-aware recommendation using sparse geo-social networking data. In: Proceedings of the 20th international conference on advances in geographic information systems—SIGSPATIAL'12, pp 199–208
5. Ye M, Yin P, Lee W-C, Lee D-L (2011) Exploiting geographical influence for collaborative point-of-interest recommendation. In: Proceedings of the 34th international ACM SIGIR conference on research and development in information—SIGIR'11, pp 325–334
6. Zheng VW, Zheng Y, Xie X, Yang Q (2010) Collaborative location and activity recommendations with GPS history data. In: Proceedings of the 19th international conference on World Wide Web—WWW'10, pp 1029–1038
7. Zheng VW, Zheng Y, Xie X, Yang Q (2012) Towards mobile intelligence: learning from GPS history data for collaborative recommendation. Artif Intell 184–185:17–37
8. Scellato S, Noulas A, Mascolo C (2011) Exploiting place features in link prediction on location-based social networks. In: KDD'11 Proceedings of the 17th ACM SIGKDD international conference on knowledge discovery and data mining, pp 1046–1054
9. Zheng Y, Zhang L, Ma Z, Xie X, Ma W (2011) Recommending friends and locations based on individual location history. ACM Trans Web, 5(1)

10. Xiao X, Zheng Y, Luo Q, Xie X (2010) Finding similar users using category-based location history. In: Proceedings of the 18th SIGSPATIAL international conference on advances in geographic information systems, GIS'10, no. 49, pp 442–445
11. Zheng Y, Wang L, Zhang R (2008) GeoLife: managing and understanding your past life over maps. In: 9th International conference on mobile data management, MDM'08, no. 49, pp 211–212
12. Zheng Y, Chen Y, Xie X, Ma W (2009) GeoLife2. 0: a location-based social networking service. In: 10th International conference on mobile data management: systems, services and middleware, MDM'09, no. 49, pp 357–358
13. Zheng Y, Xie X, Ma W (2010) GeoLife: a collaborative social networking service among user, location and trajectory. IEEE Data Eng Bull 49:32–40
14. Zheng Y, Xie X (2011) Learning travel recommendations from user-generated GPS traces ACM Trans Intell Syst Technol 2(1)
15. Liu H, Wei L-Y, Zheng Y, Schneider M, Peng W-C (2011) Route discovery from mining uncertain trajectories. In 2011 IEEE 11th international conference on data mining workshops, pp 1239–1242
16. Wei L-Y, Zheng Y, Peng W-C (2012) Constructing popular routes from uncertain trajectories. In Proceedings of the 18th ACM SIGKDD international conference on knowledge discovery and data mining—KDD'12, pp 195–203
17. Lian V, Xie X, Zhang F, Yuan NJ, Zhou T, Rui Y, Data B (2015) Mining location-based social networks: a predictive perspective. IEEE Data Eng Bull
18. Scellato S, Mascolo C (2012) Where online friends meet: social communities in location-based networks. In: 6th International AAAI conference weblogs social media, ICWSM
19. Brown C, Nicosia V, Scellato S (2013) Social and place-focused communities in location-based online social networks. Eur Phys pp 1–11
20. Brown C, Noulas A (2013) A place-focused model for social networks in cities. In: International conference on social computing, pp 75–80
21. Rösler R, Liebig T (2013) Using data from location based social networks for urban activity clustering. In: Geographic information science at the heart of Europe, Springer, pp 55–72
22. Cranshaw J, Hong JI, Sadeh N (2012) The livehoods project: utilizing social media to understand the dynamics of a city. In: ICWSM'12
23. Zhang A, Noulas A (2013) Hoodsquare: modeling and recommending neighborhoods in location-based social networks. Social computing pp 1–15
24. Noulas A, Scellato S, Mascolo C, Pontil M (2011) Exploiting semantic annotations for clustering geographic areas and users in location-based social networks. In: ICWSM Workshop the social mobile web 2011, Barcelona, Catalonia, Spain
25. Aubrecht C, Ungar J, Freire S (2011) Exploring the potential of volunteered geographic information for modeling spatio-temporal characteristics of urban population a case study for Lisbon Metro using foursquare check-in data. In: Potential of VGI for modeling population distribution, pp 11–13
26. Cho E, Myers S, Leskovec J (2011) Friendship and mobility: user movement in location-based social networks. In: Proceedings of the 17th ACM SIGKDD international conference on knowledge discovery and data mining, pp 1082–1090
27. Noulas A, Scellato S, Mascolo C, Pontil M, (2011) An empirical study of geographic user activity patterns in foursquare. In: Lada A. Adamic, Ricardo A. Baeza-Yates, Scott Counts (eds) ICWSM, The AAAI Press
28. Noulas A, Scellato S, Lambiotte R, Pontil M, Mascolo C (2012) A tale of many cities: universal patterns in human urban mobility. PLoS ONE 7(5):e37027
29. Cheng Z, Caverlee J, Lee K, Sui D (2011) Exploring millions of footprints in location sharing services. In: The 5th international AAAI conference on weblogs and social media, ICWSM
30. Wu W, Wang J (2015) Exploring city social interaction ties in the big data era: evidence based on location-based social media data from China. In: 55th Congress of the European Regional Science Association: world renaissance: changing roles for people and places, pp 25–28

31. Sarwat M, Eldawy A, Mokbel MF, Riedl J (2013) PLUTUS: leveraging location-based social networks to recommend potential customers to venues. In: 2013 IEEE 14th International conference mobile data management, pp 26–35
32. Karamshuk D, Noulas V, Scellato S (2013) Geo-Spotting : mining online location-based services for optimal retail store placement, pp 1–22
33. Foursquare [Online]. Available: https://foursquare.com/
34. Fousquare Categories [Online]. Available: https://developer.foursquare.com/docs/explore#req=venues/categories
35. Nina S, Lam N (1983) Spatial interpolation methods: a review. Am Cartogr 10(2):129–150
36. Burrough PA (1986) Principles of geographical information systems for land resources assessment. Oxford University press
37. McLaughlin D, Reid LB, Li SG, Hyman J (1993) A stochastic method for characterising groundwater contamination. Groundwater 31(2):237–49
38. Tobler AWR (1970) A computer movie simulating urban growth in the Detroit region. Econ Geogr 46:234–240
39. Shepard D (1968) A two-dimensional interpolation function for irregularly-spaced data. In The 1968 ACM National Conference, pp 517–524
40. MapQuest API [Online]. Available: http://developer.mapquest.com/
41. Li H, Calder CA, Cressie N (2007) Beyond Moran's I: testing for spatial dependence based on the spatial autoregressive model. Geogr Anal 39(4):357–375
42. A L (1988) Spatial econometrics: methods and models. Dordrecht: Kluwer Academic Publishers

Intelligent Prediction of Firm Innovation Activity—The Case of Czech Smart Cities

Petr Hajek and Jan Stejskal

1 Introduction

Innovation activity is one of the key factors currently affecting the competitiveness of entrepreneurs. In the last ten years, new production factors have become a significant source of competitive advantage—knowledge, the ability to learn and creativity. All these factors necessarily lead to the increase of innovation activity and the creation of innovations. This is possible to achieve by involving economic entities in cooperative chains. Then, the so-called knowledge spillover effects become a side effect of any type of described cooperation with a knowledge base. Smart cities provide the most suitable environments of open and user-driven innovation, where resources can be shared with the aim of establishing urban and regional innovation ecosystems [1].

Nonlinear models of innovation have recently been introduced to take interactive and recursive terms into account. However, far too little attention has been paid to predicting innovation activity using nonlinear models [2]. We aim to fill this gap and employ ensembles of decision trees, demonstrating that significantly more accurate predictions can be achieved. This is an extended version of [3]. Here we develop a general model for intelligent prediction of firm innovation activity in the context of a smart city. In addition, we perform a comparative study across various meta-learning classifiers, including boosting, rotation forest, dagging, and bagging.

P. Hajek (✉)
Faculty of Economics and Administration, Institute of System Engineering and Informatics, University of Pardubice, Studentska 95, CZ53210 Pardubice, Czech Republic
e-mail: petr.hajek@upce.cz

J. Stejskal
Faculty of Economics and Administration, Institute of Economic Sciences, University of Pardubice, Studentska 95, CZ53210 Pardubice, Czech Republic
e-mail: jan.stejskal@upce.cz

L. Ismail and L. Zhang (eds.), *Information Innovation Technology in Smart Cities*,
https://doi.org/10.1007/978-981-10-1741-4_9

The remainder of this paper is structured in the following way. Section 2 briefly reviews related literature. Section 3 describes the prediction model and data. Section 4 shows the experiment results and Sect. 5 provides concluding remarks and implications for innovation management.

2 Related Literature Background

Innovative activities are currently one of the crucial sources of competitive advantage in every developed economy. Knowledge and technological processes are determinants of all smart innovation activities [4]. This follows from endogenous growth models [5]. However, these determinants do not develop only innovative activities, but also the regions (environments) in which the firms are located. Their abilities to innovate depend on spatial and social proximity. It was shown that in the spatial context this implies that local growth depends on the amount of technological activity which is carried out locally and on the ability to exploit external technological achievements through information spill-overs [6]. It has been demonstrated that the existence of spill-over effects supports a growing number of innovations in the form of patents [7–10].

Other studies suggest that innovative activities depend on their ability to obtain information and knowledge, ability to be an effective part of the knowledge network and to be able to cooperate. Numerous studies (reviewed in [11]) confirm that spatial proximity facilitates learning processes. These processes are often influenced by knowledge spill-overs effects and by acquisition of crucial knowledge (sticky knowledge). Capello and Lenzi [11], in this context, draws attention to two issues. The first concerns whether the knowledge spill-over effects are more intense in intra-industry or inter-industry exchange of knowledge [12, 13]. On this question, there is no clear answer. Various studies come to different answers. The second problem is the spatial range of knowledge spill-overs (importance of relational capital in innovation activity). Ellison and Glaeser [14], Porter [15], Von Hipple [16] are the representatives of the first "industrial dynamic approach" and Camagni [17] is main representative of the second "spatial-relational approach".

As a specific kind of spill-over effect that was defined is the knowledge transfer between firms and universities [18]. Universities are a source of new knowledge, applications and knowledge production. Due to frequent public subsidies of the research, the positive externalities are formatted in greater extent. Mainly the businesses are consumers of this type of externality (there are two basic forms: passive, which includes journal, research papers, conference, students practices and trainings; and active, which includes training for employees, engaging in knowledge networks and cooperative chains). These forms of cooperation generated so called technological externalities, and if there is a transfer of knowledge (especially tacit knowledge), we can observe also knowledge effects called as knowledge spill-over. If the firms are recipients of externality, they can transform the new knowledge into their economic processes, economic value and increase their competitive advantage. If the universities and R&Ds are the recipients of the externality, the new knowledge can strengthen and extend their knowledge base, research and technology absorption and ability to generate new knowledge in future.

Some scholars pointed out in the conclusions of their researches that the synergistic effects (resulted from cooperation) can significantly increase the efficiency of knowledge processes. It must be admitted that these effects are strictly localized and take place only there where the knowledge production occurs in the form of innovation output. Capello [19] added that the kind of learning mechanisms envisaged in this theory which enhance innovative creativity is a collective learning. This learning process (in this case) is complemented by socializing, creative knowledge from external sources (outside the firm but within region). Knowledge, having thus arisen, can be labelled in accordance with Buchanan's theory—the club goods; subjects of the knowledge chain or regional innovation system are members of this club.

The spill-over effects can occur when the necessary milieu where collective learning takes place exists. High mobility of well skilled and motivated forces (but only in a given area) is one of the determinants of collective learning. This can help accelerate the knowledge transfer and help set the atmosphere of trust. Thus, the relationships can be quickly strengthened. The lock-in problem can be one of the risks in this situation. The stable group of subjects belonging to knowledge (production) chain is the second determinant. In the cooperation chain the knowledge transfer and acquisition are realized in unawares (tacit) way. The actors of transfer processes are: firms and customers, suppliers, public entities, agents, universities, R&Ds etc. We can see the formation of a special kind of demand from customers and consumers towards innovations. These recipients are willing to partly participate on new knowledge creation. The social milieu is the third determinant. It creates an environment in which knowledge processes take place. It depends on morality and general confidence in the society [20]. This social environment is sometimes considered as part of an innovative milieu, environment for innovation, which forms the basis for efficient production of the innovation.

The last 20 years are the innovative milieu often confronted with the emerging knowledge platforms in regions—regional innovation systems—which are made by political decisions, pursue the political aims and are often financed with public funds [21]. Many studies have shown that systems supporting the formation of human capital, interpersonal networks, specialized and skilled labour markets, local governance systems; therefore they are highly selective in spatial terms and require ad hoc local policy interventions to be adequately supported [22]. Thanks to the smart specialization approach, the inadequacy of a 'one-size-fits-all' policy for innovation at regional level is decisively transferred from the scientific literature into the institutional debate [20].

Smart innovation realized in spatial context (cities or regions) requires components to rank smart cities, but to create a framework that can be used to characterize how to envision a smart city and design initiatives, which advance this vision by implementing shared services, and navigating their emerging challenges [23].

Drivers that enhance the innovative capability of smart industries are (1) management and organization, (2) technology, (3) governance, (4) policy, (5) people and communities, (6) the economy, (7) built infrastructure, and (8) the natural environment. An extensive review of studies of individual determinants is shown in [21].

3 Model Design and Data Processing

According to the discussion above, the model of the intelligent prediction of firm innovation activity can be formally described in the following way:

$$Y = f(X_1, X_2, X_3, X_4, X_5, X_6, X_7, X_8, X_9, X_{10}), \tag{1}$$

where X_1 (internal knowledge spillovers), X_2 (market knowledge spillovers), X_3 (university and research lab knowledge spillovers), X_4 (other external knowledge spillovers), X_5 (internal R&D expenditure), X_6 (local and regional financial support), X_7 (EU financial support), X_8 (collaboration on innovations), X_9 (the presence of a university having the same discipline as the firm in the city) and X_{10} (sales—representing firm size) are causal predictors of Y (innovation activity). All variables were numeric except for the binary variables X_8 (0—no collaboration, 1—collaboration with enterprises/universities on innovation) and X_9 (0—no university in the city; 1—a university in the city). All above mentioned activities are realized at the local (city) level and they are considered key characteristics of smart cities in human conceptualization (also known as learning or knowledge city [23]). In this concept, education, learning and knowledge have central importance to smart city. In fact, it is heavily related to knowledge economy, stressing innovation as a main instrument used to nurture the knowledge and drive the knowledge-based urban development [24].

In our empirical analysis, we used primary data from the CIS—Community Innovation Survey—obtained from the Czech Statistical Office of the Czech Republic (source of data), see http://ec.europa.eu/eurostat/web/microdata/community-innovation-survey for the description of the data. The data are part of the EU science and technology statistics. The CIS questionnaire provides data broken down by the type of innovators, economic activities and size classes. The same questionnaire was used for all EU Members States [25]. The research was focused only on the firm level in the Czech Republic during the period of 2008–2010 (the surveys are carried out with two years' frequency). The questionnaire combined sample (stratified random sampling) and exhaustive surveys with a response rate greater than 60%. For our analysis overall, we gathered data for a total of 5151 firms with more than 10 employees. In accordance with the goal of this paper, we selected 523 companies, i.e., only companies from the chemical industry, for our data group—specifically, countries covering NACE (Statistical classification of economic activities in the European Community) categories 20–23

(20—Manufacture of chemicals and chemical products, 21—Manufacture of basic pharmaceutical products and pharmaceutical preparations, 22—Manufacture of rubber and plastic products, and 23—Manufacture of other non-metallic mineral products). We chose chemical industry because it is considered to be science driven, leading to highly innovative results. Chemical industry also claims a significant percent of the Czech Republic's manufacturing output and is, therefore, regarded as a key industry in smart specialization strategies in Czech regions and cities.

The collected data were related to both the input variables $X_1 \ldots X_{10}$ and output variable Y. Regarding knowledge spillovers, the companies were asked to assign importance (on a scale of 0–3, i.e. 0—not used, 1—low, 2—medium, 3—high) to communication sources (Question 6.1 of the CIS questionnaire): (1) internal, (2) market (suppliers, customers, competitors and consultants), (3) institutional (universities and research institutes) and (4) other sources (scientific journals, conferences, professional associations and the Internet). For those with more than one source, we averaged the values.

Internal R&D expenditure was calculated as the sum of expenditure for: (1) in-house R&D, (2) purchase of external R&D, (3) acquisition of machinery, equipment, and software, and (4) acquisition of external knowledge (Question 5.2 of the CIS questionnaire). Regarding the external financial support, the firms were asked whether they received any public financial support for innovation activities from the selected levels of government (Question 5.3 of the CIS questionnaire). We utilized two most frequent levels of financial support, from local or regional authorities (1 for government and local/regional support, 0.5 for either government or local/regional support, and 0 for no support), and the EU (1 for EU support and participation in the EU 7th Framework Programme, 0.5 for EU support only, 0 for no support) respectively. Question 6.2 was used to measure the collaboration on innovations. The firms were asked whether they collaborated on any of their innovation activities with other firms or institutions (1 for yes, 0 for no).

The company's innovation activity was determined according to whether the company introduced a new or significantly improved product (goods or services) or process (a production process, distribution method or supporting activity) onto the market (Questions 2.1 and 3.1 of the CIS questionnaire, 1 for yes and 0 for no). Out of the 523 companies, 276 companies (52.8%) were innovative and 247 (47.2%) were non-innovative. Before further processing, we rescaled all input variables to fit the [0, 1] range using normalization procedure: $X_{norm} = (X - X_{min})/(X_{max} - X_{min})$. About 29.9% of the firms had missing data. This ratio was similar for both discrete (X_8) and continuous variables ($X_1 \ldots X_7$). Data for X_9 (discrete) and X_{10} (continuous) variables were complete. To replace the missing data, we employed a multiple imputation scheme based on the fully conditional specification method [26]. Basic descriptive statistics of the final dataset is provided in Table 1. We further tested the differences between non-innovative and innovative companies using Student's paired t-test [27] for continuous ($X_1 \ldots X_7$, X_{10}) and Pearson's χ^2 test [28] for categorical data (X_8 and X_9).

Table 1 Normalized importance of input variables

Input variable	Non-innovative	Innovative	*t*-statistics
X_1	0.681	0.815	5.856[a]
X_2	0.538	0.563	1.406
X_3	0.225	0.232	0.368
X_4	0.379	0.457	4.695[a]
X_5	0.059	0.169	6.049[a]
X_6	0.051	0.125	4.771[a]
X_7	0.142	0.128	−0.627
X_8	241 (No)/6 (Yes)	127 (No)/149 (Yes)	$\chi^2 = 180.0$[a]
X_9	53 (No)/223 (Yes)	70 (No)/177 (Yes)	$\chi^2 = 23.7$[a]
X_{10}	0.047	0.167	7.615[a]

[a]Significantly different at $p = 0.05$

4 Empirical Experiments

The prediction of innovation activity was conducted using an ensemble of decision trees. In this method, a number of independent decision trees are generated to make the overall prediction. More specifically, we employed the bagging procedure, which forms multiple bootstrap replicates of training data and uses them as new training sets [29, 30]. Thus, higher accuracy can be achieved than with single decision trees [31]. This was also true for our data set. In addition, we compared the results using the boosting technique which, in contrast, generates a series of dependent trees [32]. To avoid overfitting, all experiments were performed using 10-fold cross-validation. The single decision tree was pruned to minimum cross-validation error, the minimum size of the node to split was set at 10 and the maximum tree level was set at 10. The boosting algorithm was set as follows: pruning trees in a series of a minimum of 10, the maximum number of trees in a series = 400, the depth of individual trees = 5, the minimum size of the node to split = 10, the proportion of rows for each tree = 0.5 and the influence trimming factor = 0.01. Rotation forest was trained using the following values of parameters: minimum and maximum size of the group = 3, the number of iterations = 10, the percentage of instances to be removed = 50, and projection filter = PCA (principal component analysis). Dagging was performed with 10 folds used for splitting the training sets into smaller chunks for the base classifier. Finally, bagging was performed using 200 trees in the forest, the minimum size of the node to split was set at 2 and the maximum tree levels were set at 50. The setting of the parameters was obtained using grid search strategy. All experiments were performed in the DTREG software.

The prediction performance was measured by the commonly used 2-class criteria (here class 1 for innovative, class 0 for non-innovative). Note that innovative firms can be classified as either innovative (true positive) or non-innovative (false negative). Similarly, non-innovative firms can be classified as either non-innovative (true negative) or innovative (false positive). We used the following criteria based on these classifications:

$$\text{accuracy} = (\text{TP} + \text{TN})/(\text{TP} + \text{FP} + \text{FN} + \text{TN}), \quad (2)$$

$$\text{sensitivity} = \text{TP}/(\text{TP} + \text{FN}), \quad (3)$$

$$\text{specificity} = \text{TN}/(\text{TN} + \text{FP}), \quad (4)$$

$$\text{positive predictive value} = (\text{PPV}) = \text{TP}/(\text{TP} + \text{FP}), \quad (5)$$

$$\text{negative predictive value} = (\text{NPV}) = \text{TN}/(\text{TN} + \text{FN}), \quad (6)$$

where TP is true positive, TN is true negative, FP is false positive and FN is false negative. Note that alternative measures can be calculated, such as an F-measure that combines sensitivity, specificity and many other factors.

5 Results

The results of the prediction in Table 2 show that all methods perform well in both classes (innovative/non-innovative chemical firms).

The highest accuracy (93.04%) was achieved by the bagging algorithm, which also outperformed other methods in terms of the remaining prediction performance indicators, except for specificity. This was also confirmed by an additional measurement, the area under the ROC curve (AUC) (combining true positive rate (sensitivity) and false positive rate), see Figs. 1 and 2. Although boosting performed better in class 0 (non-innovative chemical firms, see “Specificity” in Table 2), overall prediction performance was better for the bagging procedure (ROC = 0.970), which also predicted class 1 with an accuracy of 93.84%. Taken together, bagging performed the best among the methods compared; therefore, we further examined the effects of the input variables using this model.

Table 2 Prediction performance of decision tree models

Measure	Algorithm				
	Single	Boosting	Rotation forest	Dagging	Bagging
Accuracy [%]	89.74	89.91	90.39	82.01	93.04[b]
Sensitivity [%]	88.77	87.27	90.06	85.80	93.84
Specificity [%]	90.74	92.59	90.71	78.64	92.22
PPV[a] [%]	90.74	92.31	90.01	78.90	93.61
NPV [%]	88.77	87.72	90.06	85.80	93.05

[a]PPV is positive predictive value and NPV is negative predictive value
[b]Significantly higher accuracy at $p = 0.05$ using a paired student’s t-test

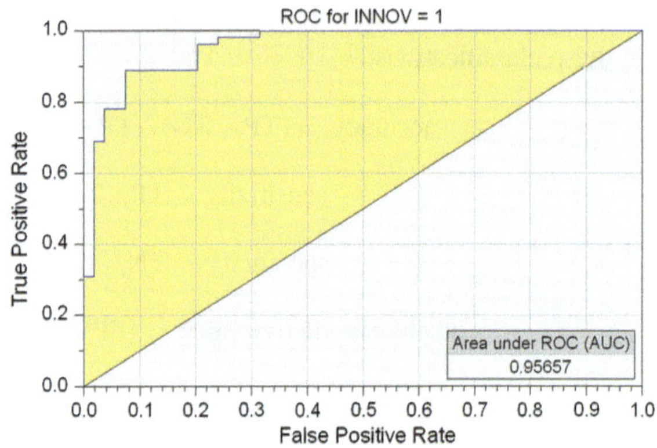

Fig. 1 ROC curves for boosting algorithm

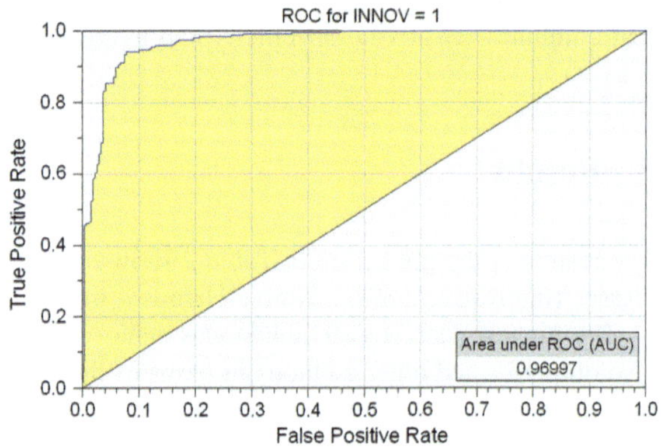

Fig. 2 ROC curves for bagging algorithm

The appendix shows the single decision tree, providing 89.76% accuracy. Although this single model was outperformed in our experiments, it provides some useful insight into the decision-making mechanisms of the firms under investigation.

The model presented shows that when companies cooperate, they then create innovations at a probability rate of greater than 96%. When companies do not cooperate, their innovation ability depends on internal spillover effects. When they are able to acquire and use internal information for generating innovation in large amounts, it was demonstrated that these activities lead unequivocally to the creation of innovation. When only a low ability to generate internal spillover effects is seen in non-cooperating companies, there is only a negligible probability that they will be able to innovate.

Non-cooperating companies with a high rate of internal spillover effects are then additionally influenced by their ability to acquire information and findings from institutional sources (universities, R&D organizations). It was demonstrated that these sources are not effectively used in the investigated group of chemical companies, thus they do not contribute to a significant increase in their innovation capabilities. When these companies acquire information and knowledge from other sources (conferences, publications, etc.), it contributes to their innovation capability more than institutional sources.

To obtain more in-depth insight into the importance of each input attribute, we performed sensitivity analysis on the model by calculating the relative importance of the input attributes using a range of 0–100. The reason for this is that some of the input attributes may seem to be unimportant in the decision trees due to the fact that there may exist a slightly better (usually correlated) splitting attribute used in the corresponding node. Thus, the importance of some input attributes may be masked. Therefore, attribute importance was computed by adding up the improvement in classification gained by each split that used a certain predictor. Thus, we were able to identify the relative importance of input attributes as opposed to in the regression tree, where splits closer to the root of the tree are typically more important.

Table 3 shows that internal knowledge spillovers were the most important determinant of the chemical firms' innovation activity during the period monitored. Other knowledge spillover effects were also important, in particular those generated by other sources (scientific journals, conferences, etc.).

Further, R&D intensity (proxied by internal expenditure), collaboration on innovation and firm size were also important determinants with a relative importance >65. In contrast, financial support from both the government and the EU was a relatively weak predictor. Finally, the weakest influence was associated with the presence of a chemical university in the region.

Wang and Chien [2] reported that neural networks outperform traditional linear models in innovation performance prediction. Therefore, we performed further experiments to compare the results of the ensembles of decision trees with Naïve Bayes (NB), neural networks and support vector machines, respectively. Specifically, we used two most common neural network architectures are represented by multilayer perceptron (MLP) and radial basis function (RBF) neural networks. Support vector machine (SVM), on the other hand, is the most commonly applied machine learning method. In contrast to empirical risk minimization by

Table 3 Normalized importance of input variables

Input variable	Importance
X_1 Internal knowledge spillovers	100.00
X_2 Market knowledge spillovers	51.74
X_3 University and research lab knowledge spillovers	63.11
X_4 Other external knowledge spillovers	81.49
X_5 Internal R&D expenditure	70.53
X_6 Local and regional financial support	32.32
X_7 EU financial support	31.98
X_8 Collaboration on innovations	65.35
X_9 The presence of a chemical university in the region	16.83
X_{10} Sales—representing firm size	76.48

Table 4 Prediction performance of decision tree models

Measure	Algorithm				
	NB	MLP	RBF	SVM	DT bagging
Accuracy [%]	69.51	78.20	84.80	83.88	93.04[a]
Sensitivity [%]	87.29	77.54	82.97	78.99	93.84
Specificity [%]	53.62	78.89	86.67	88.89	92.22
PPV [%]	62.89	78.97	86.42	87.90	93.61
NPV [%]	87.29	77.45	83.27	80.54	93.05
ROC	0.805	0.814	0.937	0.907	0.970

[a]Significantly higher accuracy at $p = 0.05$ using a paired student's t-test

neural networks, SVM represents a structural risk minimization approach which has shown better generalization performance.

MLP employed here was a feed-forward neural network with one hidden layer of neurons operating with sigmoid activation functions. MLPs were trained using conjugate gradient algorithm, where the number of neurons in the hidden layer was detected automatically from interval {2, 3, …, 20} with the step set to 1 and maximum steps without change set to 4. Maximum number of iterations was 10000 and iterations without improvement were set to 100. Again, grid search procedure was used to find the best settings of the parameters of the MLP (RBF neural network and SVM).

In the RBF neural network, Gaussian activation functions were used by neurons in the hidden layer. The optimum number of neurons and other training parameters of the RBF neural networks were detected using a genetic algorithm (with population = 200 and generations = 20), where the maximum number of neurons in the hidden layer was set to 100, and the radius of RBF ranged from 0.01 to 400.

The SVM method was trained by the sequential minimal optimization (SMO) [33] using RBF kernel function with parameters found using grid search algorithm, where complexity parameter C ranging from 0.1 to 50000, radius of RBF function ranging from 0.001 to 20, and 10 intervals were set for the grid search algorithm. Linear and polynomial kernel functions were also tested without improvement.

Table 4 shows that the bagging algorithm using decision trees significantly outperformed neural networks and SVM, respectively.

6 Conclusion

This report analyzes the prediction of the innovation activity of chemical firms using an ensemble of decision trees. It offers evidence of the importance of individual determinants for the creation of innovation. On the basis of the analysis, it is possible to state that internal knowledge spillovers are the crucial determinant of the innovation activity of chemical companies. Therefore, we argue that internal knowledge spillovers are among the factors with the greatest influence on the creation of spillover effects; next, in order of their relative importance to internal knowledge spillover effects, are: external knowledge spillovers, sales (representing firm size) and internal R&D expenditure.

As one of its outputs, our paper offers a proposal for evaluating the effectiveness of public financial support expended on innovation activity [34]. Specifically, we demonstrated the low influence of any type of financial stimulus and support (from EU funds to local budgets). Their importance represents only a third of those mentioned above. This finding corroborates the arguments against public financial support expressed by Arrow [35] (according to the classical market failure argument, firms will not invest (enough) in R&D, because the benefits of innovation activities cannot be fully reaped due to incomplete appropriability and knowledge spillovers between firms). Even despite this, all OECD countries are currently spending significant amounts of public money on programs intended to stimulate innovation activity [36]. However, not much is known about the significance and long-term effects of public financial support.

Other evidence was obtained from the constructed decision tree, in which the determinants of cooperation and the sources of key knowledge determining a company's innovation activity are modeled. It was demonstrated that, to a certain degree, internal spillover effects can replace cooperation on the creation of innovation, which is sometimes impossible or out of the question in practice. For such cases, it is necessary to maximize the percent of internal spillovers. However, the acquisition of knowledge directly from the knowledge sector is not an effective method; rather, it is more effective to acquire necessary knowledge from other sources.

It should be noted that, although we were able to obtain significantly higher accuracy by bagging decision trees, this was achieved at the cost of complexity, because 200 trees were generated in the forest. Therefore, in the future, we suggest using not only alternative machine learning and soft computing techniques such as evolutionary rule-based systems, fuzzy rule-based systems, etc., but we also encourage further studies using the bias-variance trade-off to study the prediction of innovation activity. Finally, we recommend including additional input variables associated with both business cluster initiatives [37] and regional innovation systems [38, 39].

Acknowledgements This article was created as part of the resolution of the research task No. 14-02836S entitled "Modelling of knowledge spill-over effects in the context of regional and local development", financially supported by the Grant Agency of the Czech Republic.

Appendix: Single Decision Tree

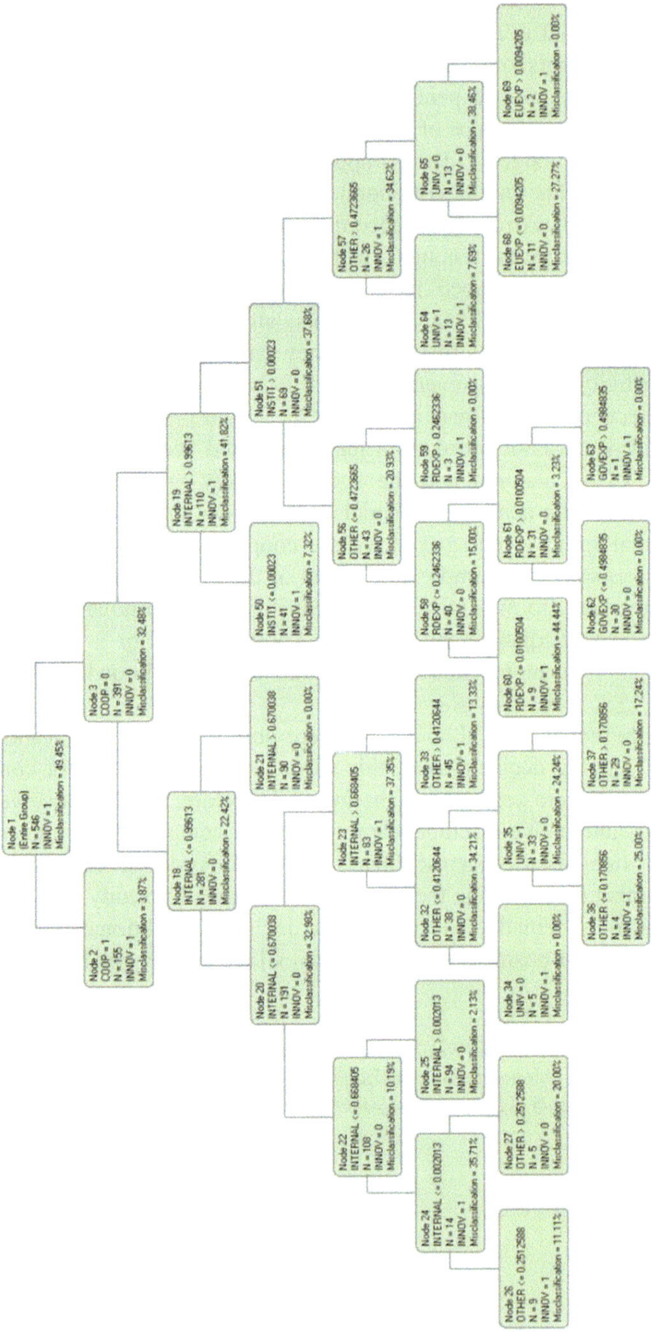

References

1. Schaffers H, Komninos N, Pallot M, Trousse B, Nilsson M, Oliveira A (2011) Smart cities and the future internet: towards cooperation frameworks for open innovation. Future Internet Assembly 6656:431–446
2. Wang TY, Chien SC (2006) Forecasting innovation performance via neural networks—a case of Taiwanese manufacturing industry. Technovation 26(5):635–643
3. Hajek P, Stejskal J (2015) Predicting the innovation activity of chemical firms using an ensemble of decision trees. In: Proceedings of the 11th international conference on innovations in information technology (IIT), IEEE, pp 35–39
4. Chourabi H, Nam T, Walker S, Gil-Garcia JR, Mellouli S, Nahon K, Scholl HJ (2012) Understanding smart cities: an integrative framework. In: System science (HICSS), pp 2289–2297
5. Romer PM (1990) Endogenous technological change. J Polit Econ 98:71–102
6. Moreno R, Paci R, Usai S (2005) Spatial spillovers and innovation activity in European regions. Environ Plan A 37(10):1793–1812
7. Glaeser E, Kallal H, Scheinkman J, Shleifer A (1992) Growth of cities. J Polit Econ 100:1126–1152
8. Nieto MJ, Santamaria L (2007) The importance of diverse collaborative networks for the novelty of product innovation. Technovation 27(6):367–377
9. Frenz M, Ietto-Gillies G (2009) The impact on innovation performance of different sources of knowledge: evidence from the UK Community Innovation Survey. Res Policy 38(7):1125–1135
10. Gallego J, Rubalcaba L, Suárez C (2013) Knowledge for innovation in Europe: the role of external knowledge on firms' cooperation strategies. J Bus Res 66(10):2034–2041
11. Capello R, Lenzi C (2012) Knowledge, innovation and economic growth: spatial heterogeneity in Europe. Growth Change 43(4):697–698
12. Abramovsky L, Kremp E, López A, Schmidt T, Simpson H (2009) Understanding co-operative innovative activity: evidence from four European countries. Econ Innov New Technol 18(3):243–265
13. Becker W, Dietz J (2004) R&D cooperation and innovation activities of firms—evidence for the German manufacturing industry. Res Policy 33(2):209–223
14. Ellison G, Glaeser E (1999) The geographic concentration of industry: does natural advantage explain agglomeration? Am Econ Rev 89:311–316
15. Porter M (1990) The competitive advantage of nations. Free Press, New York
16. Von Hipple E (1994) Sticky information and the locus of problem solving: implications for innovation. Manage Sci 40:429–439
17. Camagni R (1999) The city as a milieu: applying the GREMI approach to urban evolution. Révue d'Economie Régionale et Urbaine 3:591–606
18. Anselin L, Varga A, Acs Z (1997) Local geographic spillovers between university research and high technology innovations. J Urban Econ 42(3):422–448
19. Capello R (2002) Spatial and sectoral characteristics of relational capital in innovation activity. Eur Plan Stud 10(2):177–200
20. Camagni R, Capello R (2013) Regional innovation patterns and the EU regional policy reform: toward smart innovation policies. Growth Change 44(2):355–389
21. Hottenrott H, Lopes-Bento C (2014) (International) R&D Collaboration and SMEs: the effectiveness of targeted public R&D support schemes. Res Policy 43(6):1055–1066
22. Camagni R, Maillat D (1995) Milieux innovateurs. Théories et Politiques, Economica, Paris
23. Nam T, Pardo TA (2011) Smart city as urban innovation: focusing on management, policy, and context. In: Proceedings of the 5th international conference on theory and practice of electronic governance. ACM, pp 185–194
24. Hajkova V, Hajek P (2014) Efficiency of knowledge bases in urban population and economic growth–evidence from European cities. Cities 40:11–22

25. Tether B (2001) Identifying innovation, innovators and innovative behaviours: a critical assessment of the community innovation survey (CIS). Centre for Research on Innovation and Competition, University of Manchester
26. Van Buuren S (2007) Multiple imputation of discrete and continuous data by fully conditional specification. Stat Methods Med Res 16(3):219–242
27. Zimmerman DW (1997) Teacher's corner: a note on interpretation of the paired-samples t test. J Educ Behav Stat 22(3):349–360
28. Cochran WG (1952) The χ^2 test of goodness of fit. Ann Math Stat 23(3):315–345
29. Breiman L (1996) Bagging predictors. Mach Learn 24(2):123–140
30. Hajek P, Olej V (2014) Predicting firms' credit ratings using ensembles of artificial immune systems and machine learning—an over-sampling approach. In: Iliadis L, Maglogiannis I, Papadopoulos H (eds) Artificial intelligence applications and innovations. Springer, Berlin, pp 29–38
31. Dietterich TG (2000) Ensemble methods in machine learning. Multiple classifier systems. Lect Notes Comput Sci 1857:1–15
32. Freund Y, Schapire RE (1995) A desicion-theoretic generalization of on-line learning and an application to boosting. Computational learning theory. Lect Notes Comput Sci 904:23–37
33. Platt JC (1999) Fast training of support vector machines using sequential minimal optimization. In: Advances in Kernel methods. MIT Press, Cambridge
34. Stejskal J, Hajek P (2015) The influence of public expenditure on innovation activity in Czech manufacturing industry. In: Proceedings of the 25th international business information management association conference—innovation vision 2020: from regional development sustainability to global economic growth, IBIMA 2015, pp 1820–1827
35. Arrow KJ (1962) Economic welfare and the allocation of resources for invention. In: Nelson R (ed) The rate and direction of inventive activity. Princeton University Press, Princeton, pp 609–625
36. Klette TJ, Mren J, Griliches Z (2000) Do subsidies to commercial R&D reduce market failures? Microeconometric evaluation studies. Res Policy 29(4):471–495
37. Stejskal J, Hajek P (2012) Competitive advantage analysis: a novel method for industrial clusters identification. J Bus Econ Manage 13(2):344–365
38. Matatkova K, Stejskal J (2013) Descriptive analysis of the regional innovation system-novel method for public administration authorities. Transylv Rev Adm Sci 39:91–107
39. Hajek P, Henriques R, Hajkova V (2014) Visualising components of regional innovation systems using self-organizing maps—evidence from European regions. Technol Forecast Soc Chang 84:197–214

Part III
Accelerating the Evolution of Smart Cities with the Internet of Things and Internet of People

Emotion Identification in Twitter Messages for Smart City Applications

Dario Stojanovski, Gjorgji Strezoski, Gjorgji Madjarov and Ivica Dimitrovski

1 Introduction

Social media have gained increasing popularity over the recent years with the number of users on social networks and microblogging platforms growing rapidly. As of 2016, Twitter has over 310 million monthly active users and generates over 500 million messages per day.[1] Twitter messages are also known as tweets and have 140 character limitation. Tweets contain informal language, users use a lot of abbreviations, URLs, emoticons and Twitter specific symbols, such as hashtags and targets (user mentions).

Emotion identification and sentiment analysis have recently spiked the interest of both, academia and industry with the exponential growth of social media. Detecting users' reaction towards certain products and services can provide valuable insight for companies offering them. Additionally, it can be used to get information of the public opinion against different topics and events and many other potential use-cases. As a result there are many proposed methods for solving this task.

Twitter emotion analysis can have many potential smart city applications. In previous work [1], we presented a use-case scenario where we apply our emotion

[1]https://about.twitter.com/company.

D. Stojanovski (✉) · G. Strezoski · G. Madjarov · I. Dimitrovski
Faculty of Computer Science and Engineering, Ss. Cyril and Methodius University, Rugjer Boshkovikj 16, 1000 Skopje, Republic of Macedonia
e-mail: stojanovski.dario@gmail.com

G. Strezoski
e-mail: strezoski.g@gmail.com

G. Madjarov
e-mail: gjorgji.madjarov@finki.ukim.mk

I. Dimitrovski
e-mail: ivica.dimitrovski@finki.ukim.mk

L. Ismail and L. Zhang (eds.), *Information Innovation Technology in Smart Cities*,
https://doi.org/10.1007/978-981-10-1741-4_10

identification model to Twitter messages from the 2014 FIFA World Cup in Brazil. We provided an extensive analysis of emotional distribution detected for the duration of the matches being considered in our work. Emotion identification in sports tweets can be utilized for different applications. An emotion identification system can detect if a game has caused a lot of anger or other negative emotions amongst the fans, so the official organizers can be warned to take extra security measures after and during the match. Additionally, results from such system can be used to identify future potentially critical matches in terms of security.

Another potential smart city application of Twitter emotion analysis is in relation to local public services. Several works study how Twitter is used in this context. Sobaci and Karkin [2] present how mayors in Turkey use the social network to offer better public service. In Agostino [3] and Flores and Rezende [4] on the other hand, they analyze how citizens use Twitter to engage in public life and decisions. They showcase that the platform is heavily used by both, the government and the citizens, and can be utilized to improve communication between them in order to offer better services.

In the work presented in this paper, we showcase a deep learning system for emotion identification in Twitter messages. For this purpose, we use a convolutional neural network with multiple filters and varying window sizes which has not been adequately studied for the task of emotion detection. The approach is founded on the work of Kim [5] that reported state-of-the-art results in 4 out of 7 sentence classification tasks. We leverage pre-trained word embeddings, obtained by unsupervised learning on a large set of Twitter messages [6]. The network is trained on automatically labeled tweets in respect to 7 emotions. Additionally, we present a use-case scenario for smart city applications that leverages local government related messages. We apply our model for emotion identification in tweets related to local government projects and municipality issues.

The rest of the paper is organized as follows. Section 2 outlines current approaches to emotion identification. In Sect. 3, we present the proposed system architecture consisted of a convolutional neural network and the necessary pre-training. We give a detailed overview of the used dataset, the conducted experiments and the achieved results in Sect. 4. In Sect. 5, we present a use-case where we apply our system for emotion identification in tweets related to local government. Finally, we conclude our work in Sect. 6.

2 Related Work

There has been a lot of work done in the field of emotion identification. Current approaches mainly are based on unigrams, bigrams, Part-of-Speech tags and other hand-crafted features and machine learning algorithms such as Support Vector Machines (SVM), Naive Bayes and maximum entropy.

Appropriate labeling of tweets with corresponding emotions still poses a challenge in the field. Roberts et al. [7] used a manually annotated set of tweets with 7 labels for emotion. Their approach for classification consisted of hand-crafted features such as unigrams, bigrams, indicators of exclamation and question marks, WordNet hypernyms and several other features and a SVM classifier. Balabantaray et al. [8] also used manually labeled tweets and developed a system for emotion classification using hand-crafted features such as Part-of-Speech tags, unigrams, bigrams and others and the Word-net Affect emotion lexicon. Their work presents an extensive analysis of the effect different combinations of features have on performance.

Purver and Battersby [9] and Wang et al. [10] on the other hand, use distant supervision for automatic emotion annotation. By using well-known indicators of emotional content they were able to create noisily labeled datasets. Tweets were annotated by the presence of emotion-related hashtags. Wang et al. [10] applied additional heuristics to improve on the quality of the acquired dataset. They only considered messages where the emotion-related hashtag is at the end of the tweet, removed messages containing URLs and quotations and discarded tweets with more than 3 hashtags, non-English messages and retweets. Furthermore, the quality of the dataset was evaluated by randomly sampling a small number of tweets for manual inspection by two annotators. They received 95.08% precision on the development and 93.16% on the test set, where a tweet was manually checked whether the assigned label is relevant to the conveyed emotion. The dataset is publicly available[2] and we build our work on portion of the data.

Sintsova et al. [11] used the Amazon Mechanical Turk (AMT) to build a human-based lexicon. Using the annotators' emotionally labeled tweets, they constructed a linguistic resource for emotion classification. Their approach is able to capture up to 20 distinct fine-grained emotions.

Unlike the previously depicted approaches that use extensive feature engineering which can be both, time-consuming and produce over-specified and incomplete features, our approach is based on a deep learning technique. Deep learning approaches handle the feature extraction task automatically, potentially providing for more robust and adaptable models. Most of the current work focuses on utilizing convolutional neural networks. Collobert et al. [12] proposed a unified neural network architecture that can be applied to various Natural Language Processing (NLP) tasks including sentiment analysis. dos Santos and Gatti [13] proposed a CNN for identifying sentiment analysis by exploiting character-level, word-level and sentence-level information. Kim [5] on the other hand, proposed an approach for sentence classification including sentiment analysis using a CNN with multiple filters and feature maps. This work also showed that continuously updating pre-trained word embeddings provides for better performance. In our work, we build on the aforementioned deep learning techniques. However, these approaches have been used for sentiment analysis, which is limited to classifying tweets with three labels, positive, negative and neutral. In this paper, we use convolutional neural network for a finer grained classification into 7 emotions.

[2] http://knoesis.org/projects/emotion.

3 System Architecture

The approach we used for emotion identification in this work is a convolutional neural network architecture which is depicted in detail in Fig. 1. The model is consisted of a simple network with one convolutional layer and a softmax output layer. Each token in a tweet is represented by a word embedding or word representation generated by a neural language model [6]. These features are fed to the convolutional layer of the network. The proposed system is not dependent on hand-crafted features and manually created lexicons. Consequently, the approach is more robust than traditional NLP techniques and more adaptable when applied to different tasks and domains.

3.1 Pre-training

In order to clean noise from the tweets, we applied several pre-processing steps. We replaced each occurrence of a user mention with a generic token and lowercased all

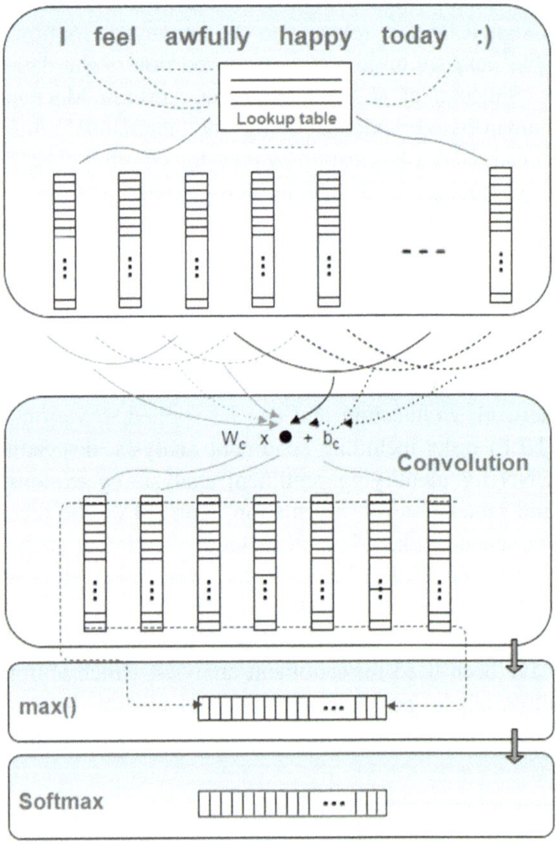

Fig. 1 Convolutional neural network architecture

words. Additionally, we removed all HTML entities and punctuation except for exclamation and question marks and stripped hashtag symbols, but we kept emoticons. Moreover, all elongated words were shortened to maximum 3 character repetitions.

We leverage word representations or embeddings, learned on large corpus of unlabeled textual data using neural language models. The word embeddings are a continuous representation of the words themselves. These embeddings capture syntactic and semantic regularities of words and have recently been used in many tasks in NLP. Our system works by first, constructing a lookup table, where each word is mapped to an appropriate feature vector or word embedding. In this work, we do not do the pre-training of word embeddings ourselves as there are several already available ones, which are used to initialize the lookup table.

Since we are dealing with Twitter messages, which usually contain a lot of informal language, slang and abbreviations, using word embeddings trained on corpus where more formal language is used may not be suitable. Using word vectors trained on Twitter data is more fitting in our task as we assume there will be less missing tokens in the lookup table and the representations will be more meaningful in this context. Therefore, we leverage 200 dimensional GloVe embeddings [6] trained on 2 billion tweets (20 billion tokens). For words that are not present in the vocabulary of word vectors, we use random initialization.

However, since the training is done in an unsupervised manner, there is no sentiment or emotion regularities encoded in the embeddings. As a result, words such as "bad" and "good", that likely appeared in similar context in the corpus are neighboring words based on cosine similarity. We use available word embeddings and by back-propagation, during network training, update them in order to adapt to the specific task at hand. The intuition behind this approach is that by back-propagating the classification errors, emotion regularities are encoded into the word representations. Upon finishing network training, "good" was no longer one of the most similar words to "bad" and vice versa. Additionally, this approach enables for building a more meaningful representation for words that are not present in the lookup table and for which random initialization is used.

3.2 Convolutional Neural Network

In this work, we utilize a convolutional neural network for classification of tweets into 7 emotion classes. Our approach is based on the work of Kim [5] which was used for different sentence classification tasks including sentiment analysis. CNNs with pooling operation deal naturally with variable length sentences and also, to some extent, take into account the ordering of the words and the context each word appears in. For simplicity, we consider that each tweet represents one sentence.

The network is trained by mapping tweets to an appropriate feature representation and supplying them to the convolutional layer. Each word or token of an input tweet, with the appropriate padding at the beginning and end of it, is mapped to an appropriate word representation. Padding length is defined as $h/2$ where h is the window size of the filter. Words are mapped from the aforementioned lookup table $L \in R^{k\times|V|}$, where k is the dimension of the word vectors and V is a vocabulary of the words in the lookup table. Each word or token is projected to a vector $w_i \in R^k$. After the mapping, a tweet is represented as a concatenation of the word embeddings

$$x = \{w_1, w_2, \ldots, w_n\}.$$

The obtained feature representation of the tweet is then supplied to the convolutional layer. In this step, we apply multiple filters with varying windows sizes h. We use rectified linear units in the convolutional layer and windows of size 3, 4 and 5. As tweets are short texts with limited character count, having such window sizes is adequate and using larger ones would not be beneficial. Filters are applied to every possible window of words in the tweet and a feature map is produced as a result. For each of the filters, a weight matrix W_c and a bias term b_c are learned. The weight matrix is used to extract local features around each word window. The convolution operation can be formally expressed as:

$$x'_i = f(W_c \cdot x_{i:i+h-1} + b_c),$$

where $f(\cdot)$ is the activation function and $x_{i:i+h-1}$ is the concatenation of word vectors from position i to position $i+h-1$. The generated feature map is then passed through a max-over-time pooling layer:

$$x' = \max\{x'_1, x'_2, \ldots, x'_{n-h+1}\},$$

which outputs a fixed sized vector where the size is a hyper-parameter to be determined by the user. In our case, we set the size of this vector to 100 and this hyper-parameter corresponds to the number of hidden units in the convolutional layer. By doing so, we extract the most important features for each feature map.

The output of the pooling operation for each of the convolution operations with varying window sizes is concatenated. Predictions are generated using a softmax regression classifier. The concatenated features from the max-over-time pooling layer are passed to a fully connected softmax layer whose output is the probability distribution over the labels.

Deep neural networks suffer from overfitting due to the high number of parameters that need to be learned. In order to counteract this issue, we use dropout regularization which essentially randomly drops a portion of hidden units (sets to zero) during training. As a result, the network prevents co-adaption between the hidden units. The proportion of units to be dropped is hyper-parameter to be determined by the user. The network is trained using stochastic gradient descent over shuffled mini-batches.

4 Experiments

4.1 Dataset

In order to train our model, we utilize an already available annotated set provided in the work of Wang et al. [10]. Tweets in this dataset are annotated with 7 basic emotions *love*, *joy*, *surprise*, *anger*, *sadness*, *fear* and *thankfulness*. The dataset was generated by extrapolating a set of keywords of the 7 basic human emotions and their lexical variants to represent a single category of human emotions. Then, they queried the Twitter API for tweets containing any of the keywords in the form of a hashtag. The dataset is not related to any specific topic or domain.

However, due to Twitter privacy policy, only the IDs of the tweets were available for download, not the content itself. Using the Twitter API, we collected the annotated messages, but because of changed privacy settings or deletion, a significant portion of the messages was not available. Out of 1,991,184 tweets for training, 247,798 for development and 250,000 for test, we were able to retrieve 1,347,959, 168,003 and 169,114, respectively. Nonetheless, this is still a representative dataset. Moreover, the distribution of emotions in the tweets was similar to that of the original set. We applied the same heuristics that are pertaining to the removal of the hashtags indicative of the emotion from the Twitter message.

4.2 Experimental Setup and Results

We reused several parameters which were used in the work of Kim [5], mini-batch size of 50, l_2 constraint of 3, rectified linear units for the convolutional layer and filter windows of 3, 4 and 5. We set the learning rate to 0.02 and the dropout parameter to 0.7. These parameters, along with the decision to use rectified linear units over hyperbolic tangent was done by doing a grid search using the 1000 samples training set.

Due to technical limitations, we did not utilized the full capacity of the dataset. We tested our model with 2000 Twitter messages while for the development set we used 1000 messages. Both sets were generated by randomly sampling from the retrieved tweets. We trained the model with 1000 and 10,000 training samples, in order to observe the gains from a larger training set. Employing the above mentioned parameters we were able to achieve 50.12% with 1000 training samples and 55.77% with 10,000. Wang et al. [10] reported 43.41 and 52.92% respectively. Our model achieves higher accuracy in both cases on our reduced set.

Results for each label are presented in Table 1. For the three most popular emotions, *joy* (28%), *sadness* (24%) and *anger* (22%), we observe the highest F1-score of 64.95, 55.48 and 58.71% respectively. Both precision and recall are

Table 1 Precision, recall and F1 score for each emotion label

Emotion	Precision (%)	Recall (%)	F1 (%)
Joy	59.59	71.38	64.95
Sadness	51.06	60.74	55.48
Anger	59.28	58.16	58.71
Love	44.68	34.29	38.8
Fear	52.78	16.52	25.16
Thankfulness	71.01	41.89	52.69
Fear	0	0	0

relatively high in comparison to the other emotions. Precision and recall for *joy* are 59.59 and 71.38%, 51.06 and 60.74% for *sadness* and 59.28 and 58.16% for *anger*. For the less popular ones, *love* (12%), *fear* (5%), *thankfulness* (5%) precision is relatively high, 44.68, 52.78 and 71.01% respectively, but recall is significantly lower compared with the top 3 emotions, 34.29, 16.52 and 41.89% respectively. The F1-score for each of these emotions are 38.8, 25.16 and 52.69% accordingly. The imbalance of class distribution in the dataset, leads to a classifier that will rarely classify a sample with uncommon labels. Since *surprise* accounts for only 1% of the training data, on the randomly sampled test set in our work, the classifier did not classify correctly any test example with *surprise*.

5 Emotion Identification in Local Government Tweets

In this paper, we present a use-case of emotion identification for local government purposes. We outline a system that can potentially be used by local governments to get real-time feedback from the local community in relation to ongoing and proposed projects. A showcase of the proposed system is presented in Fig. 2. We implemented the system for several major cities in USA such as New York, Los Angeles, San Francisco etc. The system is implemented in the Django web framework.

First, we manually identify official local government Twitter accounts for each corresponding city. This also includes the official accounts of the city's mayor and accounts for the 311 number which provide access to information and non-emergency municipal services. For example, for New York, we retrieve tweets from *@nycgov*, *@NYCMayorsOffice*, *@BilldeBlasio* and *@nyc311*. For the retrieval of Twitter messages, we utilize the Twitter Streaming API which enables access to tweets posted in real-time. Using the Streaming API, we collect tweets posted by these users. All messages mentioning any of the defined accounts are retrieved as well. In this way, we do not depend solely on what local governments tweet, but what the local community is concerned with too.

However, it doesn't necessarily mean that all tweets related to local government will mention some of the official accounts. As a result, in order to provide for bigger coverage, we also download geo-referenced tweets posted within some of the cities

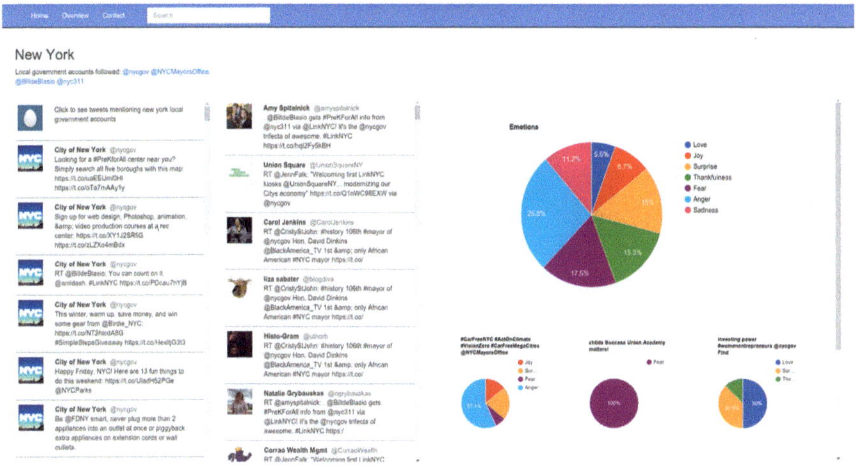

Fig. 2 Main page overview of our proposed system

under observation. This is also enabled by the Twitter Streaming API. Supplying the API with a bounding box defined by a set of coordinates, it will return every tweet posted within those boundaries. Tweets can be geo-referenced by either containing exact coordinates or by being embedded with a Place object. Places are Twitter specific objects which are defined by their name and bounding box among other information. They can relate to different geographic objects, from narrow areas such as points of interest to wider ones, such as entire cities. Tweets containing Place objects whose bounding box intersects with those of our interest are returned by the API.

Then, these tweets are compared against tweets posted by the local government accounts in order to identify which ones are related to actual local government issues. The matching between tweets is done by computing the cosine similarity of the TF-IDF representation of the Twitter messages. We can potentially, narrow the search space by only retrieving tweets containing at least two keywords from local government posted tweets. In this way, we get a high probability of getting related tweets, although additional filtering would still be required.

The system enables overview of local government related Twitter activity for New York, Boston, Washington DC, Detroit, Los Angeles and San Francisco. The model for emotion identification is trained on tweets from Wang et al. [10], but on a balanced distribution of emotion labels in order to counteract the issue of predicting underrepresented emotions in the training set. The size of the training set is 10,000 samples.

When presented with the system, the user can choose one of the multiple cities which are currently being considered in our work. Upon selecting, the user is presented with the overview screen. On the left, we list all local government tweets. This provides a quick look to what is the local government currently working on, ongoing projects and current issues. The user has the option to list all tweets

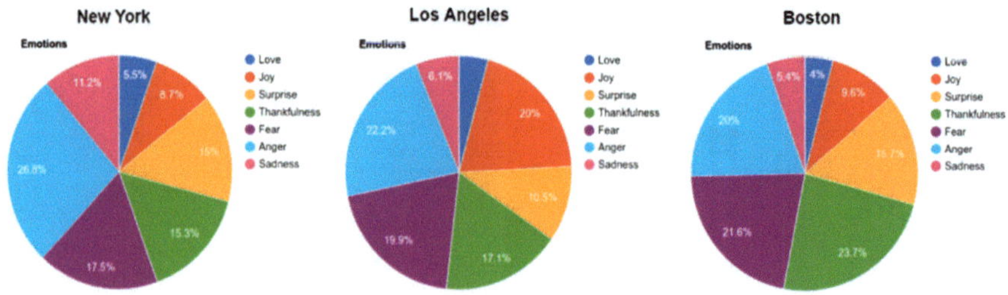

Fig. 3 Emotion distribution in local government related tweets

mentioning any of the local government accounts or choose a specific tweet and take a look at all relevant messages to that specific tweet. These messages appear in the middle column. The system presents the content of each tweet, as well as the full and screen name of the user with the accompanying profile image. On the right part of the screen, a summary of the tweets in the middle column is given. We present an overview of the identified emotions in the relevant tweets which enables users to have a closer look at the local community's satisfaction with their city. If the user has chosen to list all tweets mentioning local government accounts, he is also presented with an emotional distribution by topic, in order to get even more meaningful insight. Topics are generated by applying hierarchical agglomerative clustering over the tweets. For this purpose, we utilize the *fastcluster*[3] library for hierarchical clustering. For scalability, we only consider words that have a frequency over a predefined threshold. We cut the dendrogram at a specific threshold value as well to extract the generated clusters.

In Fig. 3, the emotion distribution for New York, Los Angeles and Boston is presented for the tweets mentioning any of the corresponding local government accounts. We can observe that certain emotions are more dominant in some cities over others. For example, *joy* in San Francisco, *anger* in New York, *thankfulness* in Boston are more present in relation to the other cities. We leave more detailed analysis of the observed emotions for future work.

6 Conclusion

In this paper, we presented a convolutional neural network for emotion identification in Twitter messages and we applied it to monitoring emotions in tweets related to public local services. The model was evaluated on a set of hashtag labeled tweets with 7 distinct emotions. Using the presented architecture, we achieved improvements over current state-of-the-art performance with the dataset on reduced training set. Our approach obtains an accuracy of 50.12 and 55.77% with a training

[3]http://danifold.net/fastcluster.html.

set of 1000 and 10,000 samples. We trained the model on a set with a balanced distribution of emotion labels and we applied it to tweets related to public local services. We present a system that retrieves tweets from official local government accounts for several cities and related Twitter messages. We showcase how such system can be utilized by the local government and the general public to get insight into the current projects and problems and improve the communication between both parties.

Acknowledgements We would like to acknowledge the support of the European Commission through the project MAESTRA—Learning from Massive, Incompletely annotated, and Structured Data (Grant number ICT-2013-612944).

References

1. Stojanovski D, Strezoski G, Madjarov G, Dimitrovski I (2015) Emotion identification in FIFA world cup tweets using convolutional neural network. In: 11th International conference on innovations in information technology (IIT), 2015. IEEE, pp 52–57
2. Sobaci MZ, Karkin N (2013) The use of twitter by mayors in Turkey: tweets for better public services? Gov Inf Q 30(4):417–425 (Elsevier)
3. Agostino D (2013) Using social media to engage citizens: a study of Italian municipalities. Public Relat Rev 39(3):232–234 (Elsevier)
4. Flores CC, Rezende DA (2013) Strategic digital city: techno-social network twitter as communication channel for popular participation in city comprehensive plans. In: Proceedings of the nineteenth Americas conference on information systems, Chicago, Illinois
5. Kim Y (2014) Convolutional neural networks for sentence classification. In: Proceedings of the 2014 conference on empirical methods in natural language processing (EMNLP). Association for Computational Linguistics, pp 1746–1751
6. Pennington J, Socher R, Manning CD (2014) Glove: global vectors for word representation. In: Proceedings of the empirical methods in natural language processing (EMNLP 2014) 12:1532–1543
7. Roberts K, Roach MA, Johnson J, Guthrie J, Harabagiu SM (2012) EmpaTweet: annotating and detecting emotions on twitter. LREC 3806–3813
8. Balabantaray RC, Mohammad M, Sharma N (2012) Multi-class twitter emotion classification: a new approach. Int J Appl Inf Syst 4:48–53
9. Purver M, Battersby S (2012) Experimenting with distant supervision for emotion classification. In: Proceedings of the 13th conference of the European chapter of the Association for Computational Linguistics. Association for Computational Linguistics, pp 482–491
10. Wang W, Chen L, Thirunarayan K, Sheth AP (2012) Harnessing Twitter "big data" for automatic emotion identification. In: Privacy, security, risk and trust (PASSAT), 2012 international conference on and 2012 international conference on social computing (SocialCom). IEEE. 587–592
11. Sintsova V, Musat C-C, Pu FP (2013) Fine-grained emotion recognition in olympic tweets based on human computation. 4th Workshop on computational approaches to subjectivity, sentiment and social media analysis

12. Collobert R, Weston J, Bottou L, Karlen M, Kavukcuoglu K, Kuksa P (2011) Natural language processing (almost) from scratch. J Mach Learn Res (JMLR.org) 12:2493–2537
13. dos Santos C, Gatti M (2014) Deep convolutional neural networks for sentiment analysis of short texts. In: Proceedings of COLING 2014, the 25th international conference on computational linguistics: technical papers. Dublin City University and Association for Computational Linguistics, pp 69–78

MedWeight Smart Community: A Social Approach

Giannis Meletakis, Rania Hatzi, Panagiotis Katsivelis, Mara Nikolaidou, Dimosthenis Anagnostopoulos, Costas A. Anastasiou, Eleni Karfopoulou and Mary Yannakoulia

1 Introduction

Information and Communication Technology (ICT) is literally changing every aspect of our life [1]. This entire new age has affected a great number of different domains providing new widely accepted tools and visions for everyday communication and collaboration among participants. A great impact on this growing area has been accomplished from on-line social networks, as they have been established as a prominent model for communication and interaction between individuals, as well as among members of communities or organizations. There is no questioning

G. Meletakis (✉) · R. Hatzi · P. Katsivelis · M. Nikolaidou · D. Anagnostopoulos
Department of Informatics and Telematics, Harokopio University, Athens, Greece
e-mail: meletakis@hua.gr

R. Hatzi
e-mail: raniah@hua.gr

P. Katsivelis
e-mail: pkatsiv@hua.gr

M. Nikolaidou
e-mail: mara@hua.gr

D. Anagnostopoulos
e-mail: dimosthe@hua.gr

C.A. Anastasiou · E. Karfopoulou · M. Yannakoulia
Department of Nutrition and Dietetics, Harokopio University, Athens, Greece
e-mail: acostas@hua.gr

E. Karfopoulou
e-mail: ekarfop@hua.gr

M. Yannakoulia
e-mail: myianna@hua.gr

L. Ismail and L. Zhang (eds.), *Information Innovation Technology in Smart Cities*,
https://doi.org/10.1007/978-981-10-1741-4_11

of on-line social network success, as social networks have been playing a significant role in the development of human society over decades [2].

Despite the human social nature, the concepts of community development and community participation took shape in the 1950s [3]. Nowadays, it is assumed that citizen participation is a desired and necessary part of community development activities. Participation means that people are closely involved in the processes that affect their lives. As mentioned in [4] citizen's participation is the process that can meaningfully tie programs to people. Furthermore, sense of community is a feeling that members have of belonging, a feeling that members matter to one another and to the group, and a shared faith that members needs will be met through their commitment to be together [5]. There is no doubt that promoting the participation of community members is the main pillar in any modern community development program [6].

Communities around the world are responding to the participants needs by discovering new ways of using information and communication technologies (ICT) for economic, social and cultural development [1] establishing a new concept called "Smart Communities". In Smart Communities Guidebook, developed by California Institute for Smart Communities (1997) at San Diego State University, the concept of Smart Community is presented as a community in which government, business and residents understand the potential of information technology, and make a conscious decision to use that technology to transform life and work in their region in significant and positive ways [7]. In present, the Smart Community concept is known and used all over the world under different names and in different circumstances [1]. In any case, it is considered an imperative constituent of the smart city environment. Could modern social network technology serve the concept of smart communities? May current popular social networks promote the creation and support of smart communities? Are there any limitations in social network interaction model hindering smart community support?

These are the questions we are dealing with in this paper in our effort to promote a smart community targeting weight maintenance support, called MedWeight. Social network technology was employed to support MedWeight community. The requirements imposed by such a community on social network technology and the necessary extensions to the social network interaction model in order to support smart communities were discussed in [8]. Based on these extensions, Medweight social network platform was developed. In this chapter, we focus on MedWeight community perspective and the experience obtained by community members during the development and usage the MedWeight platform. The basic challenges on choosing smart community direction are discussed, requirement imposed are explained and key issues on its adaptation for the interaction between nutrition experts and community members trying to maintain their weight are explored.

The rest of the paper is organized as follows. Background and motivation for Medweight community leaders on choosing the establishment of smart community technology to support volunteers for weight maintenance is explained in Sect. 2. Section 3 outlines technological key challenges in supporting MedWeight smart community and corresponding extensions proposed to the social network model.

MedWeight social network, developed to support the community, from the participant perspective, is presented in Sect. 4. Conclusions and future work reside in Sect. 5.

2 Background—Motivation

It is a common true that the Internet and Web technologies have helped many of us communicate more easily and effectively. Web 2.0 applications and platforms, such as wikis, blogs and social networks, enable people to communicate directly and without space or time limitations. Although social technology does not affect the interdependency between people, it strongly enhances the probability that people who are interdependent, in the sense that they could share common interests, learn about each other and eventually start a relationship. In any case, enhancing accessibility may have a positive impact in maintaining, enriching and building communities [9]. Thus, social technology may definitely contribute in promoting the Smart Community concept.

Social support, as perceived in the context of smart communities, is defined as the resources or aids exchanged between individuals through interpersonal ties [10]. This is one of the key benefits that users perceive from on-line social networking [11]. As indicated in [12], Social Network users perceived a greater level of emotional support and companionship than did general Internet users at a level that was almost equivalent to the amount that married or cohabiting Americans normally perceive from their live-in partners. The same view appeared in [13] where the results shown that the positive affect felt by social network users after on-line social networking was positively associated with perceived companionship support, appraisal support, and life satisfaction. In the same study [13] was noticed that it is the quality of interaction that matters in establishing social support and psychological well-being, but not the frequency or amount of social networking use.

Social technology has empowered patients to share their information and experiences and also to gain access to others information [14]. As indicated in [15] the proliferation of the Internet for acquiring information on health and developing e-health has gained a lot of attention in recent years. Data in the European Union show that when searching the internet for a specific injury, disease, illness or condition, 36% of the responders searched for testimonials or experiences from other patients and 10% looked for emotional support in dealing with the specific health issue [16].

According to [17] the key factors of on-line health support communities high popularity are:

- any time support overcoming time boundaries
- anywhere support overcoming distance boundaries that might be associated with traditional face-to-face support provision

- establishment of "safe" environment for individuals with stigmatizing or disfiguring conditions to obtain support
- anonymization makes it easier for individuals to discuss sensitive or embarrassing topics, and may increase honesty, intimacy and self-disclosure
- larger broad of experiences and opinions may be offered than face-to-face support groups

The majority of existing social health related applications are targeting on-line health care and support communities. They consist mainly of medical blogs and micro-blogs [18], wikis [19] and media sharing sites [20]. Social networking sites are also available [21]. Social technology may also bring a new dimension to health care as it offers a medium to be used by the public, patients, and health professionals to communicate about health issues with the possibility of potentially improving health outcomes [22]. There are several examples of social media applications targeting evaluation and reporting of real-time diseases, catalyzing outreach during (public) health campaigns and recruitment of patients to on-line studies and in clinical trials [23].

Despite the fact that most efforts in weight management have been focused on weight loss, it is now evident that weight loss maintenance in formerly obese/overweight individuals represents possibly the greatest challenge, both in terms of physiology and behavior, in overall weight control. However, what is the impact of technology? Could social technology contribute to establish a smart community helping its members in their effort to lose weight and maintain weight loss [24]? Who should be part of this community? Should weight management experts also participate to provide professional advice or should the community only persist of people interested in weight management?

As indicated in [25], the contribution of social technology seams helpful, since the participants of the study were almost five times more likely to perceive Encouragement support for their weight loss efforts if they used the social media tools at least once a week. In addition, this study [26] showed that an Internet weight maintenance program could sustain comparable long-term weight loss compared with a similar program conducted in person and over the phone. On the opposite side, in [27], the arm with social media demonstrated no difference in perceived support compared to in-person therapy and it also had the highest rates of attrition.

A recent review of the existing studies suggests that social networks may be effective in improving nutrition knowledge or key behaviors in weight control, such as eating behavior or physical activity, thus promoting body weight change and overall feeling of well-being [28]. In fact, it has been shown that web based interventions may be equally effective with face-to-face interventions in changing dietary behaviors, such as fruit and vegetable consumption, or physical activity levels [29] or maintaining weight loss [30]. Furthermore, when targeting specific populations, such as adolescents, the available data suggest that computer-based interventions may be superior in improving nutrition knowledge, as compared to traditional means of education. The essential components of a successful web-based intervention for weight control have not been fully established. As weight loss or maintenance is a matter that depends mainly on behavior change, it has been shown that web based interventions incorporating components of behavioral theory, counseling or self-monitoring may be superior, compared to education alone [30].

To this end, in the following we explore the potential of using social networking technology to build a smart community for weight management.

3 MedWeight Project

The MedWeight project was established as part of a research study related to weight maintenance by the Department of Nutrition and Dietetics at Harokopio University of Athens (http://medweight.hua.gr/en/index.php) two years ago. A similar study for weight maintenance exists in the National Weight Control Registry, US (NWCR) [31]. Its main aim is to collect information on common behaviours among its members that predict long-term weight maintenance status. Assessment is performed mainly through paper work that is send to the potential participants by mail. Our ambition was to advance this perspective by including not only assessment but also interactions between all users, i.e. health professionals and patients [32]; and to advance the assessment procedures by utilizing social networking technology.

MedWeight community consists of more than 1000 volunteers involved in the study and Nutrition experts and researchers, advising them. The goal of this community is to help volunteers maintain their weight and follow-up people encountering weight problems. Both other volunteers and nutrition experts are assisting them in this effort.

Volunteers include both successful losers and weight loss regainers. Their interaction may be of interest, as the assistance they provide to each other is achieved mainly through communication between them. Helpful information and guidance by the nutrition experts related to issues of interest in MedWeight community is also provided to volunteers. The prospect of participant's mutual support for health issues through direct communication was the main motivation for establishing Med-Weight community.

As such, the community should explore all available technology to help its members serve their purpose. As in face-to-face support groups, the community should consist of people interesting in maintaining their weight and experts helping them. In contrast to what is normally the case in existing social networks, where all participants are treated as equal and have exactly the same capabilities and rights, in this smart community environment participants should be able to differentiate their behaviour in the community based on their role, as in the real world, when attending a support group for example.

The main feature, differentiating it from other similar efforts on health issues, is the fact that both volunteers and nutrition experts participate in the community, having different roles and capabilities as in real life. In existing on-line communities either patients or doctors participate, having exactly the same privileges and capabilities (as for example patientslikeme.com for patients or twitterdoctors.net for medical professionals). Such a feature enables the volunteers to behave in on-line MedWeight community as they would a real-world support community, receiving similar services and perceiving their support community the same way as they would in the real-world. Thus, MedWeight obtains the characteristics of a true smart community, as prescribed in [7].

4 Supporting Smart Communities Through Social Networking Technology

The typical social network model, as supported by popular social networks, such as Facebook, Twitter and Google+, dictates that all participants are (a) described by the same characteristics, (b) belong in the same category and (c) are related to others with one unique relation type (e.g. friendship in Facebook or Follower in Twitter). Available social networking platforms do not support the participant's discrimination in different types, being unable to disseminate the produced information according to their category.

In the following, we briefly present the extended social network interaction model proposed in [8] to explore requirements imposed by MedWeight smart community. The proposed extensions are summarized in Fig. 1 and analytically discussed in the following. In the figure, proposed entities are depicted as cyan rectangles. Our vision is that our proposed model and corresponding social networking platform could be utilized from other communities as well [33].

These are thc main characteristics added to the typical social network model:

- Role definition support to reflect the role each participant plays in the community
- Relation definition support to reflect the different relation developed between of community members having different roles
- Information dissemination advancement based on roles and relations
- Group management advancement based on roles and relations
- Application execution based on roles and relations

4.1 Roles, Relations and Groups

Realizing that there was a major need to differentiate between participants to reflect their role and capabilities/responsibilities in the community, the concept of Role was introduced. It enables to manage information dissemination among participants, indicate responsibilities and enables different ways of describing participants

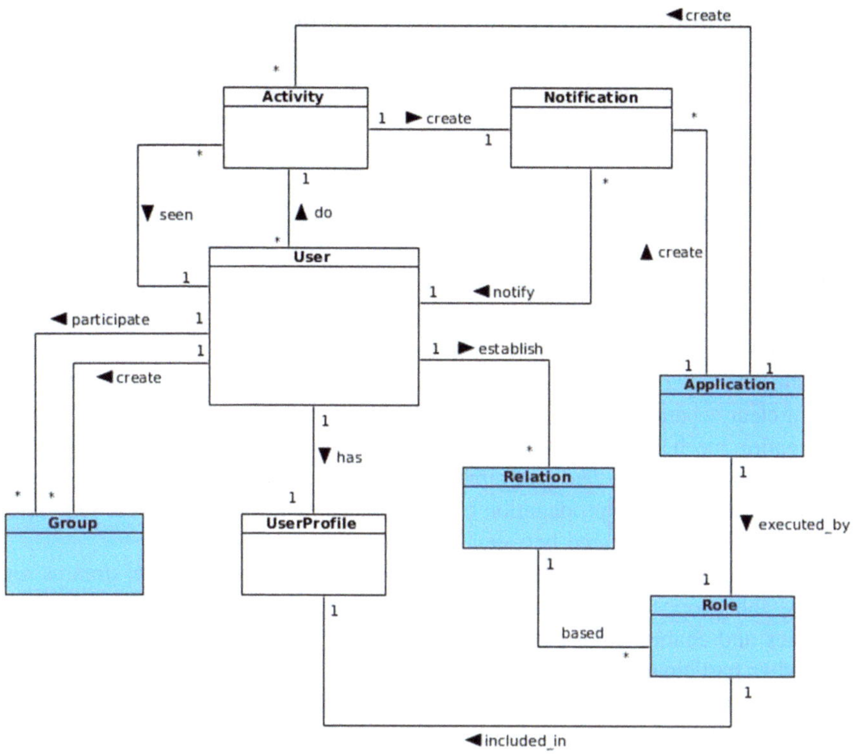

Fig. 1 Extended social network model

(for example the profile data of a Nutrition Expert may vary from those belonging to a volunteer).

Establishing relations with others is one of the main options given to social network participants. There are two general types of relations: mutual (bidirectional) and one-way (unidirectional) [34]. Our proposed model could support the dynamic creation of relations of both types between participants, based on the predefined user roles. Relations can be either unidirectional, indicating that a community member receives information from another member, or bidirectional, indicating that the members interact.

The combination of roles, relations and streams does not fully facilitate fine-grained content propagation; therefore, a more elaborate mechanism for content delivery is proposed, through groups, as supported in the typical social networking interaction model.

4.2 Information Dissemination and Application Execution

The most common operation that a participant performs in a social network is publishing content, which can be of a variety of types, such as links, texts, files, multimedia etc. Published information is propagated in the form of a stream to all participants related to the publishing entity, who receive notifications and updates about the publication, urging them to review it and possibly contribute to it, as dictated by the notion of collaborative content in Web 2.0 [35].

In communities, specific streams should be defined based on participant roles and relations. Apart from the member relations, the social aspect of the community should not be dismissed; therefore, each member may develop a social relation with any other member of the community, regardless of their roles in it. At the same time, a clear separation between them should be maintained, thus a more complex propagation mechanism is introduced incorporating more than one discrete streams. The combination of discrete participant roles, multiple streams, extended relations and rules governing the propagation of content successfully achieves the separation between information shared between community members.

In addition to sharing content and notifications through discrete streams and groups, the proposed social network model supports the provision of specific activities and enables its participants to complete specific actions in collaboration with other participants.

Actions may be provided by cooperating applications executed in a specific participant profile. In order to ask for services rather than information from another participant, a more sophisticated communication mechanism is required, facilitating information exchange between applications executed on different participant profiles.

5 MedWeight Smart Community

The MedWeight Smart Community was built, based on the extended social network model, to support MedWeight Project. It is currently deployed using Python and Django (https://www.djangoproject.com) web application framework, while the user interface, in this phase, supports only the Greek language.

There are two distinct roles and two relations supported in MedWeight community:

- *Volunteer*: a person who takes part in the study and wants to benefit from Med-Weight community to maintain weight. Most likely volunteers have been in a diet and would benefit from expert advice to maintain their weight.
- *Dietitian*: an expert scientist that provides advice, services and feedback to participants of the role Volunteer.
- *Consultant Relation*: a unidirectional relation from a volunteer to a dietitian, which enables volunteers to use dietitians to obtain expert advice.

- *Fellow Relation*: a bidirectional relation, defined between volunteers, which enables them to share experiences and information related to Med-Weight community.

Roles and relations are defined through MedWeight administration platform, as depicted in Fig. 2.

The interaction between MedWeight participants is performed mainly by publishing content either in their profile or in interest groups they may create. They may declare its visibility, as in any social network. Furthermore, they may execute applications, as discussed in the following.

The main view of a Volunteer participant profile is shown in Fig. 3. Please have in mind that MedWeight platform has a Greek language interface, as it targets Greek audience. Thus, explanation comments are embedded in the figure to make it understandable by an international audience.

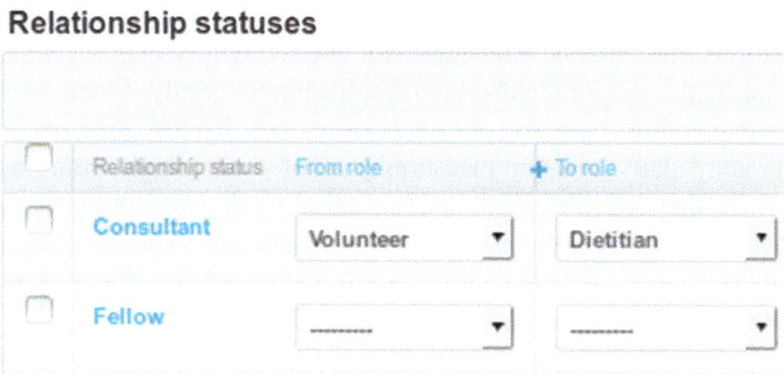

Fig. 2 Defining relations using MedWeight administration interface

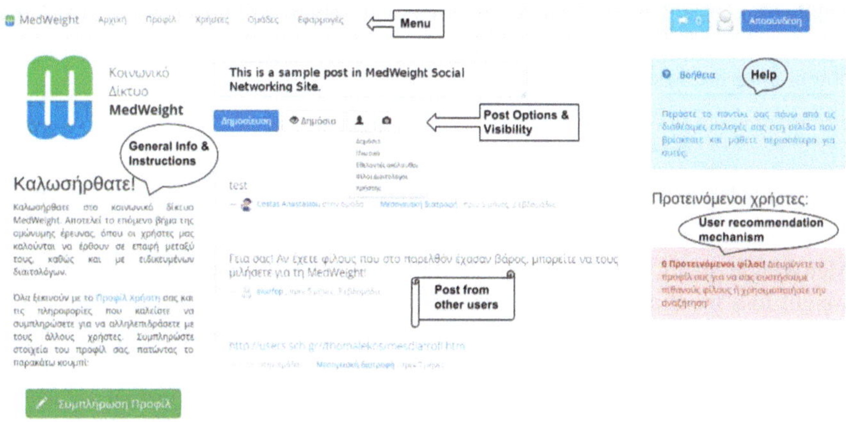

Fig. 3 MedWeight participant profile view

The content that is published by MedWeight community members in their profile is visible to all other community members related to them, either by consultant or fellow relation. In order to succeed the optimal dissemination of information, there are several given options to users related to 'visibility' of the publication such as private, public, visible to a specific participant, visible to all of them, as shown in Fig. 3.

Through the corresponding option in the right, bottom part of Fig. 3 (red rectangle), a recommendation mechanism was designed in order to match volunteers with similarities in order to promote their interaction. This process pursues to detect other volunteers with same info in their profile. Volunteers may search members of the community based on specific criteria they may combine.

Groups may be created and managed by both Volunteers and Dietitians. Based on the creator role, Medweight platform applies different group access policy. Groups created and managed by volunteers enable posting by all participants. Groups created by Dietitians have a more restricted policy, as in those groups only "expert" opinions should by posted. Only dietitians, authorized by the creator of the group, post content in it, although all Medweight participants may read this content.

Applications may also be executed with MedWeight Social Network platform. As an example, the *weight maintenance* application is briefly presented. It involves a private interaction between the volunteer and his/her dietitian consultant. Volunteers may daily register measurements of their weight, running such an application in their profile. With each measurement, the application calculates certain dietetic factors, such as Body Mass Indicator. The dietitian consultant monitors these factors, when accepting to act as a consultant. If any of these factors have exceeded a certain limit, a notification is issued to dietitian profile. Consequently, the dietitian can provide personalized feedback and expert advice to the volunteer, though the application, also running in his/her profile.

Medweight platform is used by half of the 1000 volunteers involved in the Medweight study and nutrition experts and researchers advising them, during the last 10 months. The platform was well accepted by participants, which had no problem using it. Detailed data on its impact in helping volunteers maintain weight loss is part of an on-going research and will be available on a pilot basis, after completing the first year of its usage.

6 Conclusion

MedWeight smart community was build to support volunteers trying to maintain weight loss by allowing them to be members of a community composed by both other volunteers and nutrition experts, taking into consideration the way support groups are formed in the real-world. To support MedWeight smart community, a corresponding social network platform was built, extending the typical social network model to support roles, relations and complex content dissemination policies.

Future work includes the extension of MedWeight Smart Community to provide a variety of applications allowing participants to use external services and the support other communities as well.

Acknowledgements The research presented in the paper is part of the work conducted in the project Med-Weight Study: Lifestyles for Weight Loss Maintenance funded by Coca-Cola Foundation. We would like to give special thanks to all the volunteers and dietitians participating in MedWeight Smart Community.

References

1. Lindskog H (2004) Smart communities initiatives. In: Proceedings of the 3rd ISOneWorld conference, pp 14–16
2. Xia F, Ma J (2011) Building smart communities with cyber-physical systems. In: Proceedings of 1st international symposium on from digital footprints to social and community intelligence (pp 1–6). ACM (September)
3. Murthy CA, Chowdhury N (1996) In search of optimal clusters using genetic algorithms. Pattern Recogn Lett 17(8):825–832
4. Spiegel HB (1969) Citizen participation in urban development, vol 2. Cases and programs
5. McMillan D (1976) Sense of community: an attempt at definition. George Peabody College for Teachers
6. Zomorrodian AH, Gill SS, Samaha AA, Ahmad N (2013) Quantitative models for participation evaluation in community development: a theoretical review. World Appl Sci J 25(2):314–322
7. Canada I (1998) Smart communities: report of the panel on smart communities. Information Distribution Centre, Communications Branch, Industry Canada, Ottawa
8. Meletakis G et al (2015) Building MedWeight smart community using social networking technology. In: 11th international conference on innovations in information technology (IIT), IEEE, pp 332–337
9. De Vos H (2004) Community and human social nature in contemporary society. Analyse & Kritik 26(1):7–29
10. Cohen S, Hoberman HM (1983) Positive events and social supports as buffers of life change stress. J Appl Soc Psychol 13(2):99–125. [Online] Available: http://dx.doi.org/10.1111/j.1559-1816.1983.tb02325.x
11. Park N, Kee KF, Valenzuela S (2009) Being immersed in social networking environment: facebook groups, uses and gratifications, and social outcomes. CyberPsychol Behav 12 (6):729–733
12. Hampton K, Goulet LS, Rainie L, Purcell K (2011) Social networking sites and our lives. Retrieved 12 July 2011 from, 2011
13. Oh HJ, Ozkaya E, LaRose R (2014) How does online social networking enhance life satisfaction? The relationships among online supportive interaction, affect, perceived social support, sense of community, and life satisfaction. Comput Hum Behav 30:69–78
14. Hajli MN (2014) Developing online health communities through digital media. Int J Inf Manage 34(2):311–314
15. Rains SA, Karmikel CD (2009) Health information-seeking and perceptions of website credibility: examining web-use orientation, message characteristics, and structural features of websites. Comput Hum Behav 25(2):544–553
16. Eurobarometer F (2014) European citizens digital health literacy. A report to the European Commission, 2014

17. Coulson NS (2013) How do online patient support communities affect the experience of inflammatory bowel disease? An online survey. JRSM Short Rep 4(8):2042533313478004
18. Kovic I, Lulic I, Brumini G (2008) Examining the medical blogosphere: an online survey of medical bloggers. J Med Internet Res 10(3)
19. Heilman JM, Kemmann E, Bonert M, Chatterjee A, Ragar B, Beards GM, Iberri DJ, Harvey M, Thomas B, Stomp W et al (2011) Wikipedia: a key tool for global public health promotion. J Med Internet Res 13(1)
20. Adlassnig K et al (2009) An analysis of personal medical information disclosed in youtube videos created by patients with multiple sclerosis. In: Medical Informatics in a United and Healthy Europe: proceedings of MIE 2009, the XXII international congress of the European federation for medical informatics, vol 150. IOS Press, 2009, p 292
21. Bender JL, Jimenez-Marroquin M-C, Jadad AR (2011) Seeking support on facebook: a content analysis of breast cancer groups. J Med Internet Res 13(1)
22. Moorhead SA, Hazlett DE, Harrison L, Carroll JK, Irwin A, Hoving C (2013) A new dimension of health care: systematic review of the uses, benefits, and limitations of social media for health communication. J Med Internet Res 15(4)
23. Grajales FJ III, Sheps S, Ho K, Novak-Lauscher H, Eysenbach G (2014) Social media: a review and tutorial of applications in medicine and health care. J Med Internet Res 16(2):e13
24. Dais A, Nikolaidou M, Alexopoulou N, Anagnostopoulos D (2008) Introducing a public agency networking platform towards supporting connected governance. In: Electronic government. Springer, Berlin Heidelberg, pp 375–387
25. Hwang KO, Etchegaray JM, Sciamanna CN, Bernstam EV, Thomas EJ (2014) Structural social support predicts functional social support in an online weight loss programme. Health Expect 17(3):345–352
26. Harvey-Berino J, Pintauro S, Buzzell P, Gold EC (2004) Effect of internet support on the long-term maintenance of weight loss. Obesity Res 12(2):320–329. [Online] Available: http://dx.doi.org/10.1038/oby.2004.40
27. Cussler EC, Teixeira PJ, Going SB, Houtkooper LB, Metcalfe LL, Blew RM, Ricketts JR, Lohman J, Stanford VA, Lohman TG (2008) Maintenance of weight loss in overweight middle-aged women through the internet. Obesity 16(5):1052–1060. [Online] Available: http://dx.doi.org/10.1038/oby.2008.19
28. Maher CA, Lewis LK, Ferrar K, Marshall S, De Bourdeaudhuij I, Vandelanotte C (2014) Are health behavior change interventions that use online social networks effective? A systematic review. J Med Internet Res 16(2):e40
29. Neuenschwander LM, Abbott A, Mobley AR (2013) Comparison of a web-based vs in-person nutrition education program for low-income adults. J Acad Nutr Diet 113(1):120–126
30. Neve M, Morgan PJ, Jones P, Collins C (2010) Effectiveness of web-based interventions in achieving weight loss and weight loss maintenance in overweight and obese adults: a systematic review with meta-analysis. Obes Rev 11(4):306–321
31. Wing RR, Phelan S (2005) Long-term weight loss maintenance. Am J Clin Nutr 82(1): 222S–225S
32. Karfopoulou E, Anastasiou CA, Hill JO, Yannakoulia M (2014) The medweight study: design and preliminary results. Mediterr J Nutr Metab 7(3):201–210
33. Hatzi O, Meletakis G, Katsivelis P, Kapouranis A, Nikolaidou M, Anagnostopoulos D (2014) Extending the social network interaction model to facilitate collaboration through service provision. In: Enterprise, business-process and information systems modeling. Springer, Berlin, Heidelberg, pp 94–108
34. Li Y, Zhang Z-L, Bao J (2012) Mutual or unrequited love: identifying stable clusters in social networks with uni and bi-directional links. In: Algorithms and models for the web graph. Springer, 2012, pp 113–125
35. Andersen P (2007) What is Web 2.0? Ideas, technologies and implications for education. JISC Bristol, UK 1(1)

Semi-distributed Demand Response Solutions for Smart Homes

Rim Kaddah, Daniel Kofman, Fabien Mathieu and Michal Pióro

1 Introduction

The growing deployment of intermittent renewable energy sources at different scales (from bulk to micro generation) advocates for the design of advanced Demand Response (DR) solutions to maintain the stability of the power grid and to optimize the usage of resources.

DR takes advantage of demand flexibility, but its performance depends on the granularity of visibility and demand control. The Internet of Things (IoT) paradigm enables implementing DR at the finest granularity (individual appliances), and deploying IoT-based solutions becomes feasible, both from the technological and economical points of view.

The introduction of capacity markets in several countries has provided incentives for: the flexibility end users could provide through DR mechanisms; the deployment of flexible generators (for which the energy cost is higher than the average).

R. Kaddah (✉) · D. Kofman
Telecom Paristech, 23 Avenue d'Italie, Paris, France
e-mail: rim.kaddah@telecom-paristech.fr

D. Kofman
e-mail: daniel.kofman@telecom-paristech.fr

F. Mathieu
Nokia Bell Labs, Route de Villejust, Nozay, France
e-mail: fabien.mathieu@nokia.com

M. Pióro
Institute of Telecommunications, Warsaw University of Technology, Warsaw, Poland
e-mail: michal.pioro@eit.lth.se

M. Pióro
Department of Electrical and Information Technology, Lund University, Lund, Sweden

L. Ismail and L. Zhang (eds.), *Information Innovation Technology in Smart Cities*,
https://doi.org/10.1007/978-981-10-1741-4_12

In this chapter, we focus on DR solutions for keeping power consumption below a certain known capacity limit for a well-defined period. A possible application is for utility companies, which are interested in limiting the cost of the capacity certificates they have to acquire in the capacity market (for securing supply). Such a cost reduction is facilitated by keeping power consumption below known thresholds.

In [1], the authors propose and analyze several IoT-based DR mechanisms. They show that fine-grained visibility and control on a set of households at an aggregation point enables to maximize user's perceived utility. However, this approach may cause scalability as well as privacy problems. On the other hand, they consider two levels control systems where a central controller allocates available capacity to households based on some static information (e.g., type of contract). Then, local controllers leverage IoT benefits for local optimization, without any feedback to the central controller. The drawback of such approach is that it may reduce the total utility perceived by the users.

Our main contribution is a proposition and evaluation of an intermediate approach, based on two-level systems with partial feedback from the local controllers to the central entity, where the feedback sent has little impact on privacy. The proposed solution enforces fairness by considering two levels of utility for each appliance (i.e., vital and comfort). We compare the performance of the proposed scheme with the two cases studied in [1] (fully centralized solution and two level system with no feedback). The results are analyzed for homogeneous and heterogeneous scenarios. We show that for both cases, the proposed algorithm outperforms the scheme with no feedback. Moreover, it runs in a limited number of feedback iterations, which ensures good scalability and limited requirements in terms of communication.

The chapter is organized as follows: Sect. 2 presents the related work. The system model and allocation schemes are introduced in Sects. 3 and 4, respectively. Section 5 studies the performance of the proposed control scheme and compares it with two benchmark control approaches through a numerical analysis of the model. Conclusions and future work are presented in Sect. 6.

2 Related Work

The idea of using vertically distributed control schemes with no or limited feedback is quite natural. Indeed, the literature contains a significant amount of proposals based on hierarchical control schemes (with feedback) in the context of limiting consumption capacity to a certain desired value (e.g., [2–5]).

However, these proposals usually address the problem by taking the dual of system capacity constraint. These dual variables can be seen as prices for time slots (e.g., [2–4]). Thus, these schemes are usually presented as DR schemes based on pricing. In contrast, our approach is based on direct control of the appliances.

The proposals in [3–5] are examples of schemes that are designed for residential consumers and that can take into account flexibility of generic appliances.

Authors in [3] propose a customer reward scheme that encourages users to accept direct control of loads. They propose a time-greedy algorithm (maximizes utility slot by slot) based on the utility that each appliance declares for each slot. As discussed in the previous chapter, instantaneous value of an appliance has to be carefully evaluated to capture the real benefit from using this appliance (e.g., heating system). Authors in [4] propose a dynamic pricing scheme based on a distributed algorithm to compute optimal prices and demand schedules. Closer to our proposal is the direct control scheme presented in [5] which is very similar to [4] if prices are interpreted as control signals. The authors in [5] propose to solve a problem similar to ours, but in their approach, intermediate solutions can violate the constraints, so that convergence of the algorithm is required (like all other schemes based on dual decomposition) to produce a feasible allocation. The authors do not discuss scalability and communication requirements in terms of the number of iterations required. They also assume concave utility functions. Moreover, the proposed scheme still requires disclosure of extensive information to the central entity (i.e., home consumption profile), so the approach is not adapted to reduce privacy issues.

In the present work, we target to better deal with privacy while guaranteeing the fulfillment of capacity constraints even during intermediate computation. We achieve that by building on the DR framework introduced in [1].

3 System Model

We consider an aggregator in charge of allocating power to a set of H households under a total capacity constraint $C(t)$. t represents a time slot. We suppose that during a defined time period (measured in slots), in absence of control, predicted demand would exceed available capacity. We call this period a DR period. We denote by DE_a and DE_h the functional groups in charge of decision taking (Decision Entities) at the aggregator side and at the user h side (one per home), respectively. DE_a is in charge of allocating power to each household (C_{ht}), under the total power constraint. For each house h, DE_h has two main roles: collecting information on variables monitored at user premises (state of appliances, local temperature, etc.); enforcing control decisions received from DE_a (e.g. by controlling the appliances). More details will be given in Sect. 4 when introducing the considered allocation schemes.

A utility function is defined for each controlled appliance to express the impact of its operation on user's satisfaction. We assume electrical appliances are classified among A classes. Appliances of the same class have similar usage purposes (e.g., heating) but may have different operation constraints. Appliance of class a at home h operates within a given power range $\left[P_m^a(h), P_M^a(h)\right]$.

Following [1], a specific utility function is modeled for each class of appliances based on usage patterns, criticality, users' preferences and exogenous variables (e.g. external temperature). The utility of an appliance is expressed as a function of its consumption or of some monitored variables (see Sect. 5 for an example).

In the present work, we introduce two levels of utility per appliance, vital and comfort. The first one expresses high priority targets of high impact on users' wellbeing and the second one expresses less essential preferences.

For notation, we write utilities as vital/comfort pairs: $U^a_{ht} = \left(U^a_{v_{ht}}, U^a_{c_{ht}}\right)$ denotes utility of appliance a at time t for home h. Control decisions are based on the lexicographical order comparison of utility values: higher vital value is always preferred regardless of the comfort value. Formally, for two utilities U^a_{ht} and U'^a_{ht}, we say $U^a_{ht} > U'^a_{ht}$ iff $U^a_{v_{ht}} > U'^a_{v_{ht}}$ or ($U^a_{v_{ht}} = U'^a_{v_{ht}}$ and $U^a_{c_{ht}} > U'^a_{c_{ht}}$).

Utilities can be summed using element-wise addition.

The maximal values of utilities depend on the home, type of appliance and time: they represent how the importance of appliances is modulated depending on the preferences and service agreement of the users. We assume that each house has a subscribed power limit $L(h)$ sufficient to achieve a maximal utility.

The optimization problem considered in this chapter consists in maximizing the total utility (using the lexicographic total order) of users under system constraints. Fairness is introduced through the vital/comfort separation: no comfort power is allocated to any house if some vital need can be covered instead. We do not directly focus on revenues but expect that reaching maximal users' utility leads to maximal gains for all involved players. Utility companies can provide better services for a given total allocated power, which should translate into higher revenues, or reduce the expenses in the capacity market for a given level of service, which should reduce total costs. End users can save money due to attractive prices they get for participating to the service and adjusting energy consumption to their predefined policies. Notation is summarized in Table 1.

4 Allocation Schemes

We present here two reference schemes that will be used for benchmarking purposes, along with our proposed solution.

4.1 Benchmark Schemes

The two following schemes were proposed in [1].

Table 1 Table of notation

System parameters and exogenous variables	
H	Number of homes
A	Number of appliance classes
$P_m^a(h)\,/\,P_M^a(h)$	Minimal/maximal power required by appliance a in home h
$C(t)$	Available power capacity at time slot t
$L(h)$	Subscribed power for home h
t_M	DR period duration in time slots
$T_m(h)\,/\,T_M(h)$	Minimal/maximal acceptable indoor temperature for home h
$T_0(h)\,/\,T_P(h)$	Initial/preferred indoor temperature for home h
$T_e(t)$	Exterior temperature at time t
$F(h), G(h)$	Coefficients for temperature dynamics in home h
Control variables and controlled variables	
$U_{ht}^a = \left(U_{V_{ht}}^a, U_{C_{ht}}^a\right)$	Utility (vital, comfort) of appliance a in home h at time t
X_{ht}^a	Power consumed by appliance a in home h at time t
x_{ht}^a	Activity indicator of appliance a in home h at time t (0 or 1)
T_{ht}	Temperature of home h at time t
C_{ht}	Capacity limit allocated for home h at time t
g_{ht}	Greedient of the utility function at point C_{ht} for home h at time t

4.1.1 Global Maximum Utility

The centralized global optimization is formulated by Eq. (1a, b, c).

$$\max_{X_{ht}^a, x_{ht}^a} \sum_{t=1}^{t_M} \sum_{h=1}^{H} \sum_{a=1}^{A} U_{ht}^a \tag{1a}$$

s.t.

$$\sum_{h=1}^{H} \sum_{a=1}^{A} X_{ht}^a \le C(t), \ \forall t \tag{1b}$$

$$P_m^a(h) x_{ht}^a \le X_{ht}^a \le P_M^a(h) x_{ht}^a, \quad \forall t, \forall h, \forall a \tag{1c}$$

$$x_{ht}^a \in \{0, 1\}, \quad \forall t, \forall h, \forall a \tag{1d}$$

Equation (1a, b, c) can be solved if all information about appliances and their utility functions are transmitted by the home repartitors DE_h to the aggregator DE_a, which can then compute an optimal global solution and notify the repartitors accordingly.

Decision variables in this case are variables x_{ht}^a and X_{ht}^a. Binary variables x_{ht}^a correspond to turning ON (i.e., $x_{ht}^a = 1$) or OFF (i.e., $x_{ht}^a = 0$) appliance a at home h on time slot t. If appliance is turned ON, power allocation X^a_{ht} can take values between a minimum value $P^a_m(h)$ and a maximum value $P^a_M(h)$ (see Eq. 1c).

While being optimal with respect to the utilities (by design), this allocation, called *GM*, has two major drawbacks. First, it requires computing the solution of a complex problem, which may raise scalability issues. Second, information harvesting may cause privacy issues that can affect the acceptance of the control scheme by users. Thus, it may be preferable to store information locally at homes with local intelligence. This leads to the following scheme.

4.1.2 Local Maximum Utility

This control scheme, denoted *LM*, considers only one-way communication from DE_a to DE_h (no feedback from DE_h to DE_a). Decisions are made at both levels.

First, DE_a allocates power to homes proportionally to their subscribed power, so the power allocated to home h is $C_{ht} = \frac{L(h)}{\sum_i L(i)} C(t)$.

Then, at each home h, DE_h decides the corresponding allocation per appliance by solving the restriction of (1a, b, c) to h, using C_{ht} instead of $C(t)$.

By design, *LM* is scalable (only local problems are solved) and private information disclosure is kept to a minimum. The drawback is that the corresponding allocation may be far from optimal [1].

4.2 Greedient Approach

We now propose a two-way scheme that aims at achieving a trade-off between performance, scalability and privacy.

To reach privacy and scalability goals with limited feedback, we propose a simple primal decomposition of the global *GM* problem into a master problem, described in Eq. (2a, b, c), and subproblems, described in Eq. (3a, b).

Master problem

$$\max \sum_{h=1}^{H} U_h \tag{2a}$$

$$\sum_{h=1}^{H} C_{ht} = C(t), \forall t \tag{2b}$$

$$C_{ht} \geq 0, \quad \forall h, \forall t \tag{2c}$$

Subproblems
For each home h, the following MILP is solved:

$$U_h = \max \sum_{t=1}^{t_M} \sum_{a=1}^{A} U_{ht}^{a} \tag{3a}$$

$$\sum_{a=1}^{A} X_{ht}^{a} \leq C_{ht}, \ \forall t \tag{3b}$$

If the C_{ht} are known, the subproblems (3a, b) can be solved like in the *LM* scheme. The main issue is the master problem (2a, b, c): how to shape an optimal per-home allocation while keeping the full characteristics of appliances private?

To treat this problem, we propose a new heuristic called the Sub-Greedient method (*SG*). This heuristic is inspired by the Sub-Gradient method [6], but is adapted to take into account the specificities of our model. In particular, we introduce the notion of *Greedient*, inspired by the gradient method and the metric used to sort items in the knapsack greedy approximation algorithm.[1] Greedients will be used instead of more traditional (sub)-gradient approaches to estimate the utility meso-slope of a given house.

We briefly describe the main steps of *SG*:

- *SG* needs to be bootstrapped with an initial power allocation.
- DE_a transmits to each home DE_h the current allocation proposal $C_{ht} \forall t$. DE_h then solves the corresponding subproblem 3a, b). It sends back the total utility U_h feasible, along with the Greedient associated to the current solution.
- Using the values reported by homes, DE_a then tries to propose a better solution.
- The process iterates for up to K_{MAX} iterations, and return the best solution found.

We now give the additional details necessary to have a full view of the solution.

4.2.1 Initial Allocation

Following [1], we use a round-robin strategy for the first allocation (before the first feedback): we allocate to some houses up to their power limit until the available capacity $C(t)$ is reached; we cycle with time the houses that are powered. The interest for *SG* of such an initial allocation (e.g. compared *LM*) is that it breaks possible symmetries between homes and gives an initial diversity that will help finding good Greedients.

[1]The term *discrete gradient* could be used instead of this neologism. However, the greedient, which will be formally defined below, differs from the usual definition of a discrete gradient [7].

4.2.2 Greedient

We define the greedient g_{ht} as the best possible ratio between utility and capacity improvements of home h at time t. Formally, if $U'_h(\Delta C_t)$ represents the best feasible utility for home h if its current allocation is increased by ΔC_t at time t, we have

$$g_{\mathrm{ht}} := \max_{\Delta C_t > 0} \frac{U'_h(\Delta C_t) - U_h}{\Delta C_t}.$$

To compute g_{ht}, we define the greedient g^a_{ht} of an appliance a as follows: for a given allocation C_{ht}, $C^0_{ht} \geq 0$ represent the capacity unused by house h at time t in the optimal allocation. $U^{a'}_h(\Delta C_t)$ represents the maximum utility for appliance a if an additional capacity of up to ΔC_t is added its current consumption. Then we have

$$g^{\mathrm{a}}_{\mathrm{ht}} := \max_{\Delta C_t > 0} \frac{U^{a'}_h(C^0_{ht} + \Delta C_t) - U^a_h}{\Delta C_t}.$$

The greedient of a home is the greedient of its best appliance : $g_{ht} = \max_a g^{\mathrm{a}}_{\mathrm{ht}}$.

Note that if we suppose that the utility functions have a diminishing return property, which is the case for our numerical analysis, the greedient of an appliance is equivalent to the gradient of the utility function when $C^0_{ht} = 0$ and continuous variation of power is allowed: for these situations, the best efficiency is observed for $\Delta C_t \to 0$. The only difference (under diminishing return assumption) is when allowed allocations are discrete: the greedient will consider to the next allowed value while the gradient will report 0.

Remark The improvement advertised by the greedient is only valid for a specific capacity increase, which is not disclosed to DE_a to prevent the central entity to infer the characteristics of users based on their inputs. As a result, the greedient hints at the potential interest of investing additional capacity to a given home, but it is not reliable. This is the price we choose to pay to limit privacy issues.

4.2.3 Finding Better Solutions

To update the current solution at the kth iteration, DE_a does the following:

- It first computes values $\alpha_k g_{ht}$ $\forall h$ $\forall t$. These values represent potential increase of C_{ht}. The values of α_k, called the *step size*, are discussed below.
- It then adjusts the new values of C_{ht}. based on these values, while staying positive and fitting the capacity constraints.

For the adjustment phase, it is important to deal with cases where allocation update $\alpha_k g_{ht}$ is larger than available capacity $C(t)$ or even maximum subscribed power $L(h)$ of home h, so we first cap $\alpha_k g_{ht}$ at the minimum between power limit of the smallest home $(L_m := \min_h L(h))^2$ and system capacity $C(t)$. We therefore define $\beta_{kht} = \min(\alpha_k g_{ht}, L_m, C(t))$.

Then for each t, we remove some positive common value λ_t to the C_{ht} to keep the sum of the allocations equal to the total capacity $C(t)$. To avoid houses with low C_{ht} to be badly impacted (in particular to avoid negative allocations that will be impossible to enforce), a subset I_t of the houses will be "protected" so that their values cannot decrease. In details, we do the following, starting with $I_t = \emptyset$:

- We compute λ_t such that the values

$$C'_{ht} = \begin{cases} C_{ht} + \max\{\beta_{kht} - \lambda_t, 0\} \text{ if } h \in I_t, \\ C_{ht} + \beta_{kht} - \lambda_t \text{ otherwise,} \end{cases} \tag{4}$$

  sum to $C(t)$. See [8, 9] for more details.
- We protect (e.g. add to I_t) all houses that get a negative value C'_{ht}.
- We iterate the steps above until all C'_{ht} from Eq. (4) are positive. DE_a then proposes C'_{ht} as a new solution to investigate.

Remark The solution described here applies to a 2-level hierarchy (DE_a, DE_h), but it can be generalized to M levels to take into account different aggregation points on a hierarchical distribution network: considering an aggregation point m at a certain level, the greedient for m is the maximal greedient of its children. The adjustment phase can take into account capacity constraints of m, such as static power limits at each level of the hierarchical distribution network.

Also note that the proposed scheme does not require all houses to communicate simultaneously: it can run asynchronously. In fact, as soon as at least two homes respond, a local reallocation can be made: we just need to restrict the problem to the corresponding subset of homes, using their current cumulated allocation as capacity limit.

4.2.4 Choosing the Step Size

The step size α_k for each iteration k is a crucial parameter to speed up resolution. Intuitively, large values of α_k make the allocation update (dictated by $\alpha_k g_{ht}$) useful for high consumption appliances, while lower values are more adapted to low consumption appliances.

Among the step size sequences proposed for subgradient methods, we consider for our performance analysis the two following ones (see [6]):

A diminishing non-summable step size rule of the form $\alpha_k = \frac{a_1}{\sqrt{k}}$.

[2] We chose the capacity of the smallest home instead of the capacity of the current home to avoid a *masking* effect where the demands of larger homes cloud the demands of smaller homes.

A constant step length rule of the form $\alpha_k = \frac{a_2}{\|g_{ht}\|_2}$, where $\|g_{ht}\|_2$ is the euclidean norm of the vector of all greedients.

The value of parameter a_1 (resp. a_2) is currently manually adjusted to provide the best result, but we believe that an automatic estimation of the best value given the static parameters of a given use case is a promising lead for future work.

5 Numerical Analysis

We now evaluate the performance of our proposed solution for a specific use case.

5.1 Parameters and Settings

We consider three typical types of appliances ($A = 3$): lighting ($a = 1$), heating (index $a = 2$) and washing machines (index $a = 3$). For these appliances, user's perceived utility respectively depends on: instantaneous power consumption; exogenous variables (temperature); the completion of a program. Utility functions for these appliances have a vital and a comfort component. For lighting, vital light utility is fully obtained as soon as the minimal light power $P_m^1(h)$ is reached, while comfort utility linearly grows from $P_m^1(h)$ to $P_M^1(h)$ (see Fig. 1). For heating, vital utility linearly grows until the minimum tolerable temperature $T_m(h) := 15°\text{C}$ is reached, while comfort utility linearly grows from $T_m(h)$ to the preferred temperature $T_P(h) := 22°\text{C}$ (see Fig. 2). For washing machines, an operation of duration $D(h)$ needs to be scheduled between an earliest start time $t_s(h)$ and a deadline $t_d(h)$. Once started, an operation cannot be interrupted. Vital utility function is maximal whenever the operation is successfully scheduled, while comfort utility depends on the execution time, e.g. the sooner the better for this use case (see Fig. 3).

To study the performance of the control schemes for several values of capacity, we choose the following system parameters:

- The size of the system is $H = 100$ houses.
- We select a slot duration of 5 min.
- The DR period is set to $t_M = 100$ slots ($\cong 8$ h).
- We suppose a constant external temperature $T_e(t) = 10°\text{C}\ \forall t$ and an initial temperature $T_0(h) = 22°\text{C}\ \forall h$.
- We suppose the same maximal utility values for all appliances, homes and time, arbitrary set to 1.
- Temperature in homes evolves according to a simplified conductance/capacity model that leads to the following dynamics:

Fig. 1 Utility of light power

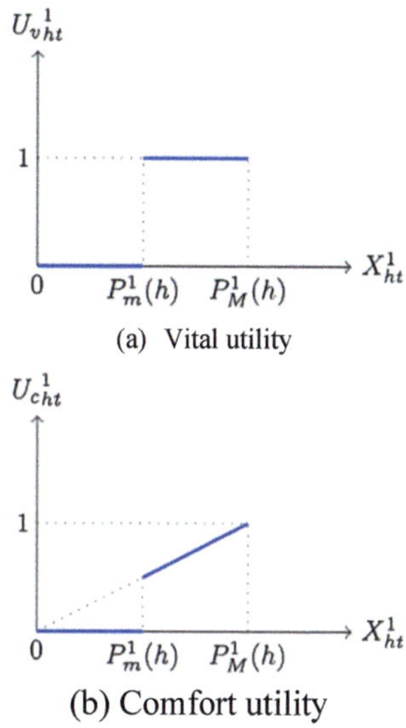

Fig. 2 Utility of T_{ht}

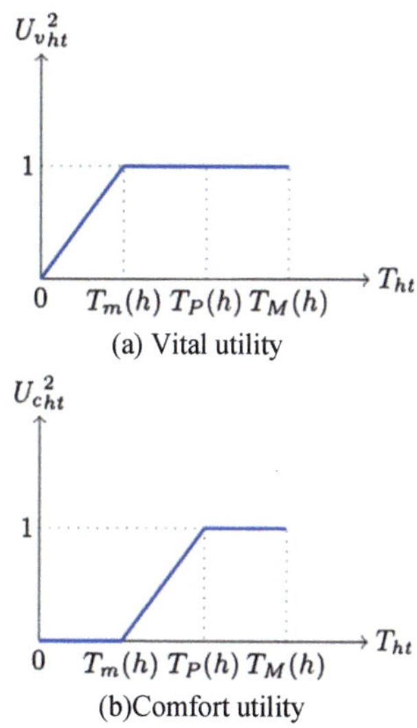

$$T_{ht} = T_{h(t-1)} + F(h)X^2_{ht} + G(h)\big(T_e(t) - T_{h(t-1)}\big).$$

- Two types of houses are considered (see Tables 2, 3, 4 and 5). Compared to class 1, class 2 has a better energetic performance (less light power required, better insulation and more efficient washing machine), resulting in a lower power limit $L(h)$).

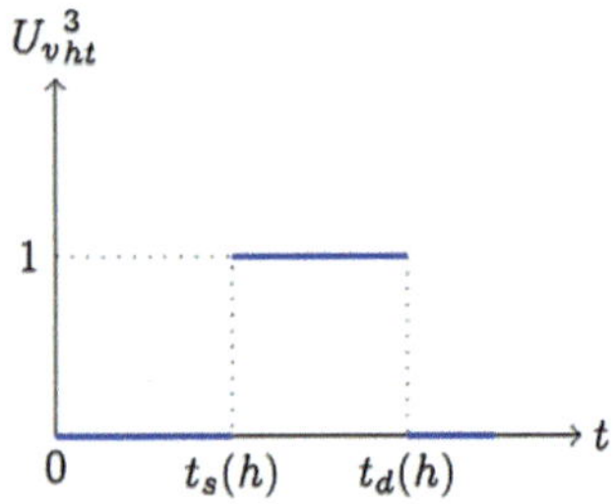

(a) Vital utility

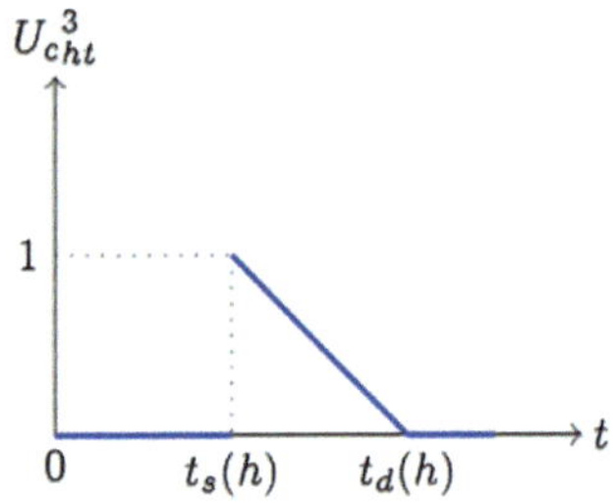

(b) Comfort utility

Fig. 3 Utility of a washing machine

Table 2 Lighting parameters

Type	$P^1_m(h)$	$P^1_M(h)$
1	50	1000
2	50	500

Table 3 Heating parameters

Type	$P^2_m(h)$	$P^2_M(h)$	F(h)	G(h)
1	1000	4000	0.0017	0.075
2	1000	2000	0.0008	0.0365

Table 4 Washing machine parameters

Type	$P^3_m(h) = P^3_M(h)$	D(h)	$t_s(h)$	$t_d(h)$
1	600	8	1	100
2	400	6	1	100

Table 5 Houses parameters

Class	Lighting type	Heating type	Washing machine type	L(h)
1	1	1	1	5600
2	2	2	2	2900

We suppose that the total available power is constant over the DR period, $C(t) = C$. We analyze the model for different values of C, ranging from low (only one type of appliances can be used) to full capacity (all appliances can be used).

While this model is simple (three types of appliances, constant values), we believe that the knowledge required to compute good solutions is sufficient to capture the trade-off between the efficiency of an allocation and the privacy of the users.

For the Sub-Greedient problem, we fix the maximum number of iterations to $K_{MAX} = 100$. Two variants are considered (cf Sect. 4.2.4): $SG-1$ uses a diminishing step ($a_1 = 1200000$) and $SG-2$ uses a constant step length ($a_2 = 6000$). Parameters a_1 and a_2 were manually tuned.

The numerical analysis of the various presented mixed integer linear problems has been carried out using IBM ILOG CPLEX [10].

In the following, we discuss two cases: homogeneous and heterogeneous. For the homogeneous case, all houses belong to class 1 and for the heterogeneous one, we suppose 50 houses of class 1 and 50 houses of class 2.

5.2 Results on the Homogeneous Case

The main results on the homogeneous case are presented in Fig. 4. It displays the relative utility per home over the DR period as a function of the available capacity $\boldsymbol{C}$, for the four supposed schemes: ***GM***, ***LM***, $\boldsymbol{SG-1}$ and $\boldsymbol{SG-2}$.

The maximal feasible utility (vital and comfort) is normalized to 1 which is reached when all appliances from all homes of a given class reach their maximal utility. Another value of interest for vital utility is 0.58, which corresponds to situations where all houses are able to achieve vital light ($P_m^1 = 50$ W) but none has the power necessary for heating ($P_m^2 = 1000$ W) so there is no control of temperature, nor washing machines are scheduled. When washing machine are scheduled in addition to lights (without heating), vital utility reaches 0.92.

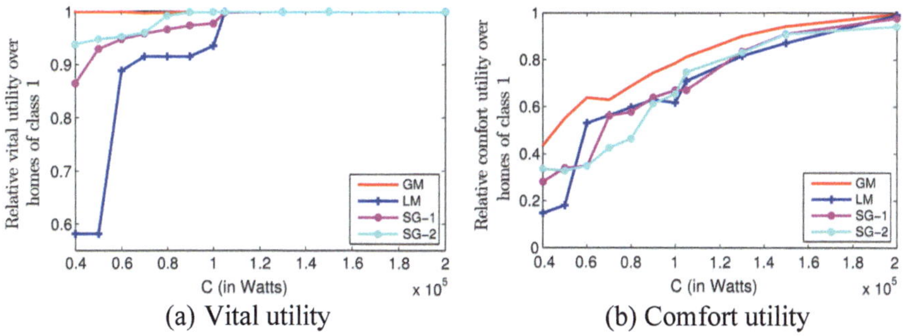

Fig. 4 Relative utility as a function of the available capacity (homogeneous case, class 1)

GM, the optimal solution, achieves maximal vital utility even for very low capacities (down to 4×10^4), thanks to its ability of finding a working rolling allocation that allows all houses to use heat for a sufficient part of the period while scheduling the other appliances. Based on *GM* results, we can measure the gap between optimal allocation and allocations obtained with *LM*, $SG-1$ and $SG-2$.

Using a static allocation, *LM* struggles for rising the vital utility above the 0.58 and 0.92 thresholds. It can schedule washing machines when $C \geq 6 \times 10^4$ ($P_m^3 = 600$ W per home). It can only start to use heat for $C = 10^5$ (1000 W per house). Maximal vital utility is reached for $C = 105 \times 10^3$ (1050 W per house) and maximal utility (vital and comfort) necessarily requires $C = 2 \times 10^5$ (2000 W per house). However, *LM* achieves descent performance results for high enough available capacity values.

Our proposal, $SG-1$ and $SG-2$, stands in-between the two opposite schemes *GM* and *LM*. Indeed, for very low capacity values ($C < 2 \times 10^4$ W), $SG-1$ and $SG-2$ perform slightly better than *LM*. Then, for capacity values up to $C = 6 \times 10^4$ W, $SG-1$ and $SG-2$ significantly over-perform *LM*. In particular, the schemes manage to activate most washing machines starting from $C = 4 \times 10^4$ W (400 W per home on average, to compare with the 600 W required to operate a washing machine). It is able to improve the vital utility of houses for values below $C = 10^5$, even if it fails to perform as good as *GM*. With respect to the comfort utility, it performs on par with *LM* even in situation where it devotes resources on heating (for vital utility) while *LM* does not.

As for the number of iterations required to reach the best solution, $SG-1$ takes around 12 iterations on average and $SG-2$ takes 21 iterations. The slowest convergence is observed for capacity $C = 2 \times 10^5$ W (maximal considered C) where $SG-1$ takes 95 iterations and $SG-2$ takes 98 iterations.

5.3 Results on the Heterogeneous Case

Figure 5 illustrates the main results for the heterogeneous case for homes of classes 1 and 2. The optimal solution given by *GM* shows that, for vital utility, the results are pretty much similar for both classes to the homogeneous case, with maximal value obtained even for low capacities (down to 4×10^4). For the comfort utility, however, *GM* leads to better values for class 2 compared to class 1. This is due to the fact that class 2 houses have better energetic performance, so once vital utility is ensured for all, it is more efficient to allocate energy to homes of class 2.

The same reason explains the poor performance of *LM*. Let us remember that the static allocation is proportional to the maximum power $L(h)$ of homes. So for a given capacity, class 1 homes get more power than class 2 ones. As a result, while performance of class 1 is satisfactory, performance of class 2 is terrible despite the better energy performance of class 2 homes. In particular, the capacity required for

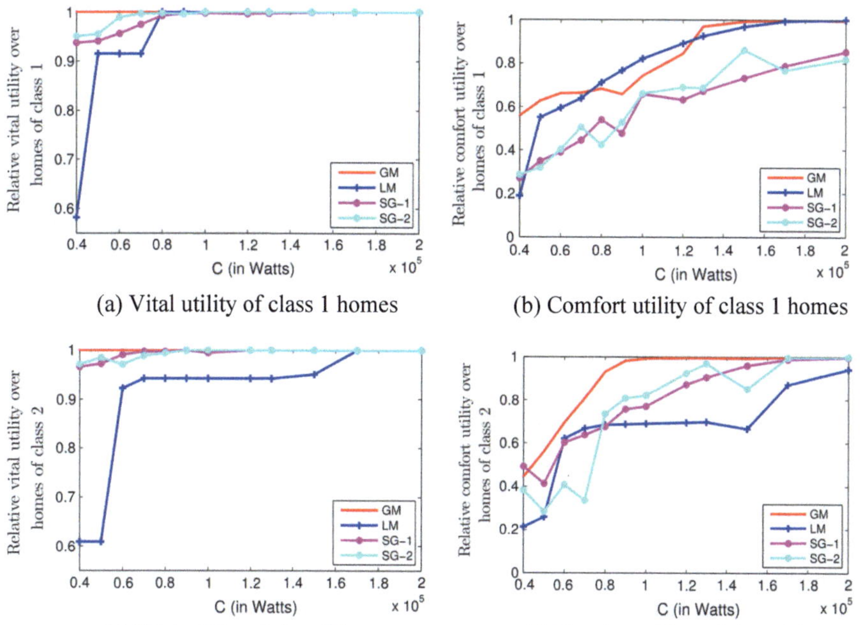

(a) Vital utility of class 1 homes

(b) Comfort utility of class 1 homes

(c) Vital utility of class 2 homes

(d) Comfort utility of class 2 homes

Fig. 5 Relative utility as a function of the available capacity (heterogeneous case, classes 1 and 2)

class 2 houses to achieve maximal vital utility is very high: $C = 1.7 \times 10^5$, which corresponds to 1700 W per house (regardless the class).

For lower capacity values, performance depend on the possibility of scheduling washing machines and heating. A global capacity $C = 5 \times 10^4$ will only allow homes of class 1 to schedule their washing machines. Actually, for this capacity value, homes of class 1 will get a power limit of 659 W whereas homes of class 2 will only get 341 W (insufficient for turning on a washing machine). For capacity values above $C = 7 \times 10^4$ (corresponding to slightly more than 450 W for each home of class 2), performance obtained corresponds to washing machines being scheduled and minimum lighting requirements being fulfilled for both classes while only homes of class 1 have their vital heating requirement.

As for $SG - 1$ and $SG - 2$, we observe that compared to the homogeneous case, the performance of our solution SG is now closer to GM than to LM. Indeed, SG is capable of providing near maximal vital utility for $C = 0.6 \times 10^5$ W.

As for the number of iterations corresponding to the last solution improvement, $SG - 1$ takes up to 3 iterations and $SG - 2$ takes 14 iterations on average.

5.4 Discussion

The results presented previously can be seen as intuitive. However, they show some interesting tradeoffs that need to be considered when proposing a DR solution. As suggested by the results, a fine grained control may not be always needed depending on the system's available capacity: its value is the highest for very low capacities. As a matter of fact, a solution based on static information can have high performance thanks to the deployment of a fine grained solution in smart-homes that can manage to efficiently schedule appliances based on user's needs when capacity is high enough.

However, producing an efficient solution based on static information is challenging especially when considerations like heterogeneity in users' needs and time-dependent constraints for appliances (e.g., minimum duration of operation) are supposed. In addition, one may imagine that available capacity will vary in time which also increases complexity of finding such a solution.

To address the lack of visibility while preserving privacy, a solution that uses limited information and is able to update allocations based on actual needs is needed. Actually, if high performance can be delivered by such a solution, a centralized approach will not be required.

The hierarchical solution proposed in this chapter is a promising one that is capable of addressing this need. It also seems to deal well with appliances introducing time dependence between time slots even if it is not a built in feature. It is capable of rendering a performant solution is a reasonable amount of iterations.

6 Conclusions

We propose an IoT-based demand response approach, named *Sub-Greedient*, that relies on a 2 level control scheme. Intelligence (decision taking) is split between a centralized component and a set of local controllers (one per home). The proposed control approach enables reaching good performance in terms of the utility perceived by the users while keeping privacy and providing scalability. Moreover, priority is provided for critical needs, which introduces some degree of fairness among households.

We show that the approach outperforms schemes where the central controller takes decisions based solely on the available total capacity and on static (contract-based) information about the households. Results for the considered use cases show that the proposed scheme requires a limited number of iterations to render effective solutions. Moreover, the proposed solution is robust as the algorithm stays inside the set of feasible allocations and can tolerate lost or delayed information.

Future work will encompass a study on the power allocation algorithms for the *Sub-Greedient* scheme considering the effect of communication impairments on the global performance and on fairness. We will analyze the cost savings under realistic cost models, looking for solutions that will target minimizing the total expenses a provider will incur in the capacity market while keeping a predefined level of service.

Acknowledgements The work presented in this chapter has been partially carried out at LINCS (www.lincs.fr)

References

1. Kaddah R, Kofman D, Pióro M (2015) Advanced demand response solutions based on fine-grained load control. In: IEEE international workshop on intelligent energy systems, San Diego (USA), October 2015, pp. 38–45
2. Deng R, Lu R, Xiao G, Chen J (2015) Fast distributed demand response with spatially-and temporally-coupled constraints in smart grid. IEEE Trans Ind Inf 11(6):1597–1606
3. Vivekananthan C, Mishra Y, Ledwich G, Li F (2014) Demand response for residential appliances via customer reward scheme. IEEE Trans Smart Grid 5(2):809–820
4. Li N, Chen L, Low SH (2011) Optimal demand response based on utility maximization in power networks. In: IEEE power and energy society general meeting, Detroit (USA), July 2011, pp 1–8
5. Shi W, Li N, Xie X, Chu C-C, Gadh R (2014) Optimal residential demand response in distribution networks. IEEE J Sel Areas Commun 32(7):1441–1450
6. Boyd S, Xiao L, Mutapcic A (2003) Subgradient methods. Lecture notes of EE392o, Stanford University (USA), Autumn Quarter
7. Bagirov AM, Karasözen B, Sezer M (2008) Discrete gradient method: derivative-free method for nonsmooth optimization. J Optim Theory Appl 137(2):317–334
8. Pióro M, Medhi D (2004) Routing, flow and capacity design in communication and computer networks. Morgan Kaufmann Publishers
9. Held M, Wolfe P, Crowder HP (1974) Validation of subgradient optimization. Math Program 6(1):62–88
10. IBM Ilog CPLEX optimizer. http://www-01.ibm.com/software/commerce/optimization/cplex-optimizer/

Part IV
Internet-of-Things, Big Data, Cloud—Security, Privacy and Reliability

Securing Smart Cities—A Big Data Challenge

Florian Gottwalt and Achim P. Karduck

1 Introduction

Senseable Cities as envisioned by from Carlo Ratti at MIT, with their promised resource efficiency and inhabitant life quality improvements, are ultimately based on the collection and insightful processing of Big Data [1]. In the last years, progress in science and industry has already led to a huge rise of the volume, variety, and velocity of the data that is generated. This resulted in the terms Big Data and Big Data Analytics—certainly a long-term development as stated by Coffman and Odlyzko [2]. This development opens new opportunities for business and society and drives innovation, such as for our envisioned Smart Cities [3, 4]. However this development also poses a challenge for data scientists and real nightmares for network security analysts. Network security analysts are not only facing a higher amount of attacks, the attacks are also getting more and more customised to their individual targets. This makes it extremely difficult to detect and prevent intrusion attempts with conventional techniques in the mass of legitimate traffic. As a consequence, this could decelerate our techno-social visions, such as Smart Cities, to become reality. For these visions to materialise, it is essential that all assets can communicate in a secure and reliable fashion at any time, plus to guarantee their evolution of their assets and infrastructures as a whole. Our focus is on the cyber-security challenges for Smart Cities. Despite the visions, research on how to address the versatile cyber security challenges in developing technological assets and infrastructures like for Smart Cities is very limited [4–6].

F. Gottwalt (✉)
University of New South Wales, Canberra, Australia
e-mail: f.gottwalt@unsw.edu.au

A.P. Karduck
Furtwangen University, Furtwangen im Schwarzwald, Germany
e-mail: karduck@hs-furtwangen.de

L. Ismail and L. Zhang (eds.), *Information Innovation Technology in Smart Cities*,
https://doi.org/10.1007/978-981-10-1741-4_13

Assuming, that a Smart City is supervised by one or more operation centres, security data has to be aggregated and analysed in these facilities. To understand and make use of this mass of data, our human abilities simply do not suffice without proper tools. Log Management alias Security Information Management (SIM) or more sophisticated systems like Security Information and Event Management (SIEM) systems are the de facto standard to take care of this challenge. However, due to their frequent deprecated architecture, they show weaknesses when it comes to Big Data.

Figure 1 illustrates a common architecture of a SIM system. In a first step generated events from security-relevant devices are normalised and consolidated into a centralised database. Simple aggregation and correlation rules are then applied to summarise the received data into a more concise summary. This aggregation is for monitoring purposes displayed in a dashboard, which at the same time allows searching through the data.

The first major challenge this architecture is facing is how to generally handle and process a huge amount of data, ideally in real time. This is not a specific issue

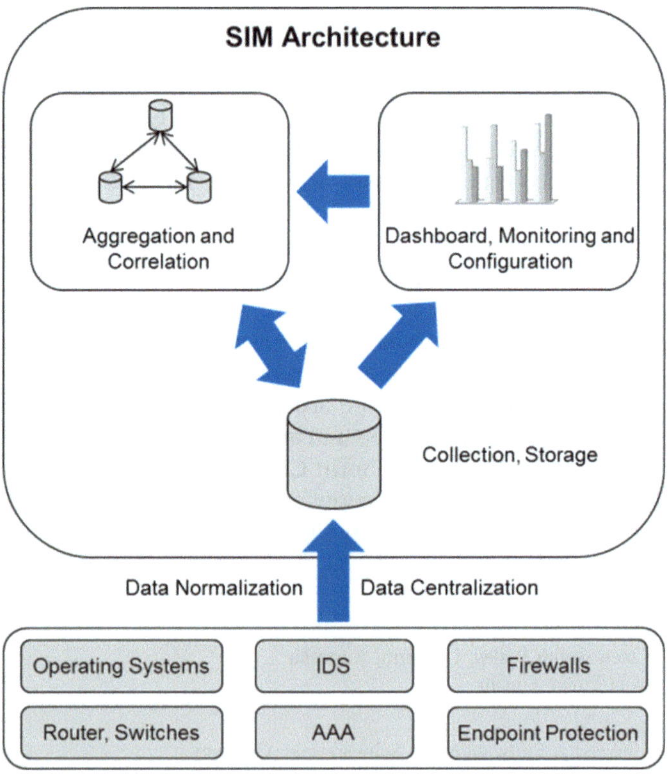

Fig. 1 SIM architecture

related to log management and thus there are already many approaches how to address that problem [7].

The second severe challenge is, after being able to process this high-volume of data, how to get insights out of it. Conventional methods, which are very often based on deterministic rules, start to struggle when it comes to a certain point of complexity and volume. The Big Data approach to find similarities and detect abnormalities in high-volume data sets is termed "data mining", here applied for the IT Ecosystem of an organisation. Much research has been done in applying data mining techniques to detect intrusions [8–10]. However, the results have shown, that this is a challenging area and alternative approaches, such as simple data summarization and visualisation of events are still very important. The application of Big Data principles, particularly to SIM environments, has not been discussed yet and only Zope et al. as well as Asanger and Hutchinson investigated into the application of data mining techniques to the closely related SIEM environments [11, 12]. In addition to the low number of studies in that area, they have focused on specific data mining algorithms and there remains the need for a general overview of enhancement possibilities.

The next section provides an overview about the status quo of research in this area and is followed by two proposed concepts on how to enhance SIM systems with Big Data principles. The concept leads to an architecture and implementation, which is applied and evaluated on real-world data.

2 Related Research

Security challenges for Smart Cities operation have only been researched very limitedly. With respect to the complexity of Smart Cities, an initiative to identify future cyber security challenges and solutions in the context of Smart Cities has to be elaborated trough collaboration and co-innovation between public-private-partnerships, including governments, companies, and social involvement of the citizens [5]. Bartoli et al. [6] have categorised security and privacy challenges coming up with Smart Cities from an architecture and Internet of Things technology perspective.

To our knowledge, no particular work has been conducted on the open questions of readiness, challenges, and required improvements of current state of the art security technologies, like SIM, with respect to our envisioned Smart Cities. Nevertheless, a lot of research has been made in related fields, like intrusion detection based on log events and data mining techniques for log summarization. Further, a large number of studies in data mining have gained attention in addressing network security incidents, including network intrusion detection [9, 10, 13]. SIM in the light of big data is timely in terms of its adoption in large organisations, as well as the Cyber-Security infrastructure of our shaping Smart Cities [14, 15].

Additionally, the success in other domains has proven that exactly these techniques are applicable to generate value in commercial areas. Due to those two points, it is surprising to experience the actual success of data mining methods for intrusion detection in operational environments. Despite the massive potential and huge academic research effort, one can find data mining techniques rarely used in commercial security solutions.

This circumstance is based on various challenges, which a Big Data enhanced SIM system is facing. The first challenge, coming with the usage of data mining techniques in SIM environments, is related to the nature of logs. Most of the successfully applied data mining techniques so far, for other fields of application, were applied on numerical data. This is also the area, where data mining algorithms were originally built for. They are built for doing mathematical operations and statistical evaluations—on numerical data. The problem coming up with data types used in log events is that they are mostly of categorical nature and hence cannot be applied directly. Another challenge to apply data mining algorithms to log collections is also based on the nature of logs—the structure. Various event structures from different devices with diverse data types make it difficult to apply techniques on an unnormalized log set. Due to that, nearly all the researched data mining approaches for intrusion detection have been applied on preprocessed/normalised data sets, like the KDD cup set [16].

To highlight potential enhancements and upcoming challenges, the functions of a SIM architecture, displayed in Fig. 1, will be investigated separately and supplemented with Big Data methods in the next section. Based on further elaboration of related research, the status quo in the area of SIEM systems will be described in more detail, followed by proposed approaches for the complementing fields of intrusion detection based on data mining and data mining for log collections.

3 Enhancing SIM Systems

The four main drawbacks of traditional SIM systems, when it comes to large amounts of data are as displayed in Fig. 2. These refer to:

- Collection and Storage
  SIM solutions often collect data to relational database architectures, which are neither designed for unstructured data nor for the scaling of data in the magnitude expected in Smart Cities [17].
- Normalization
  All data needs to be normalised before the system can make use of it. This normalisation process costs additional processing power and makes it extremely hard to get unstructured data into a valuable format.
- Aggregation
  The "intelligence" of a SIM system is very often based on deterministic rules to aggregate data and reveal suspicious events. Due to the mass of logs, those

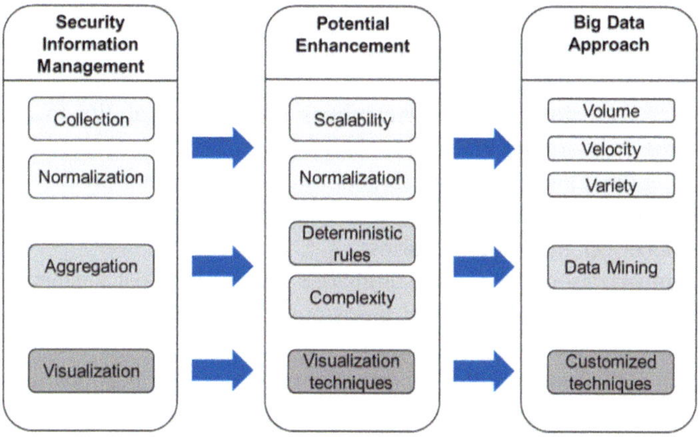

Fig. 2 Potential enhancements for SIM systems using big data approach

conventional methods are however insufficient when it comes to a certain level of complexity [18].

- Virtualization
  Common SIM systems offer reporting capabilities and dashboards which use charts, diagrams and other graphics to visualise data. However, most of the systems were not developed with a focus on visualisation and hence only basic methods are used without exploiting the full potential.

The architectural drawbacks of a SIM system are already addressed in the general definition of Big Data. The volume attribute in Big Data enables us to process huge amounts of data. The ability to work with data flowing in a very fast manner is addressed with the term velocity, and the capability to handle data, in many forms, e.g. unstructured data is covered with the attribute variety.

In addition to the architectural enhancement possibilities, there is also potential to improve the actual intelligence of a SIM system to aggregate data and reveal suspicious events.

Due to the mass of logs, conventional methods to reveal anomalous events are insufficient as most of their intelligence is based on deterministic rules. This is inadequate when it comes to extremely complex environments, like our envisioned Smart Cities, as only things, which have been defined beforehand, can be detected. The defined Big Data approach to deal with that problem is data mining. It offers wide opportunities to enhance the intelligence of a traditional SIM system. Three different ways how data mining techniques can be used to improve the aggregation process were identified:

- Unsupervised intrusion detection
- Supervised intrusion detection
- Data summarization.

Unsupervised intrusion detection methods can be applied ad hoc without any further knowledge and enable to detect abnormalities by highlighting frequent relations and infrequent occurrences.

Supervised intrusion detection techniques have the capability to classify new events on the basis of previous classified events. The events are classified into normal and anomalous events, which simplifies the analysing process by reducing the amount of events to be analysed. It can be applied on top of unsupervised techniques to cover real-time scenarios.

Data summarization techniques summarise all events to get a general overview. Although they are not able to detect intrusions directly, they reduce the number of logs to be examined drastically.

Common SIM systems offer reporting capabilities and dashboards which use charts, diagrams and other graphics to visualise data. However, most of the systems were not developed with a focus on visualisation and hence only basic methods are used without exploiting the full potential. With the rise of Big Data, visualisation tools became an essential part in finding patterns and abnormalities. By using adequate visualisation techniques, data is easier perceptible and interpretable by human beings. Interactive tools, that not only support visualisation but also data exploration, can improve the visualisation part as well.

To see the potential of the identified enhancement possibilities, the next subsection will first describe the achieved results in the area of SIEM systems, followed by proposed approaches for intrusion detection based on data mining and data summarization for log collections.

3.1 SIEM

Asanger and Hutchinson discussed unsupervised anomaly detection techniques to improve rule-based correlations in a SIEM environment. Their outcome is that adequate preparation and preprocessing steps (normalisation) are necessary to achieve valuable results. With different views on the data e.g. events per user or IP address per user, they were able to detect suspicious behaviour which was not detected by an existing SIEM solution [12]. Zope et al. gave an overview of data mining techniques and the architecture of a SIEM system. They proposed various association rule mining algorithms to filter redundant information, match association rules and generate, based on that, security events [11].

3.2 Intrusion Detection

After the SIGKDD Conference on Knowledge Discovery and Data Mining was held on the subject "Computer network intrusion detection" in the year 1999, many researchers began to investigate this problem domain. The key challenge identified at SIGKDD was to devise a predictive model capable of distinguishing between legitimate and illegitimate connections in a computer network [16]. Even though the data set created for that purpose is now 15 years old, it is still the most used data set for evaluating the effectiveness of data mining techniques in network intrusion detection. This is mainly a result of a lack of labelled data sets generated by real life data. This fact influenced the development of intrusion detection by data mining dramatically. Over the last years, researchers started to build a kind of competition of who can achieve the best results on the KDD99 data set. Researchers were using slightly modified approaches in order to achieve the best results instead of focusing on how attackers really operate. Nonetheless, it is worth mentioning that partially very good detection results were achieved by using supervised and unsupervised learning techniques [19, 20].

3.3 Data Summarization

The basic idea behind data summarization is that frequent occurring events are clustered together while abnormal events get revealed and highlighted for further investigation by a human being. Vaarandi proposed a data clustering algorithm specifically for event logs. He considered the nature of logs and used a density clustering approach to summarise frequent occurring events. Outliers, which are not assignable to a cluster, are separated for further investigation possibilities. His results are promising for detecting frequent patterns and anomalous events [21].

4 Big Data Enhanced SIM: Challenges and Concept

Unsupervised learning techniques, from the idea, are the optimal solution for detecting intrusions. However, they are facing several challenges. First of all, most unsupervised methods have a high computational complexity, which makes them hard to apply on real time scenarios. For unsupervised Clustering algorithms, the hardest challenge is to find a distance function for the determination of a distance between two categorical points. As a result of weak distance functions, most of the algorithms detected only obvious attacks with frequent occurring patterns like DOS attacks [19, 22].

Another challenge coming with the usage of clustering algorithms on log messages is that they are optimised to cluster groups together and generally perform poorly when data has outliers or less frequent patterns. Association rule algorithms are powerful, but they often cannot be applied directly to log files as they do not consist of a common format. Even if the input is in the same format it is a challenging task to obtain appropriate rules. This comes mainly from the difficulty when using audit data. Very often derived rules from audit data depend too much on that data and cannot be applied directly to a real scenario as seen in [23, 24].

Supervised intrusion detection methods require a classified input set to be able to distinguish between normal and anomalous behaviour. The success of them depends on how well these two behaviours can be distinguished. In the rapidly changing environment of network services, this is a vast challenge as requirements are changing on a daily basis and rules have to be adapted continuously. In addition, the characteristic of normal/anomalous behaviour is specific for every individual network and no audit data can be used to create this baseline [25].

The challenge with data summarization techniques is to find a good balance between data summarization and information loss. In a SIM scenario, it is important to retain all relevant information without losing a single event. An undetected incident could denote a potential serious damage for a company. As data summarization techniques are making use of common supervised and unsupervised data mining techniques the challenges to handle categorical data pertains likewise. We propose two concepts to enhance SIM systems with Big Data principles. The first concept is a reactive approach, which aims to give a general overview of the events happened as well as a forensic analysis platform. The second, proactive, approach targets to create situational awareness by highlighting anomalous events based on self-learning mechanisms while inheriting the functionality of the reactive approach.

4.1 *Forensic Approach*

The reactive approach is designed for forensic investigation purposes and outlines a SIM platform, which can be utilised to follow up events reactively, e.g. security intrusions which already took place. However, the detection of an intrusion is at first not the primary target. In an initial step, it is even more important to get an overview of what is really going on. Large amounts of events do not allow getting this overview by only looking at it. All three modules, shown in Fig. 3, offer a

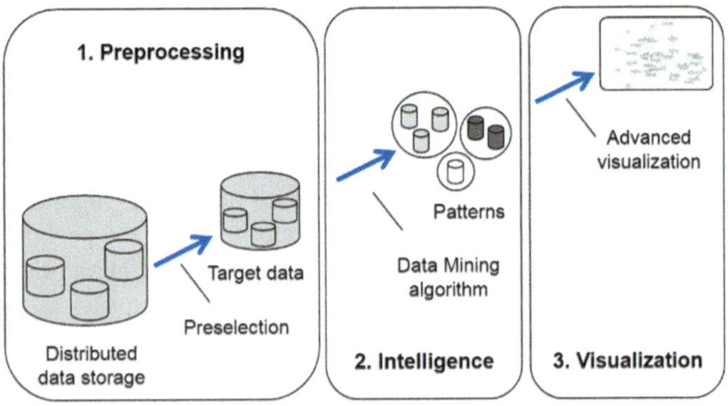

Fig. 3 Forensic approach

building block to simplify the complexity and reduce the mass of data. The pre-processing module enables to search through and preselect data with arbitrary sizes. The advantage compared to traditional SIM system is that various data types and structures can be processed without the necessity to normalise data.

The intelligence unit clusters frequent occurring events together and emphasises outliers. This enables to get an overview of what is going on without any prior configured rules and regardless of the data volume or data type. Depending on the size of input data the magnitude of the outcome after the intelligence unit can still be in a dimension which is not easy to evaluate for a human being. Due to that, optimised visualisation techniques for a specific scenario are used to further summarise the data. Visualisation to provide insightful contexts for human stakeholders is a complex topic, and graph design principles e.g. would go beyond the scope of this paper. Nevertheless, visualisation techniques have by far not fulfilled their potential in network security incident detection yet. Most of the visualisation techniques nowadays are able to display two or three dimensions of numerical data, often in a Cartesian coordinate system manner. However, log events very often contain a lot more attributes and furthermore consist mainly out of fields with categorical nature. Shiravi et al., Marty and other researchers have proposed several visualisation techniques with the purpose to visualise security relevant information more efficient [26, 27]. We have decided to use a visualisation technique termed parallel coordinates, which enables to display multidimensional data of any kind.

4.2 Proactive Approach

That a vision like Smart Cities can become reality, the detection accuracy of intrusions has to be extremely high and in addition it has to happen in real time. To achieve this, the proactive approach has to be adopted as well, which describes a self-learning SIM system. The method inherits the functionality of the reactive concept and supplements it with a supervised learning algorithm to highlight anomalous events based on previous observations. As a consequence, it provides a means to obtain situational awareness in real time. The self-learning system is comprised of two phases. A training phase, where continuously normal and anomalous behaviour is defined, and a prediction phase, where live incoming traffic is classified as normal or anomalous.

To encounter the challenges coming with a supervised learning method in a SIM scenario, the data used for training is abstracted to a higher level with the purpose, that sudden network changes are adapted slowly over time. In the training stage, displayed in Fig. 4, a model is trained with real, pooled SIM input data. This solves the dependence on audit or training data in the classification step.

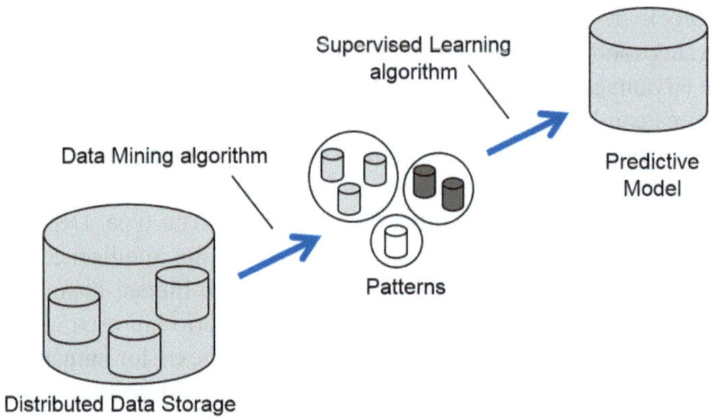

Fig. 4 Proactive approach (1)—training

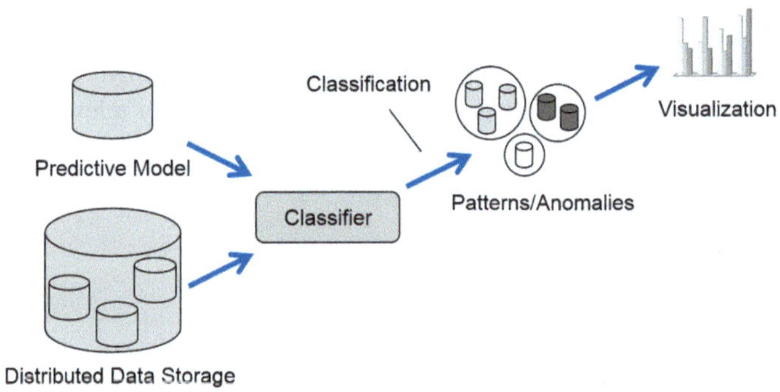

Fig. 5 Proactive approach (2)—prediction

The classification phase, displayed in Fig. 5, uses a classification algorithm to sort new incoming events into previous detected clusters or outliers. New events, which do not fit into any of these clusters, can be highlighted and visualised. After a human inspection, they can be used to further train the model and build new clusters.

5 Implementation and Evaluation

To see if the suggested approaches are applicable to a real life SIM environment, the reactive approach has been implemented completely alongside an existing logging solution of a company. Due to time limitations, the outlined proactive approach will be investigated in future work. Figure 6 shows the implementation, wherein a first step, the modules have been implemented individually without

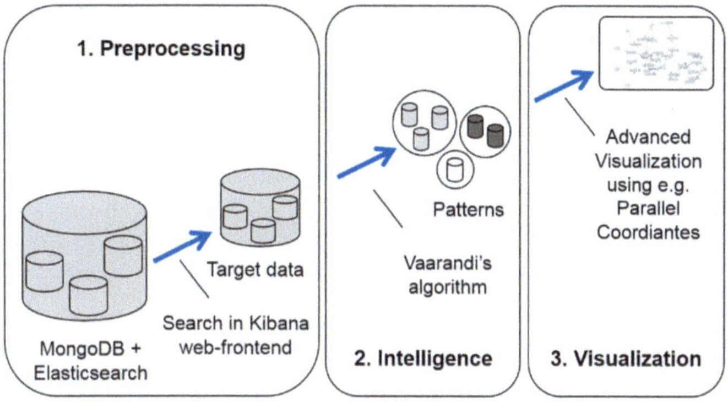

Fig. 6 Forensic approach example implementation

interfaces or automated transitions to other modules. In the pre-processing unit, at first the log events will be collected into a distributed, document-oriented database with the help of Logstash [28]. This enables to save data without normalising it and offers scalability for even large amounts of data. Elasticsearch complements the NoSQL database as a search engine to allow scalable and near real-time search on the data [29]. Kibana, a part of Elasticsearch, offers a web front-end for full-text search, correlation and simple visualisation and fulfills the pre-processing unit. No evaluations are done for this part as the solution has already proven their ability to fulfill that job in various other areas. Based on the output of the pre-processing unit, the intelligence unit, represented by a Java command line implementation of the, by Vaarandi proposed summarization algorithm, has been tested with different input sizes and diverse thresholds. Afterwards, a small python script extracts relevant features, which could then be visualised.

We've decided to implement the visualisation unit with a tool named Advizor Analyst Office, as it offers a good implementation of a parallel coordinates visualisation for our purpose [30].

The description of the real SIM data set used can be found in our previous publication on SIM in the context of Big Data [31]. The approach has been evaluated with statistical measures in [31] and in this paper we're supplementing the insights with the evaluation of the visualisation unit. The target of the intelligence unit is to summarise the data as best as possible while losing as little relevant information as possible. In [31], we've shown, that by using Varaandi's algorithm as intelligence unit, the amount of events to be analysed can be reduced to a tenth of the actual size. Yet in terms of Big Data, a tenth of Big Data is still data of a large magnitude. To further summarise the data and generate an overview, selected features have been plotted in a parallel coordinate graph.

In Fig. 7, each individual axis represents a dimension of the data set. The values of the data sets are plotted along the vertical axes respective the correct dimension. A data row is mapped by a line from left to right and could e.g. represent a firewall session. By using lines of different thickness, shape and colour even more

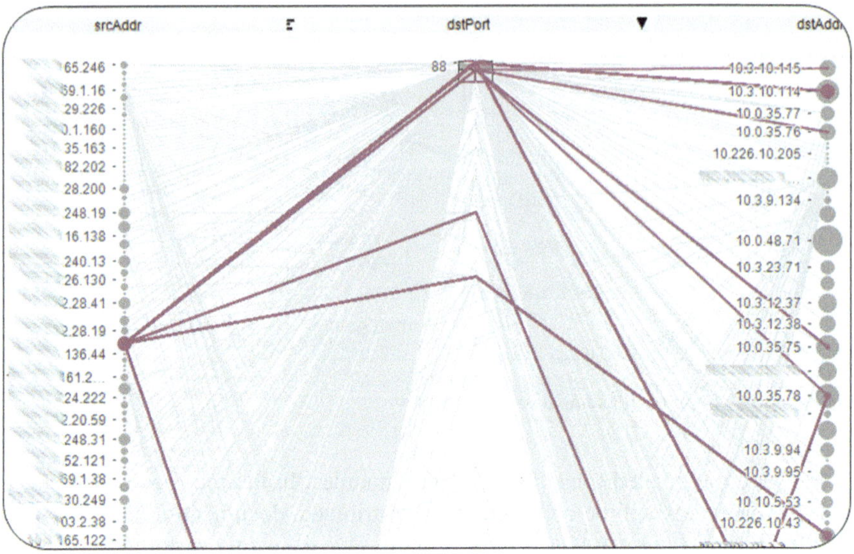

Fig. 7 Parallel coordinates graph interactive usage

dimensions can be integrated, such as e.g. the amount of occurrences of an event. Dynamic selection and arrangement of the axes, as well as colour usage, can support the detection of relationships inside the data additionally.

One disadvantage of parallel coordinate graphs is the restriction to a certain amount of data, as it can get confusing very quickly with too much data. A benchmark set by Marty recommends using maximum thousands for each dimension by up to 20 dimensions [26].

In our evaluation, the parallel coordinates graph surrounded by an interactive interface has proven to be a good method to display logs with several dimensions of various data types. For our specific use case of firewall logs, it is a good possibility to generate a general overview. Especially the interactive interface enhances the overview. Rarely used ports can be emphasised over frequently occurring well-known port numbers.

The graph has also proven to be a very good option to visualise intrusion detection system logs. An attack scenario can be displayed easily beginning with a source address, followed by an attack type and a destination address.

The main restriction for the parallel coordinate graph is the number of events displayable simultaneously. For our application around 10,000 events were the maximum amount displayable to get a meaningful picture. Depending on the number of events per second, only a small timeframe can be visualised in the graph compared to traditional techniques. A big part of Smart Cities will consist of critical infrastructures, where predominantly fixed defined processes can be found. Deviations from them can be detected easily with the visualisation unit presented.

6 Conclusion and Outlook

Our envisioned Smart Cities, being Senseable Cyber-Physical Environments of the outmost complexity, are powered by Big Data and insightful Processing. Our research addresses the fact, that our envisioned Smart Cities will only become a reality in the future if the confidentiality, integrity and availability of all assets within the digital ecosystem can be ensured at any time. The envisioned resource efficiency in terms of logistic services, including mobility, and life quality improvements for a cities citizen, can only be achieved if their Smart City operates on a reliable and respectful basis. Our paper rationalises, why traditional network security systems like SIM are not sufficient anymore to satisfy the need for an appropriate security level in a manageable way in such complex Cyber-Physical Logistic Ecosystems. However, these systems are still the de facto standard for complex environments and additionally to that; they haven't been researched extensively on the outlined upcoming challenges.

This paper has investigated the main challenges, which SIM systems are facing with the rapid growth of network traffic, and has addressed them with Big Data principles. The major problems identified were at first the general ability to process, store and analyse the amount of data, and secondly the big challenge to get insights out of it.

In order to enhance SIM for being able to handle big amounts of data in the future Smart Cities, two complementing concepts have been proposed and investigated. Firstly, the concept for the reactive approach targets to offer a forensic interface for reactive intrusion investigations and covers the main functionality, which traditional SIM systems are offering. Secondly, the concept for the complementing proactive approach uses previously seen data to predict and sort events in a proactive way. This enables to react in real time to ongoing events and supports to gain situation awareness, which is essential in critical infrastructure environments like Smart Cities. In addition, insightful processing involves the human stakeholders. Visualisation to provide human stakeholders with insightful contexts is a whole area of research in itself, and e.g. graph design principles are out of the scope and focus of this paper. Nevertheless, it has been outlined, that we decided to implement the visualisation unit with Advizor Analyst Office. For our purposes, it offers a good implementation of a parallel coordinates visualisation.

This paper has fully focused on the application in a real-life scenario and therefore evaluation of the forensic blueprint took place in a real SIM environment. This is in contrast to the limited real-life context of most other research approaches, which have been focusing on standard data sets. The obtained results indicate, that our proposed concept is generally applicable with the achievement of generating a good overview about all occurring events in a certain time frame. The fact of being independent from input data and format, while shrinking the input data to a tenth of the size with a reasonably low information loss, clearly indicates that traditional SIM systems could be enhanced in several points. The usage of adjusted visualisation techniques completes the process furthermore to an easily interpretable result for humans.

This paper contributes to a generalizable overview of enhancement possibilities for SIM systems in the context of Big Data and the complexity of the envisioned future Senseable Smart Cities. Furthermore, the presented blueprint provides a concept how to realise it. The adoption ranges from large organisations to our future Smart Habitats with their promises for resource efficiency and life quality improvements for their citizens, providing a basis safety and security to its inhabitants. On a short-term view, these concepts offer a good addition to existing SIM solutions. However, they are not yet able to replace them entirely. This is mainly a result of the core challenge located in the intelligence unit. To satisfy the requirement for various data sets with different structures and data types, the processing part was realised by a density-based data mining algorithm. However, these algorithms have the constraint of losing valuable information as all data types are considered as equal. Further enhancements, especially into the direction of replacing the summarization algorithm with a direct incident detection model, would require a normalisation process to make use of the full capabilities of data mining algorithms.

The potential of Big Data principles, especially data mining, in the area of network security incident detection, has by far not exploited its full potential. This is not a consequence of missing effort, but more a result of a general wrong research direction. Instead of focusing on the target, to increase the security level in a real working environment, most of the researchers have competed in achieving the best results on outdated standard data sets with slightly different approaches.

Due to that, on a short-term view, a combination of traditional systems like SIM and partial supplements of Big Data principles offer still the best security solutions. Nevertheless, looking at techno-social visions like the Senseable Smart Cities, where complexity and quantity of produced data will certainly further grow, only a self-learning system can offer an appropriate level of security.

References

1. Senseable City Laboratory. http://senseable.mit.edu/. Accessed 3 2016
2. Coffman KG, Odlyzko AM (2002) Internet growth: Is there a "Moore's Law" for data traffic? In: Handbook of massive data sets. Springer, pp 47–93
3. Ratti C (2014) Smart city, smart citizen. Egea, Milano
4. Nanni G (2014) Transformational Smart Cities: cyber security and resilience, Symantec Corporation
5. Securing Smart Cities Initiative. http://www.securingsmartcities.org. Accessed Mar 2016
6. Bartoli A, Hernández-Serrano J, Soriano M, Dohler M, Kountouris A, Barthel D (2011) Security and privacy in your smart city. In: Proceedings of the Barcelona smart cities congress
7. Cohen J, Dolan B, Dunlap M, Hellerstein JM, Welton C (2009) MAD skills: new analysis practices for big data. Proc VLDB Endowment 2(2):1481–1492
8. Tsai C-F, Hsu Y-F, Lin C-Y, Lin W-Y (2009) Intrusion detection by machine learning: a review. Expert Syst Appl 36(10):11994–12000. doi:10.1016/j.eswa.2009.05.029

9. Patcha A, Park J-M (2007) An overview of anomaly detection techniques: existing solutions and latest technological trends. Comput Netw 51(12):3448–3470. doi:10.1016/j.comnet.2007.02.001
10. Chandola V, Banerjee A, Kumar V (2009) Anomaly detection: a survey. ACM Comput Surv (CSUR) 41(3):15
11. Zope AR, Vidhate A, Harale N (2013) Data Mining approach in security information and event management. Int J Future Comput Commun 2(2):80–84. doi:10.7763/IJFCC.2013.V2.126
12. Asanger S, Hutchison A (2013) Experiences and challenges in enhancing security information and event management capability using unsupervised anomaly detection. In: 2013 international conference on availability, reliability and security, pp 654–661. doi:10.1109/ARES.2013.86
13. Zhong SHI, Khoshgoftaar TM, Seliya N (2007) Clustering-based network intrusion detection. Int J Reliab Qual Saf Eng 14(02):169–187. doi:10.1142/S0218539307002568
14. Smart Cities Research Center. Berkley University. http://smartcities.berkeley.edu. Accessed Aug 2014
15. Secure Smart Cities. CSIT Center for Secure Information Technologies, QUB, Belfast. http://www.csit.qub.ac.uk. Accessed Aug 2014
16. Pfahringer B (2000) Winning the KDD99 classification cup: bagged boosting. ACM SIGKDD Explor Newsl 1(2):65–66
17. Peikari C, Chuvakin A (2004) Security warrior. O'Reilly Media, Inc
18. Pinto A (2013) Defending networks with incomplete information: a machine learning approach. DEFCON 21, Las Vegas
19. Pathak V, Ananthanarayana V (2012) A novel multi-threaded K-means clustering approach for intrusion detection. In: 2012 IEEE 3rd international conference on software engineering and service science (ICSESS), IEEE, pp 757–760
20. Poojitha G, Kumar KN, Reddy PJ (2010) Intrusion detection using artificial neural network. In: 2010 international conference on computing communication and networking technologies (ICCCNT), IEEE, pp 1–7
21. Vaarandi R (2003) A data clustering algorithm for mining patterns from event logs. In: Proceedings of the 2003 IEEE workshop on IP operations and management (IPOM), pp 119–126
22. El-Semary A, Edmonds J, Gonzalez-Pino J, Papa M (2006) Applying data mining of fuzzy association rules to network intrusion detection. In: 2006 IEEE information assurance workshop, pp 100–107. doi:10.1109/IAW.2006.1652083
23. Tajbakhsh A, Rahmati M, Mirzaei A (2009) Intrusion detection using fuzzy association rules. Appl Soft Comput 9(2):462–469. doi:10.1016/j.asoc.2008.06.001
24. Tsai FS (2009) Network intrusion detection using association rules. Int J Recent Trends Eng 2 (2):1–3
25. Sommer R, Paxson V (2010) Outside the closed world: on using machine learning for network intrusion detection. In: 2010 IEEE symposium on security and privacy (SP), IEEE, pp 305–316
26. Marty R (2009) Applied security visualization. Addison-Wesley, Upper Saddle River
27. Shiravi H, Shiravi A, Ghorbani AA (2012) A survey of visualization systems for network security. IEEE Trans Visual Comput Graph 18(8):1313–1329
28. Logstash. http://logstash.net. Accessed Aug 2015
29. Elasticsearch. http://www.elasticsearch.org. Accessed Aug 2015
30. Advizor Solutions. http://www.advizorsolutions.com. Accessed Aug 2015
31. Gottwalt F, Karduck AP (2015) SIM in light of big data. In: 2015 11th international conference on innovations in information technology (IIT), IEEE, pp 326–331

Zero-Knowledge Authentication and Intrusion Detection System for Grid Computing Security

Mohammed Ennahbaoui and Hind Idrissi

1 Introduction

Grid computing [1] is a new trend in high performance computing. It involves sharing heterogeneous computers and resources, which are located in geographically distributed places belonging to different administrative domains. This technology provides large scale computation and easy access to the resources at a lower cost. According to Fig. 1, a functional architecture for the grid computing is composed of four essential entities: a *Grid User* (GU) that communicates with a *Grid Resource Broker* (GRB), which is responsible for locating the possibly available *Computational Resources* (CRs) using *Grid Information Services*. These services include the negotiation of access costs using trading services, the allocation of the jobs to the CRs (scheduling) and their execution, the monitoring and tracking of the execution progress along with the adaptation to the changes, and finally the collection of results [2].

The resolution of complex computations and data-intensive problems is among the various benefits of the grid computing, that makes it a very attractive topic in business, academic and research environments worldwide. This is mainly due to its heterogeneity, its dynamism, its reliable services supplied under conditions of scale, and especially for its reduced costs. However, grid computing stills a new paradigm that raises many security issues and conflicts in the infrastructures where it is integrated. This is mainly related to its open and distributed environment that increasingly attracts potential attackers. Among the solutions that would be efficient to address these security problems is to mutually authenticate the interacting entities with at least of knowledge, as well as to detect intruders and depict their malicious

M. Ennahbaoui (✉) · H. Idrissi
Laboratory of Mathematics, Computing and Applications, Faculty of Sciences, Mohammed V-University, Rabat, Morocco
e-mail: ennahbaoui.mohamed@gmail.com

L. Ismail and L. Zhang (eds.), *Information Innovation Technology in Smart Cities*,
https://doi.org/10.1007/978-981-10-1741-4_14

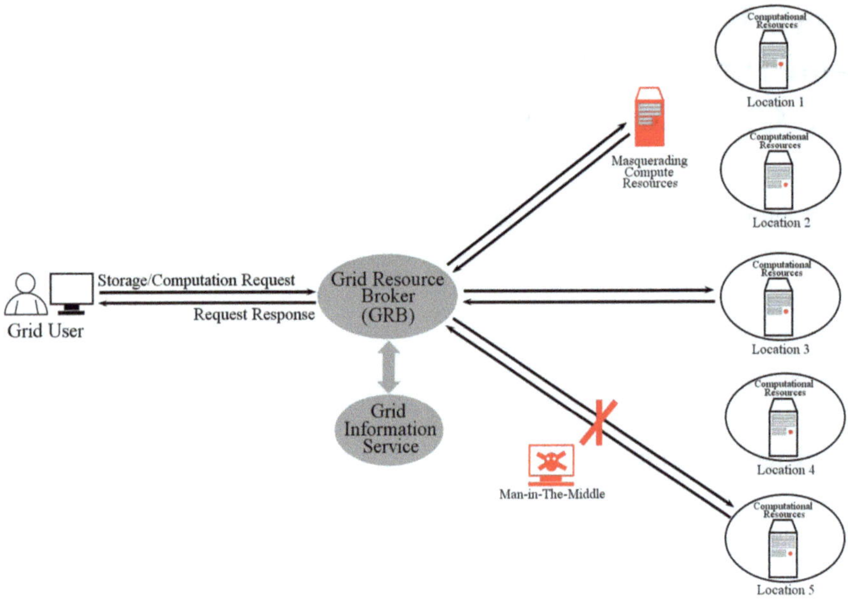

Fig. 1 Security issues raised in grid computing

activities and their behaviours susceptible to cause serious damages, through integrating an Intrusion Detection System (IDS).

In this paper, we propose a novel contribution for grid security based on two mechanisms. The first one consists in a robust and lightweight mutual authentication based on key exchange protocol and zero-knowledge proof. The second mechanism represents an IDS based on mobile agent paradigm to ensure secure, trusted and reliable communications inside the grid infrastructure, namely between the Grid Resource Broker (GRB) and its related computational resources (CRs). Actually, we are trying to conceive innovative techniques for advanced services in grid, where intelligence, pro-activity, autonomy and interoperability of mobile agents [3] are the main core. Thus, upon each received client request, the GRB is able to create multiple mobile agents charged with migrating to different computational resources (CRs), in order to perform the requested tasks in parallel.

The rest of the paper is structured as follows. Section 2 discusses the security issues in the Grid Computing Environments. Section 3 describes the proposed solution based on authentication and intrusion detection. An evaluation based on efficiency and security performances of our solution is presented in Sect. 4. Finally, a conclusion with perspectives for future work is mooted in Sect. 5.

2 Problem Formulation

The need to implant grid environments was filled through incorporating two innovative concepts: virtualization and replication. This promotes the resources coherence and fault tolerance, and facilitates the execution of jobs without need to be made compatible with an environment supported by a large resource provider. However, the shared resources and computations become not fully secure neither well monitored by the proper users. Among the security threats faced in this context, there are: Masquerading attacks and Man-In-The-Middle (MITM) attacks.

Actually, a GRB is related to several Computational Resources (CRs) with varied requirements, policies and applications, located on different control domains. Once a GU asks the GRB for the execution of a specified job, the GRB collects necessary information and decides which CRs will be involved and authorized to perform this task. Thereafter, it divides the global job into sub-tasks distributed on the named CRs, that execute required operations and send back the results. The GRB assembles the given results to get the final job result, which is forwarded to the Grid User. Hence, these interactions raise many security issues and challenges, especially when it concerns processing between the GRB and their related CRs as illustrated in Fig. 1:

1. The communication between the Grid Resource Broker (GRB) and the Computational Resources (CRs) can be intercepted by an intruder, that leads a MITM attack to recover the information flow and even more exploit it or modify the exchanged messages, in order to harm the communicating entities.
2. The Computational Resources (CRs) may be victim of masquerading attack that consists in impersonating one of its servers, in order to get in touch with the Grid Resource Broker (GRB) or other CRs without being detected. This attack may extend to the "Resource Broker" and make all the grid architecture vulnerable, which will cause huge damages.

3 Proposed Solution

In this section, we present a thorough description of our proposed solution, which combines a lightweight authentication based on zero-knowledge proof (ZKP) and an intrusion detection system (IDS) based on mobile agents.

3.1 Authentication Based on ZKP

Authentication is the first aspect that concerns the majority of infrastructures that store or distribute highly sensitive data. Thus, it becomes strongly required to

Table 1 Notations used during the authentication process

Notation	Description	Notation	Description
ID_e	Identity of entity e	g	Generator of F_p
X_e	Private key of entity e	$Encr(M)_K$	Encryption of M using K
y_e	Public key of entity e	RI	Random interval
p	Random prime	$\|\|$	Concatenation operator

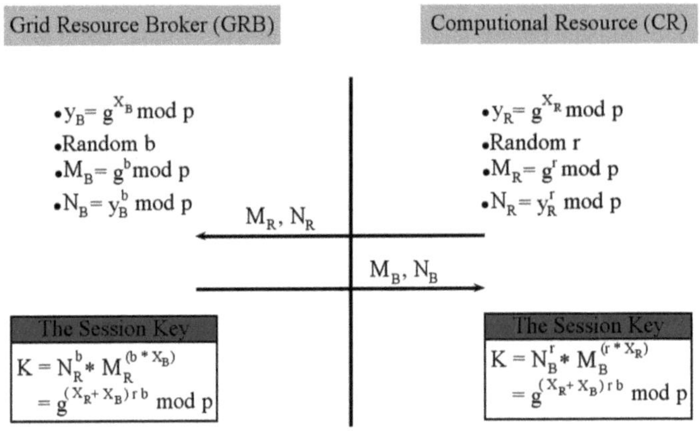

Fig. 2 Session key sharing scheme

confirm and validate the identity of a requester before providing him the access to services. There exist a wide variety of forms and schemes of authentication proposed in the literature. The zero-knowledge authentication scheme is one of these forms that would encode the identity of the requester as a hard problem, and authenticate him using zero-knowledge proof [4]. In this solution, we integrate an authentication based on ZKP, which first of all requires to exchange a session key then conducts a challenge/response mechanism. Table 1 presents the main notations used in the description of the authentication process.

3.1.1 Session Key Exchange

In order to generate a common key for the current session, we make use of an enhanced version of Diffie-Hellman key exchange protocol (DHKE), which makes it well-resistant to the Man-In-The-Middle (MITM) attacks, as illustrated in Fig. 2.

Since the first concern in DHKE is the generation of random events for modulo and primitive root computations, we make use of the robust cryptographic pseudo-random generator ISAAC + (Indirect, Shift, Accumulate, Add and Count) [5], instead of the usual DH-provider in language runtimes. After choosing a secret key and computing the public key by modulo of the prime p, each one of the grid resource broker (GRB) and the computational resource (CR) chooses an ephemeral secret (b / r) and employs it to calculate two exponential values (M and N). Those later are exchanged between the GRB and the CR, in order to compute the session key. The basic idea behind this enhanced and fixed version of DHKE is that all computations are based on incorporated ephemeral secrets, chosen in parallel by the GRB and the CR. This makes the protocol endowed with two significant properties:

- Forward Secrecy: even if the long-term private key of any party is exposed, the previous session keys cannot be discovered since the ephemeral secrets b and r for that session are unknown.
- Key Freshness: every session key is a function of ephemeral secrets so neither party can predetermine a session key's value since he would not know what the other party's ephemeral secret is going to be.

3.1.2 Challenge/Response with Commitment

Our authentication is mutual, that is to say the GRB must prove its identity to the CR and verify that the CR is the expected one. Thereby, to establish this authentication our proposed ZKP uses the challenge/response mechanism along with a commitment scheme, that allows one party to commit an encrypted message and reveal the secret key later when receiving a signal. Figure 3 shows the process of

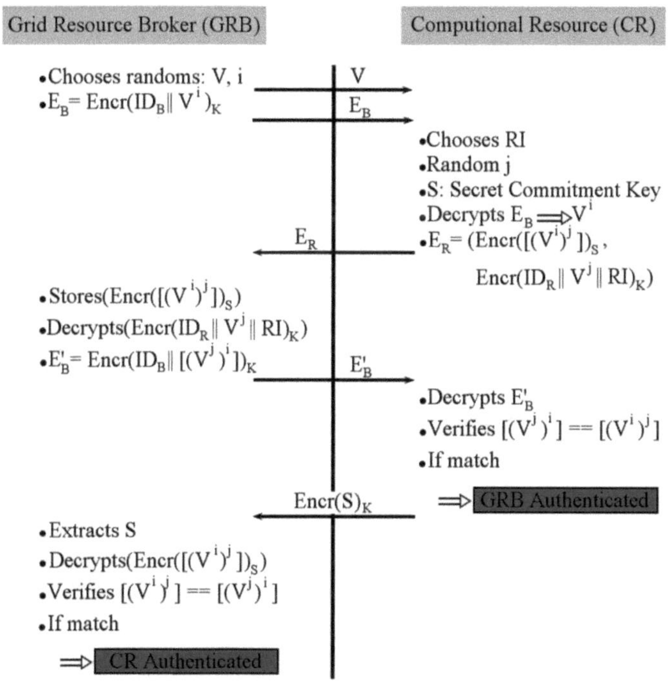

Fig. 3 The process of the zero-knowledge authentication scheme

the proposed zero-knowledge authentication scheme, which can be compressed in the following steps:

- Step 1. GRB → CR: V, E_B
  The GRB chooses a random information V to share with the CR, then generates a random integer i in order to calculate V^i. This later is concatenated with the identity of the GRB (ID_B), before being encrypted using the session key K: $E_B = Encr(ID_B||V^i)_K$. Finally, V and E_B are sent to the CR.
- Step 2. CR → GRB: E_R
  Once receiving the message of the GRB, the CR decrypts E_B and extracts the value V^i. Then, using a random integer j, the CR computes $(V^i)^j$, chooses a random interval (RI) of at most 256 bits from the beginning of this value, and encrypts the selection using a random key (S) related to the adopted commitment scheme, $Encr([(V^i)^j])_S$. Thereafter, using the received information V, the CR calculates V^j and concatenates it with its identity ID_R and with RI. The given value is then encrypted using the session key K: $Encr(ID_R||V^i||RI)_K$. At the end, the entire message $E_R = (Encr([(V^i)^j])_S, Encr(ID_R||V^i||RI)_K)$ is sent to the GRB.
- Step 3. GRB → CR: E'_B
  Upon reception of a message from the CR, the GRB extracts the first part and stores it as a commitment message. The second part of E_R is decrypted using the session key K, then the value $(V^j)^i$ is calculated. At this stage, the CR extracts the random interval RI, deduces its size in order to send the same interval size from the beginning of the calculated value, $[(V^j)^i]$. The later interval is concatenated with the ID_B before being encrypted using the session key K:
  $E'_B = Encr(ID_B||[(V^j)^i])_K$. The value E'_B is directed to the CR.
- Step 4. CR → GRB: $Encr(S)_K$
  In order to check the authenticity of thc GRB, the CR decrypts E'_B and compares the received $[(V^j)^i]$ with the value $[(V^i)^j]$ it has calculated. If both values are equal, then the CR considers the authentication of the GRB successfully performed, and sends to it the encryption of the key S employed for commitment in step 2. Else, the authentication is ceased.
- Step 5. GRB
  After decrypting the received key S using the session key K, it would be possible for the GRB to decrypt the value stored in step 3. Then, in order to verify the authenticity of CR, the GRB compares the given interval value $[(V^i)^j]$ with the calculated $[(V^j)^i]$. If both values are equal, then the CR is successfully authenticated.

3.2 IDS Based on Mobile Agents

According to the NIST [6], an IDS is a software or hardware device potentially capable to identify an attack and notify appropriate personnel immediately, which help to stop possible threats or at least prevent them from succeeding. In this part, we describe the structure and process of the proposed IDS, which is mainly based on a robust detection technique that depends on the generation of a cryptographic trace by each mobile agent during the execution of its assigned task. A mobile agent is defined as a software entity able to move from one node to another across the network, with a set of actions to perform (code), resources to deploy (data), and a state of execution. Thus, the interoperability guaranteed in mobile agents interactions makes them efficient in negotiating with grid users and allocating processing and resources to applications.

In this work, we adopt a fully-distributed IDS, where real-time detection is performed on subjects activities and attitudes. Indeed, the use of mobile agents ensures low response time, less energy consumption. This structure allows to perform fast detection, where the known and the new attacks are discovered without need to be updated. A basic and simple alert processing is provided, such that malicious behaviours are notified, stated, pursued and clustered in local variables of the agents.

Our approach proposes a solution that aims to secure the grid computing without compromising its features and properties. At the moment when a Grid User (GU) sends a task execution request to the GRB, this later appeals for the help of the *Grid Information Service*, that monitors all information about the existing CRs, in order to recognize the equipment and resources showed to be capable, available and authorized to execute specified job. Thereby, it divides the requested task into sub-tasks and determines which CRs will be involved for their execution. These sub-tasks are distributed throughout multiple mobile agents accurately created by the GRB and dispatched in parallel, as illustrated in Fig. 4. Accordingly, our solution makes use of two kinds of mobile agents:

- *Execution Agents*: each one of these agents migrates to the intended geographical area where the CRs are located, in order to execute its assigned sub-task. We note that we make use of RSA public key cryptosystem [7] such that public and private keys of 2048 bits are generated at the time of Grid creation, and updated according to the intern security policy of the system. Hence, the data and code of the agent are encrypted before its mobility using the public key of the targeted CRs, which in turn decrypts the agent when arrived and before launching its execution, using its private key.
- *Result Collector Agent*: It is charged with collecting the results provided by every execution agent, through browsing the different CRs concerned using their IP addresses. Along the itinerary of this agent, its sensitive data, code and the collected results are encrypted and decrypted using the same cryptosystem.

Along their migrations, the "Execution Agents" and the "Result Collector Agent" may be victims of several intrusion attacks, and when retrieving execution

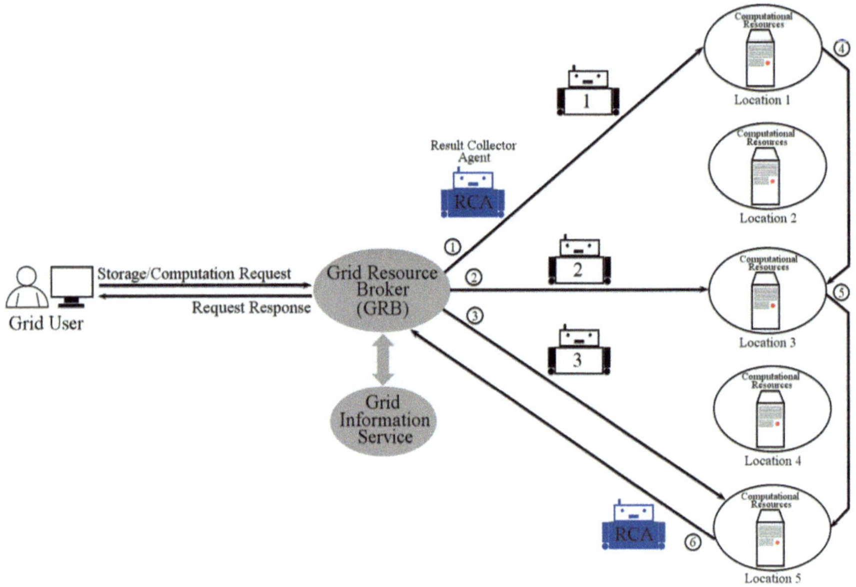

Fig. 4 Proposed solution for grid security using mobile agents

results, the GRB needs to be sure that they are not falsified neither harmful. Thus, we develop an advanced detection mechanism for Grid based on tracing technique inspired from the Cryptographic Traces proposed in [8]. Those later are generated through deep post-mortem analysis of data (called Traces) collected during the agent execution. According to this, the agent code is composed of a sequence of statements, that can be white or black, such that white statements modify the agent's execution using its internal variables only, while black statements use information received from external environment to alter the state of the program.

While moving to a CR location, a trace is produced and interpreted as a pair of a *unique identifier* (UI) randomly generated using the Java package java.util.UUID, and a *signature* (SIGN) generated using the public key of the GRB, and which comprises a set of identification information such as: the identity of the mobile agent, its owner and the CRs receiving it, the arrival time to the CRs, the agent code and the task assigned to it, in addition to a timestamp for freshness. When the "Execution Agent" performs its job, the black statements of the executed code are abstracted and encrypted along with the given results, using the GRB public key then added for signature in the trace, as viewed in the pseudo-code of Fig. 5. As it should be remarked, the global signature "SIGN" includes an inner signature (*Sign*) employed by the GRB to verify the concerned CRs. It is obtained using the secret key of the specified CRs and contains the identities of the agent and its owner, the hash value of the black statements and the results using SHA-3 [9], and the time of execution termination. As long as the "Result Collector Agent" needs to move across multiple CRs locations to collect execution results, it holds a list of the

```
Cryptographic_Trace (Mobile_Agent A, CRs Li)
ID_A     <--   A.getAgent_ID();
ID_O     <--   A.getOwner();
ID_CRs   <--   A.getReceivingCRs();
AT       <--   A.getArrivalTime(Si);
Code     <--   A.getCode();
JOB      <--   A.getRequestedJob();
GRB_PuK  <--   GRB Public Key;
Li_PrK   <--   Li Private Key ;
ts       <--   timestamp;
UI       <--   UUID.randomUUID();
if A.getClass()== "ExecutionAgent.class"
  A.Execute();
  BS      <--  A.getBlackStatementExecuted();
  Result <--   A.getExecutionResult();
  EET     <--  A.getExecutionEndTime(Si);
  Tr_Ex (Li)= < UI , SIGN_GRB_PuK ( ID_O, ID_CRs, AT, Code, JOB,
  ts, Encrypt_GRB_PuK (BS, Result), Sign_Li_PrK( ID_O, ID_A,
  SHA3(BS, Result), EET) )>;
else if A.getClass()== "CollectorAgent.class"
  if (i==1)
    Tr_Co (L1)= < UI , SIGN_GRB_PuK ( ID_O, ID_CRs, AT, Code,
    JOB, ts, Encrypt_GRB_PuK (Tr_Ex (L1)), Sign_L1_PrK( ID_O,
    ID_A, SHA3(Tr_Ex (L1)) ) )>;
  else
    Tr_Co (Li)= < UI , SIGN_GRB_PuK (Tr_Co (Li-1), ID_O,
    ID_CRs, AT, Code, JOB, Encrypt_GRB_PuK (Tr_Ex (Li)),
    Sign_Li_PrK ( ID_O, ID_A, SHA3( Tr_Ex (Li) ),
    SHA3( Tr_Co(Li-1)) ) ) >;
```

Fig. 5 Java pseudo-code of the cryptographic trace generation

locations to visit including their domains, their IP addresses and their public keys. In practice, the "Result Collector Agent" gathers the cryptographic trace of each "Execution Agent" then terminates it. In order to guarantee a chaining mechanism among the traces produced by all "Execution Agents" and also those of the "Result Collector Agent", each trace that has been generated on the previous CRs location, either while execution or collection, is also included in the signature (SIGN) on the current CRs and its hash is integrated in the inner signature. This mechanism makes all the traces dependent to each other, such that the latest trace keeps an instance of all the prior ones, which allows the inspection of agents behaviours while comparing the requested tasks with those really executed.

Once receiving the final traces from the "Result Collector Agent", the GRB proceeds to the verification of the nested traces. In order to see how this process is performed, we suppose that the GRB suspects the CRs at the location 4 of being malicious. First, the GRB decypts the global signature (SIGN) of the trace (*Tr_Co* (L_4)) provided by the "Result Collector Agent" using its private key, then checks the identity of the agent owner and its hosting CRs location. When this verification is affirmative along with the arrival time that corresponds to the common reference clock, a decryption of the inner signature (Sign) is performed using the public key of the CRs at location 4 (L_4*_PuK*). Given the second hash included in this inner signature and which contains the trace on the previous CRs location 3 (*Tr_Co* (L_3)), we compute a hash of the trace provided in the global signature, and we compare the both hashes. If they don't match, then we attest that an intrusion has been detected while collecting execution results, else the verification of the "Collector Agent" reliability is said to be affirmative.

At this stage and assuming normal conditions, the trace of the "Execution Agent" (*Tr_Ex* (L_4)), that was hosted by the CRs location 4, also needs to be checked. Thus, it is first hashed and compared to the first hash value in the inner signature of (Tr_Co (L_4)). If conforming, then the global signature "SIGN" of (*Tr_Ex* (L_4)) is decrypted using the GRB public key, the included identities and the arrival time are verified. Afterwards, the abstracted black statements (BS) along with the obtained results are decrypted, hashed and compared with the hash provided in the inner signature, that was decrypted using the public key (L_4*_PuK*). Once the hashes are equal, a time checking verification is achieved, basing on the end execution time, the arrival time and the timestamp. Finally, whether a cryptographic trace does not fulfil these defined constraints, its originator is categorized as suspicious and it is revived to clarify its situation or regenerate the trace if needed. The fact of reproducing a non-valid trace or not replying to the claim of GRB leads to recognize that CRs location is malicious, or being victim of intrusion attack.

4 Evaluation

In this section, our proposed solution undergoes a security analysis to evaluate its resistance against some well-known attacks, as well as a set of experimental investigations are conducted to prove its reliability, flexibility and effectiveness.

4.1 *Security Analysis*

4.1.1 Man-in-the-Middle Attack

The main concept of this attack is to get between the GRB and the CR in order to intercept and control their conversations. Our solution is resistant to this attack for many reasons. First of all, the authentication is mutual between the GRB and the CR. In the key exchange, we make use of ephemeral secrets in computations, so that without knowledge of these secrets and the private key of each party, an adversary can not reproduce the values M and N. Concerning the zero-knowledge proof scheme, no information about the secret is disclosed, since all exchanged data are encrypted before being sent.

4.1.2 Replay Attack

Assuming that an adversary attempts to maliciously replay the interval values: $[(V^i)^j]$ and $[(V^j)^i]$, in order to convince the GRB or the CR that they are provided by a legitimate party and then gain access to sensitive data. Our solution is resistant to this attack due to use of the commitment scheme, which promotes the revelling of the secret key after committing the message once receiving a positive signal.

4.1.3 Guessing Attack

In this attack, an adversary tries to collect many exchanged messages, in order to guess the key or the secret information. Since all exchanged messages in our solution are not sent in clear, also, only random intervals of at most 256 bits are provided, then our solution is resistant to this attack. Moreover, a wide variety of random events and secrets which are used in the scheme are changed for every communication.

4.2 *Efficiency Analysis*

For practical evaluation purposes, we simulate a grid architecture using the toolkit GridSim [10], which is based on Java language and its libraries are imported into such Java programming IDE, like Eclipse in our case. In implementation, heterogeneous types of Computational Resources (CRs) are considered, where each resource may contain a different number of machines, and each machine may have one or multiple Processing Elements (PE). The allocated machines holds different operating systems (WindowsXP, MacOs, Ubuntu), Core i7 at 2.7 GHz, and 4 Go of RAM and necessary equipped with JADE (4.3.3) [11] FIPA-compliant agent platform, which is also configured on Eclipse and charged with receiving, executing and dispatching the mobile agents. The resources capability can be defined in the form of Millions Instructions Per Second (MIPS) as per Standard Performance Evaluation Corporation (SPEC) benchmark. The characteristics of the grid architecture used for experiments are listed in Table 2.

The resources are scheduled using random scheduling algorithm, and the jobs are also selected randomly. On completion of given task, the resource may be selected for the next task. Concerning the Grid Resource Broker (GRB), we used GridSim to create Nimrod-G broker [2]. Simulation is conducted for increasing number of CRs: {10, 20, 30, 40, 50} where varying number of gridlets: {100, 200, 400, 600, 800, 1000} is scheduled. Each gridlet is contained and clustered inside a mobile agent, that adapts its execution to the specifications of the Grid environment.

Our proposed Grid-IDS based on mobile agents demonstrates very challenging performances compared to an IDS implemented basing the traditional Client/Server architecture. Starting by rating the response time performance, we have measured the overhead of security added to the system through integrating the cryptographic traces mechanism. Thus, the calculated time spent by an "Execution Agent" to produce a cryptographic trace of a job execution was about 0.093–0.116 s, while the time spent by a "Result Collector Agent" to generate multiple traces on an increasing number of CRs is shown in Table 3. These overheads present a very low percentage of the total time spent during the agent trip, where only one migration takes about 0.248 s. The use of mobile agents is proved to be more beneficial and scalable compared to Client/Server as illustrated in Fig. 6, where our Grid-IDS, endowed with a reliable pattern, can support increasing job execution requests and provides results in less time. In the same context, the network load of our integrated IDS using mobile agents seems to be lower as shown in Fig. 7.

It was worthy and important at this stage to test the capacity of our solution to detect advanced intrusions, either while execution of tasks (at the "Execution

Table 2 Characteristics of the grid architecture used for experiments

Parameter	Value	Parameter	Value
Number of resources	50	Length of job	1000 MIPS
Number of machine per resource	1	Cost per job	3G–5G
Number of PEs per machine	5	Bandwidth	4000–6000 B/S
Number of gridlets	1000	Job input size	25 MB
PE ratings	10–40 MIPS	Job output size	45 MB

Table 3 Time cost of "result collector agent" cryptographic trace generation face to the increase of CRs

Number of CRs	5	10	15	20	30	40	50
Trace generation time (s)	0.428	0.778	1.041	1.375	2.003	2.711	3.348

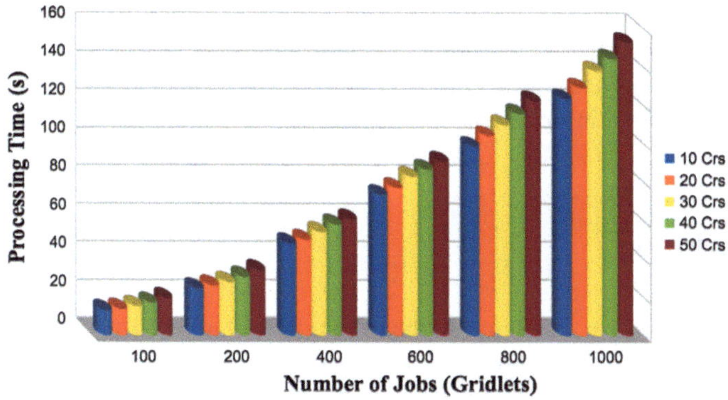

Fig. 6 Time processing performance of our grid-IDS versus the increase of jobs

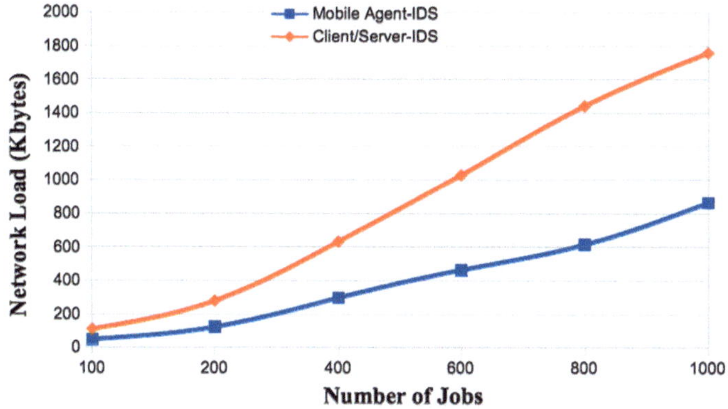

Fig. 7 Network load versus the increase of jobs while using mobile agents and client/server

Table 4 Comparison of detection performance between our IDS based on mobile agent and an IDS based on client/server

Total number of CRs		10	15	20	30	40	50
Masquerading CRs/malicious brokers		2	4	8	12	16	20
CRs detected using client/server		0	2	5	7	9	11
CRs detected using mobile agents	EA	2	4	6	8	10	14
	RCA	0	0	2	4	6	6

Agent" (EA) level) or collecting results (at the "Result Collector Agent" (RCA) level). For that purpose, we have integrated some masquerading CRs that pretend the identities prescribed in the path of both types of agents, and we have simulated some brokers as malicious intruders that try to corrupt and control them. The results given in Table 4 prove the reliability and robustness of our Grid-IDS as it was able to detect more intrusions, and specify at which levels they occurred. Experiments show that the majority of intrusions were detected in the traces of "Execution Agents", which seems to be logical since this latter carries more sensitive and attractive information.

5 Conclusion

In this paper, we give an interest to the insider threats in the Grid computing. Thus, to address these security issues we have proposed a solution based on a mutual authentication using zero-knowledge proof and an IDS that benefits from the flexibility and interoperability of mobile agents to conserve traces and proofs of the jobs execution. The practical evaluation of our solution, when compared to Client/Server architecture, demonstrates very promising results: low response time, less network load and high intrusion detection capacity. For future works, we think to associate the detection with suitable prevention policy and make use of heuristics to replace the random scheduling and optimize agent's mobility.

References

1. Foster I, Kesselman C, Tuecke S (2001) The anatomy of the grid: enabling scalable virtual organizations. Int J High Perform Comput Appl 15(2):200–222
2. Buyya R, Abramson D, Giddy JNG (2000) An architecture for a resource management and scheduling system in a global computational grid. In: Proceedings of the 4th international conference and exhibition on high performance computing in Asia-Pacific Region (HPC ASIA 2000), May 14–17, 2000, Beijing, China, IEEE CS Press, USA, 2000
3. Gavalas D, Tsekouras GE, Anagnostopoulos C (2009) A mobile agent platform for distributed network and systems management. J Syst Softw 82(2):355–371
4. Smart NP (2016) Zero-knowledge proofs. In: Cryptography made simple. Springer, pp 425–438
5. Aumasson J (2006) On the pseudo-random generator ISAAC. IACR Cryptology ePrint Archive, p 438
6. Scarfone K, Mell P (2007) Guide to intrusion detection and prevention systems (idps). NIST Spec Publ 2007(800):94
7. Boneh D, Shacham H (2002) Fast variants of RSA. CryptoBytes 5(1):1–9
8. Vigna G (1998) Cryptographic traces for mobile agents. In: Mobile agents and security, Springer, Berlin Heidelberg, pp 137–153
9. Jaffar A, Martinez JC (2013) Detail power analysis of the SHA-3 hashing algorithm candidates on xilinx spartan-3E. Int J Comput Electr Eng 5(4):410–413
10. Buyya R, Murshed M (2002) GridSim: a toolkit for the modeling and simulation of distributed resource management and scheduling for grid computing. Concurrency Comput Pract Experience 14:1175–1220
11. Bellifemine F, Poggi A, Rimassa G (2001) JADE: a FIPA2000-compliant agent development environment. In: 5th international conference on autonomous agents, ACM Montreal, pp 216–217

Part V
Advanced, Intelligent Communication Infrastructure for Smart Cities

SPMA: A Centralized Macrocellular Approach for Enhancing System Capacity of Future Smart Networks

Muhammad Usman Sheikh, Sharareh Naghdi and Jukka Lempiäinen

1 Introduction

It is forecasted that the future mobile network traffic will increase rapidly in the imminent years [1]. The exponential demand of user services with different requirements will be an issue for mobile operators. Usage of internet in daily life and excessively used online applications on the cellular phone generates a huge traffic. While there is an increase in demand for network capacity and data services, the optimization of network topology becomes important and critical. Therefore, minimizing the interference and maximizing the system capacity are the major design goals for the future wireless networks. In order to enhance the network performance, optimizing the antenna configuration i.e. tilt, height, azimuth and beamwidth are considered as baseline actions [2, 3].

One of the fundamental methods to increase the capacity of a macrocellular network is to increase the base station density also known as site densification. In site densification the base stations are placed more closer which results in decrease in intersite distance, thus more interference comes from the neighboring cells. Other possible way of increasing the system capacity is the higher order sectorization. Traditionally a single site is designed with three sectors, using 65° Half Power Beamwidth (HPBW) antenna in each sector. Six-sector site and 12-sector sites are the example of higher order sectored sites, which uses 32° and 16° HPBW antennas, respectively [4]. To limit the interference and to reduce the overlapping area among the sectors, it is recommended to use antennas with narrow beams for higher order sectorization. There are some other techniques introduced in Release 8 of 3 GPP which can address the problem of interference such as Inter-Cell Interference Coordination (ICIC) and Coordinated Multipoint (CoMP) transmissions which were further enhanced in Release 10 and Release 11 [5].

M.U. Sheikh (✉) · S. Naghdi · J. Lempiäinen
Tampere University of Technology (TUT), Tampere, Finland
e-mail: muhammad.sheikh@tut.fi

L. Ismail and L. Zhang (eds.), *Information Innovation Technology in Smart Cities*,
https://doi.org/10.1007/978-981-10-1741-4_15

The novel concept of Single Path Multiple Access (SPMA) introduced in [6] is an 'Innovative deployment paradigm' which is based on advanced antenna solutions. SPMA can be considered as an evolved and enhanced version of Space Division Multiple Access (SDMA). In SPMA, it is proposed to establish a link between the transmitter and receiver using a single path instead of using multipaths. Frequency spectrum is aggressively reused in SPMA, which can dramatically increase the system capacity by multifold. In [7], the performance of SPMA was evaluated in relatively urban and dense urban environment; however this article focuses on the SPMA performance in almost flat and suburban environment.

This article provides a comprehensive overview of research carried to discover an optimal antenna downtilt point for different antenna beamwidths in the real world scenario. This article shows the impact of adopting antennas with different beamwidths for 3 and 6-sector sites in macrocellular suburban environment. It also shows the behavior of a network in terms of quality i.e. Signal to Interference Ratio (SIR) when those sectored sites are replaced with SPMA nodes. Finally, the performance of SPMA is compared with of 3 and 6-sector sites in terms of capacity. This study was carried out using sophisticated 3D ray tracing and real world map and site data.

2 Theory

2.1 Antenna Tilting and Antenna Beamwidth

One of the most economic ways of improving the network performance with respect to coverage and capacity is to optimize the antenna configuration. Antenna tilting directly affects the coverage and thus limits the interference. It is an efficient method to control the radiation of an antenna. There are two ways of antenna tilting i.e. mechanical tilting and electrical tilting. Mechanical tilting involves the physical tilting of an antenna in downward or upward direction. Whereas the electrical titling is achieved by changing the phase of the antenna elements without physically changing the antenna position [2, 8]. In case of tower mounted antennas, mechanical downtilting has a drawback of high back lobe. An optimum downtilt angle is the function of antenna vertical beamwidth ($\theta_{VER,BW}$) and geometrical factor (θ_{GEO}) as shown in Eq. (1).

$$v = f\left(\theta_{VER,BW}, \theta_{GEO}\right), \tag{1}$$

where the geometrical factor (θ_{GEO}) can be calculated as follows.

$$\theta_{GEO} = \tan^{-1}\left(\frac{h_{BTS} - h_{MS}}{d}\right) \tag{2}$$

In Eq. (2), (h_{BTS}) and (h_{MS}) represents the height of the Base Station (BS) antenna and Mobile Station (MS) antenna, respectively, and (d) is the size of the dominance area i.e. separation between BS and MS.

Another factor which affects the coverage of the cell is the beamwidth of an antenna in the horizontal (azimuth) plane and in vertical (elevation) plane. It is denoted by the term Half Power Beamwidth (HPBW), and is calculated from the −3 dB point in the radiation pattern with respect to main lobe. Antenna with narrow HPBW also offers higher antenna gain as antenna gain is inversely proportional to the beamwidth of an antenna. Sample radiation patterns of antennas with different beamwidths are shown in Fig. 1, which are later used for simulations and research work of this paper.

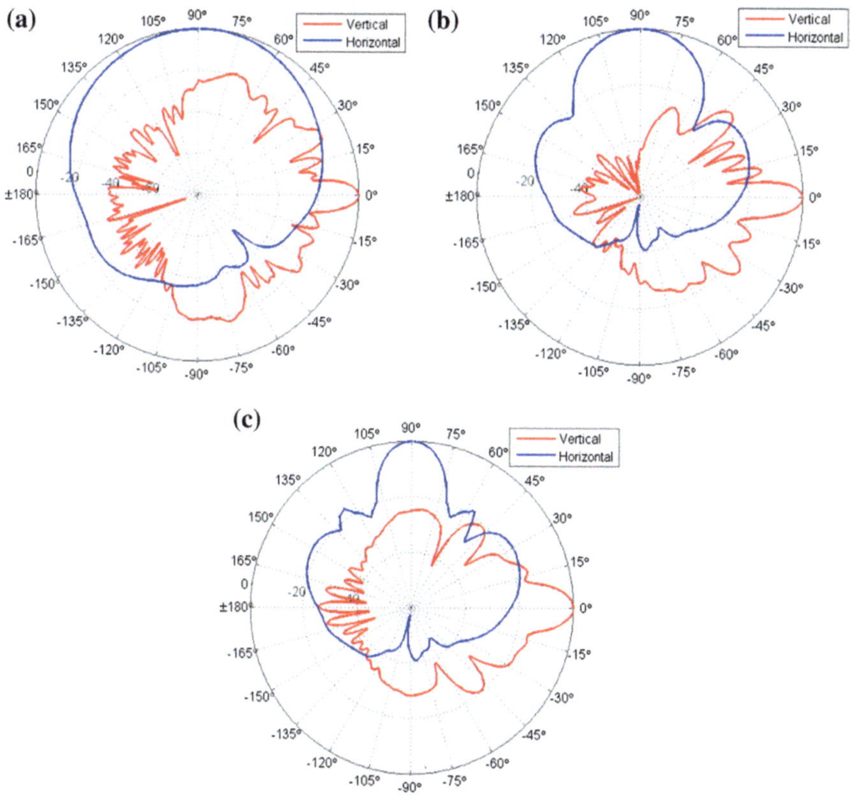

Fig. 1 Horizontal and vertical radiation patterns of antennas, **a** 65° horizontal HPBW, **b** 32° horizontal HPBW, **c** 16° horizontal HPBW

2.2 Higher Order Sectorization

In general, an outdoor macro site has multiple sectors i.e. directional antennas are used instead of an omni directional antenna. Traditionally each macro site has three sectors. Higher order sectorization extends the number of sectors to 6 or even more at an individual site. Sector densification reduces the spatial separation between the sectors which increases the overlapping area among the sectors. Therefore, in order to avoid the interference from the neighbor sectors of the same site it is recommended to use antennas with narrow antenna pattern for higher order sectored sites [4, 9].

Higher order sectorization at macrocellular level provides a solution to tackle the challenge of growing traffic demand, while reusing the existing macro site locations and spectrum in an efficient way. It is already established in [4, 9] that higher order sectorization provides capacity gain with respect to 3-sector sites. From the OPEX and CAPEX point of view, it is easier and cheaper to upgrade the existing 3-sector site to a higher order sectored site compared to adding a new site in the network.

2.3 Single Path Multiple Access (SPMA)

The novel and innovative concept of Single Path Multiple Access (SPMA) was proposed in [6], which allows the frequency resources to be reused more frequently by employing needle beams. The main target of using the needle beam is to limit the interference. Instead of using multipaths, it is proposed to find a single path to establish a link between BS and MS. The essence of the SPMA concept relies on the strong assumption that the future novel antennas would be able to form simultaneously several narrow adaptive antenna beams. These extremely narrow beams also known as "Needle beams" will have the horizontal beamwidth of around 0.5° and vertical beamwidth of around 0.2°. Each user is served and track by individual beam. SPMA can be considered as an evolved and enhanced version of Space Division Multiple Access (SDMA), where two nearby users i.e. separated by few meters can reuse the same frequency resources.

It is also expected that new electrical materials will be used in antenna manufacturing. These electrical materials include e.g. metamaterials, graphene, Carbon Nano Tube (CNT), Graphene Nano Ribbon (GNR), and cloaking [10, 11] etc. A centralized macro site approach is recommended for SPMA where a traditional base station with a finite number of sectors is replaced by a SPMA node. A single SPMA node is assumed to be capable of forming multiple narrow beams in any direction to serve multiple users. A scheduler at SPMA node should also be able to schedule different users in the time domain as well. The traditional concept of 'Cell' is no more valid for SPMA, as in the case of SPMA each user is assumed to have its own "Virtual Cell", where each user is re-using the frequency resources with a global reuse factor of one.

2.4 3D Ray Tracing

For the given parameters of base station i.e. location, height, transmission power, frequency, characteristics of antenna etc., the propagation models give the information about Quality of Service (QoS) in the area under consideration. The received signal strength (RX level) and the signal quality (SINR) are still considered as the most significant metrics for describing the network performance. The radio propagation models can be characterized by three factors, (1) Intelligence, (2) Accuracy, and (3) Processing (computational) time. The empirical and semi-empirical models consider the wave propagation only in a vertical plane which contains transmitter and receiver point [Fast 3D ray tracing]. Since those empirical models are based on measurement done in some other cities, therefore they need fine tuning and optimization to make them applicable for different propagation environment. The empirical models are computationally less complex, have short computational time, and are easy to implement. However, their prediction accuracy is limited.

There are two basic approaches for searching the propagation paths using a ray optical path finding, one is ray launching and second is ray tracing. The basis for ray tracing and ray launching model is a 3D building data of propagation environment. Ray launching approach is originated from computer graphics, and in this approach several rays are launched from a fixed transmitting antenna in all relevant directions with discrete small angular separation. At potential receiving points, the field strength is obtained by summing up the rays reaching those points. The ray propagation is stopped, if either the energy is below a minimum threshold or a given maximum number of interactions are reached. The computational time is almost independent of the number of receiving points, whereas it is linearly dependent on the number of interaction points, and is inversely proportional on the angular separation of launching rays [12, 13].

Another approach is ray tracing that basically looks for only valid paths between transmitter and receiver point with finite interactions. A well known algorithm used for ray tracing is "Image Theory". The ray tracing techniques precisely model the channel and provide fairly accurate prediction results. However, they are computationally complex and require large computation time, and the computational time for such ray tracing techniques increases exponentially with the increase in number of walls and interaction points in the considered area [14].

3 Simulation Environment, Cases and Simulation Parameters

The key assumptions and the simulation tool used to study the impact of different antenna configurations (beamwidths and tilts) in macrocellular environment are explained in this section. Also, the assumptions considered regarding the design and beamwidth of SPMA antenna is highlighted in this section.

3.1 Simulation Platform and Simulation Environment

A huge campaign of simulations was carried out using a MATLAB as a simulation platform. An indigenous 3D ray tracing tool “sAGA” was developed in MATLAB environment for doing the coverage prediction. Image theory is used in sAGA radio wave propagation tool to find the possible ray paths between the transmitter and the receiver point with the finite number reflections and diffractions using 3D building map. For the case of macrocellular environment where the antenna is above the rooftop, this tool also finds rooftop diffracted ray path along with ground reflected paths. After finding all the propagation paths with the given number of reflections and diffractions, the field strength for each propagation path is computed using propagation theory. Reflection coefficient and diffraction coefficient at each reflection and diffraction point is calculated using ray path geometry. Image theory provides the ray paths with high accuracy and precision, but the accuracy of propagation model is highly dependent upon the input building data and the number of reflections and diffractions considered. For the indoor users, the 15 dB building penetration (wall penetration) loss is used to account for outdoor to indoor propagation.

For the research work of this paper, an area of around 5.25 km $\times$ 5.25 km from Chicago, USA is selected. A real world 3D building data from the city of Chicago is used instead of hypothetical, fictive regular building grid with equal heights and spacing. Similarly, actual practical network site locations and heights are used for simulations, so that a fair comparison between the performance of 3-sector site, 6-sector site and SPMA deployment can be made. There are total ten sites with different heights, where all the sites are macro sites with antenna height clearly above the average building height. The height of the sites ranges from 20 m to 38 m. The selected area has almost flat terrain, and representing a suburban area with low height (average 6.5 m) building structures. The selected area with 3D building data and site locations can be seen in Fig. 2a.

Total 3450 RX points were homogeneously distributed with 75% of the users in an indoor environment and 25% of the users in an outdoor environment, which reflects a realistic distribution in real work. For both the outdoor and indoor users the RX height was set to 1.5 m. Figure 2b shows the distribution of users in 2D map generated by the MATLAB.

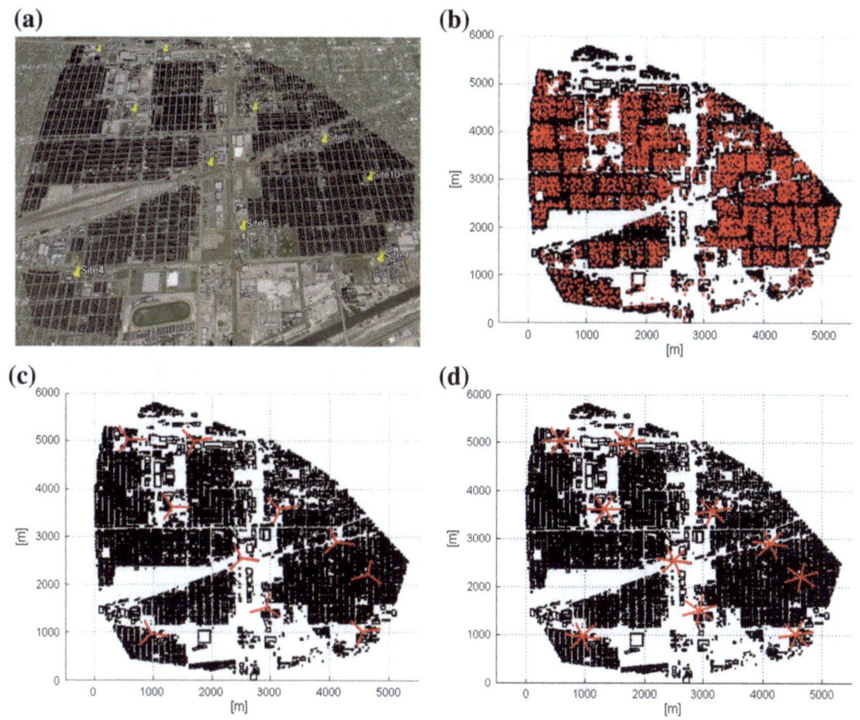

Fig. 2 **a** Location of sites, **b** User distribution, **c** Orientation of antennas for 3-sector case and, **d** Orientation of antennas for 6-sector case

3.2 *Simulation Cases*

In this paper we have studied the impact of antenna beamwidths and antenna down tilts in three sector and six sector site deployments, and compared their performance with single path multiple access technology. The following simulation cases were studied in this paper.

- **3-sector site with 65° HPBW antenna**: It is the traditional and the most commonly used case in a conventional macrocellular networks. Each site comprises of three sectors, and every sector has single 65° HPBW antenna with a maximum antenna gain of 15.39 dB in the direction of main lobe. The antenna pattern for 65° HPBW antenna in a horizontal plane along with in vertical plane is shown in Fig. 1a. For all the sites in the network, there is a spatial separation of 120° in a horizontal plane between the sectors of the same site. However, the base azimuth can be different for different sites as shown in Fig. 2c. It acts as a baseline configuration, and is used as a reference case for comparing the results with other configurations.

- **3-sector site with 32° HPBW antenna**: In this case, 32° HPBW antenna with a maximum antenna gain of 18.20 dB in the direction of main lobe is used for three sector sites. The antenna pattern for 32° HPBW antenna in a horizontal plane along with in vertical plane is shown in Fig. 1b.
- **6-sector site with 32° HPBW antenna**: In this case each site comprises of six sectors, and every sector has single 32° HPBW antenna. For all the considered sites, there is a fix spatial separation of 60° in an azimuth plane between the sectors of the same site. However, the base azimuth can be different for different sites as shown in Fig. 2d.
- **6-sector site with 16° HPBW antenna**: In this case, a narrow antenna of 16° HPBW antenna with a maximum antenna gain of 21.15 dB in the direction of main lobe is used for six sector sites. The antenna pattern for 16° HPBW antenna in a horizontal plane is shown in Fig. 1c.
- **SPMA with needle beam**: It is the special and unique case in which a SPMA node is assumed to have an extremely narrow needle beam for each serving user. The SPMA node is assumed not to be limited by the maximum number of serving users e.g. thousands of beams can be generated by single SPMA node. The needle beam is assumed to have 0.5° beamwidth in horizontal plane and 0.2° in a vertical plane. The beam is assumed to have ideal radiation pattern with flat response and no additional gain in an azimuth and elevation plane. In case of SPMA, a pencil beam is steered precisely to the serving user while keeping the user in the middle of the beam.

The general simulation parameters are provided in Table 1.

Table 1 General simulation parameters

Parameter	Unit	Value
Operating frequency band	MHz	2600
Number of sites	No.	10
Transmit power	dBm	46
TX height	m	Variable
Indoor/outdoor user height	m	1.5
Building penetration loss	dB	15
Ground permittivity Building material permittivity		10 5
Polarization		Vertical
Reflections	No.	2
Diffractions	No.	2
Rooftop diffraction		Enabled

4 Simulation Results and Analysis

The main target of the radio network planning and optimization is to provide better coverage (RX level) and enhanced quality (capacity). Received signal strength and Signal to Interference ratio (SIR) are measures of network coverage and network quality, whereas system capacity is directly proportional to SIR. There are different ways of optimizing the coverage and network quality, and antenna down tilting is one of them. Figure 3 shows the mean cell RX level for considered antenna configurations against the different antenna downtilt. The x-axis indicates the antenna downtilt in degrees and the y-axis indicates the corresponding received signal strength in dBm.

The results presented in Fig. 3 show, that initially for all the considered antenna configurations the mean RX level improves with antenna downtilting, however after reaching the optimal point with respect to providing best coverage i.e. 2° downtilt, the RX level starts to deteriorate by further increasing the antenna downtilt. At 2° downtilt, the mean RX level of −61.6 and −64.5 dBm is achieved with 6-sector and 3-sector, respectively. However, by using aggressive downtilting the mean RX level drops up to −80.5 and −83.2 dBm for 3-, and 6-sector sites respectively. Due to more number of sectors in the considered area, the six-sector site cases clearly provide better coverage compared to three-sector site cases. Even with a 16° HPBW antenna, the six-sector site provides a dominant coverage over 3-sector site. Interestingly, it can be seen that in terms of coverage (RX level) the antenna with wide radiation pattern in horizontal plane i.e. 65° HPBW for 3-sector site, and 32° HPBW for 6-sector site provides better result up to 6–7° downtilt compared to narrow antenna radiation pattern i.e. 32° HPBW and 16° HPBW for 3-, and 6-sector sites, respectively.

Figures 4a, b show the heat map of received signal level with 3-sector, and 6-sector sites, respectively. The color bar indicates the power level in dBm, and the

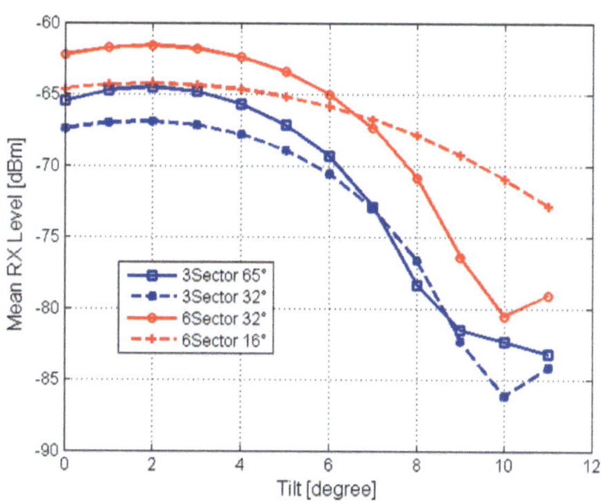

Fig. 3 Mean cell RX level for different configurations against antenna downtilt

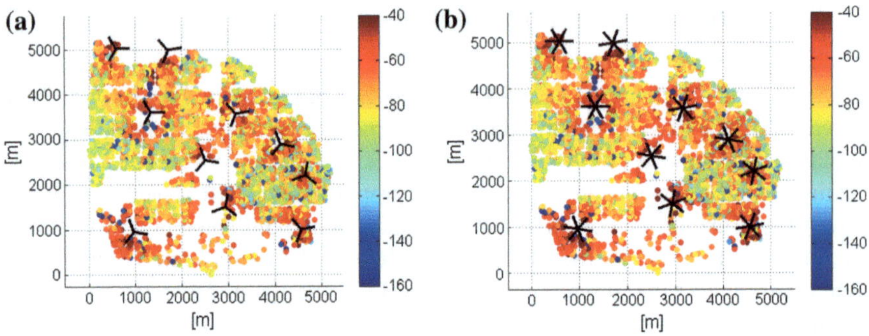

Fig. 4 **a** RX level heat map for 3-sector case, **b** RX level heat map for 6-sector case

Fig. 5 RX level heat map for SPMA case

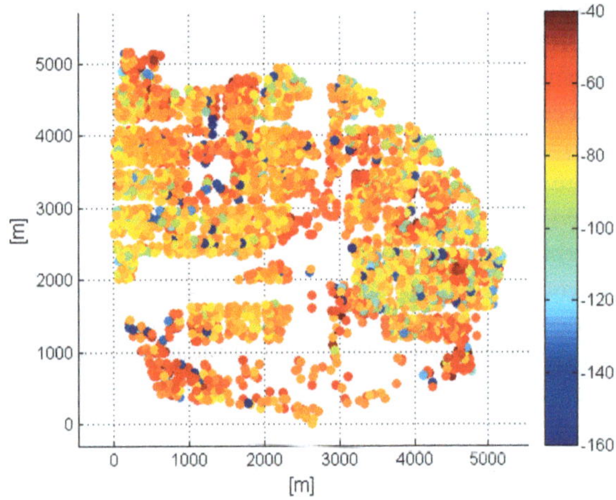

antenna locations and azimuths are marked with black marker. Due to aggressive antenna downtilting e.g. 8° for 3-sector site and 9° for 6-sector site, it can be seen that samples with strong RX level are spotted only in the closed vicinity of the sites. Signal propagation was restricted with antenna tilting, which makes the signal coverage non-homogeneous over the area under consideration. However, most of the bad samples are located on the left-center and right centre of the map, as those areas lack the dominant site. As a narrow antenna radiation pattern is used for 6-sector sites, therefore the coverage heat map has not considerably changed with 6-sector site compared to 3-sector site, and it is difficult to spot the difference in heat map with a naked eye.

Figure 5 shows the heat map of received signal level with SPMA node. The same scale is used in the color bar to make a fair comparison of SPMA coverage with 3-, and 6-sector site coverage. It is fascinating to see that fairly homogeneous coverage is provided with SPMA, and those bad sample areas earlier shown with 3-, and 6 sector sites are adequately covered with SPMA nodes. The signal strength

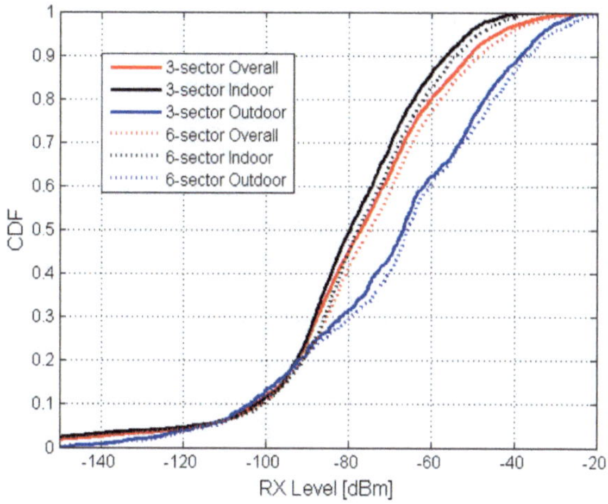

Fig. 6 CDF plot of RX level for 3-sector and 6-sector deployment

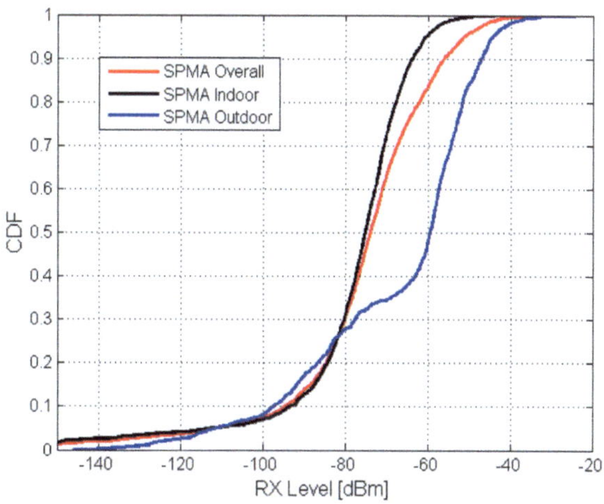

Fig. 7 CDF plot of RX level for SPMA case

of strongest samples has been reduced due to which the dark maroon or dark red color is missing in Fig. 5. However, the quite acceptable level of received signal is maintained, despite of assuming no antenna gain for SPMA.

Figure 6 shows the CDF plot of received signal level for 3- and 6-sector site deployment. In Fig. 6, separate results are presented for indoor and outdoor users along with combined results. The mean RX level of −81.1 and −79.06 dBm is achieved in an indoor environment, whereas the mean received signal level of −69.85 and −68.2 dBm is acquired in an outdoor environment, with 3-, and 6-sector sites respectively. The ten percentile of the indoor and outdoor users represents the worst or cell edge users, and they are at the level of about −104 to −100.5 dBm.

Figure 7 shows the CDF plot of received signal level for SPMA case. The mean RX level of −78.8 and −67.85 dBm is attained with SPMA for an indoor environment and outdoor environment, respectively, which is higher compared to sectored sites. Although the received power of strongest samples were reduced with SPMA, however still the better mean received signal power is achieved. Similarly, the ten percentile value for indoor and outdoor users has improved by SPMA, and is raised to −97.4 and −93.43 dBm for indoor and outdoor users, respectively. The detailed statistical analysis of received signal level in an indoor and outdoor environment for 3-sector, 6-sector and SPMA case is presented in Table 2.

For the considered simulation environment, an optimum antenna downtilt which corresponds to maximum SIR was also found for different antenna configurations. Figure 8 shows the mean SIR of cell for the cases of 3-sector and 6-sector sites with different antenna configurations.

It can be seen from the Fig. 8 that down tilting helps in improving the SIR. However, after certain point the SIR starts to degrade with increasing tilt. It was

Table 2 Statistical analysis of RX level for considered cases in different environments

Case	RX level 10 percentile [dBm]	RX level 90 percentile [dBm]	RX level median [dBm]	RX level mean [dBm]
3-sector indoor	−102.03	−56.57	−79.69	−81.11
3-sector outdoor	−103.75	−38.51	−66.75	−69.84
3-sector overall	−102.53	−50.98	−77.10	−78.33
6-sector indoor	−100.68	−54.11	−77.73	−79.06
6-sector outdoor	−104.14	−37.07	−65.95	−68.19
6-sector overall	−101.21	−48.71	−74.96	−76.38
SPMA indoor	−93.49	−63.91	−75.28	−78.79
SPMA outdoor	−97.40	−47.16	−59.62	−67.84
SPMA overall	−94.53	−56.19	−74.01	−76.09

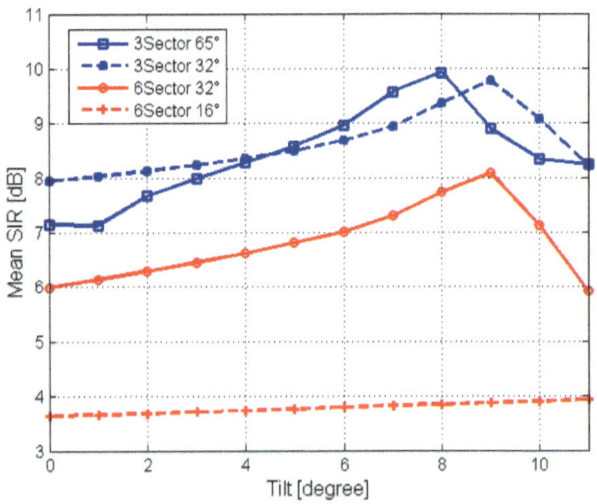

Fig. 8 Mean cell SIR for different configurations against tilt

found that among the considered cases the maximum SIR of 9.92 dB was achieved by 3-sector site with 65° HPBW antenna at 8° downtilt. Reducing the beamwidth of antenna from 65° to 32° did not help in improving the SIR, and maximum of 9.78 dB SIR was provided by 3-sector site with 32° antenna at 9° downtilt. It is interesting to see that higher order sectorization lowers the mean SIR at cell level. Even by using narrow 32° HPBW antenna pattern for 6-sector site, the maximum achievable SIR was found to be 8.08 dB with 9° downtilt, which is around 2 dB less compared to 3-sector site. However, the SIR performance was degraded significantly with 16° HPBW antenna. Now onward the analysis and results presented in this paper for the case of 3-sector sites are with respect to 65° BW antenna at 8° downtilt, and for the case of 6-sector sites are with respect to 32° antenna at 9° downtilt.

Figure 9 shows the SIR heat map for the case of 3-sector sites. It shows that in the near region of all site locations there are samples with high SIR values (greater than or equal to 18 dB). It re-affirms that in the clear dominance of cell area there is more probability of having high SIR values. In the left central part, as there is no site covering that area therefore user experiences bad signal quality. Whereas, in the right part there is more density of sites which results in better SIR. As there is no nearby interferer for Site4 in the left lower part of the area, therefore it witnesses a large number of samples with good SIR. In traditional cellular approach, cell border area has always been a problematic area due to sever interference coming from the neighbor cells. Users with bad link quality are clearly evident at the cell border area in Fig. 9.

Figure 10 shows SIR heat map for the case of 6-sector sites. In the case of 6-sector site deployment, it is important to note here that the maximum achievable SIR has reduced to around 15 dB compared to +20 dB in 3-sector site deployment. These results show that in suburban environment with macro site, the higher order sectorization increases inter-sector interference due to large visibility area of sector, which results in lower SIR. Although, 6-sector sites were deployed with narrow

Fig. 9 SIR heat map for 3-sector case

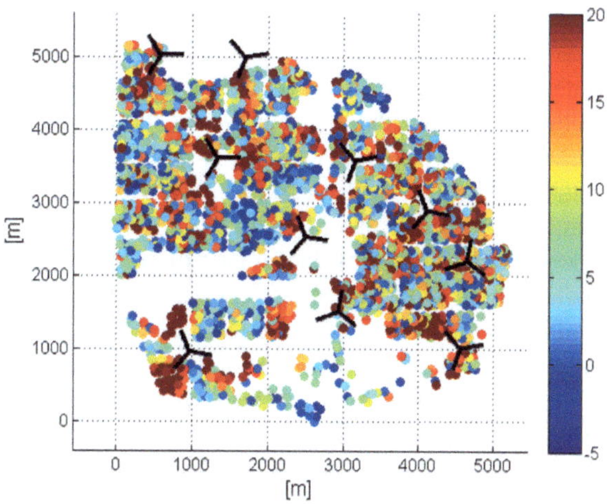

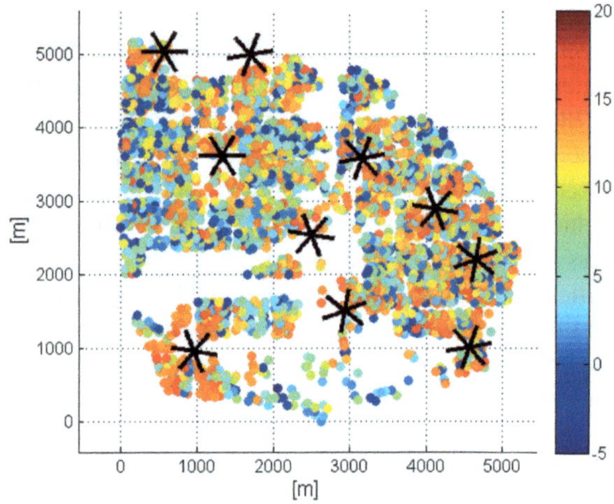

Fig. 10 SIR heat map for 6-sector case

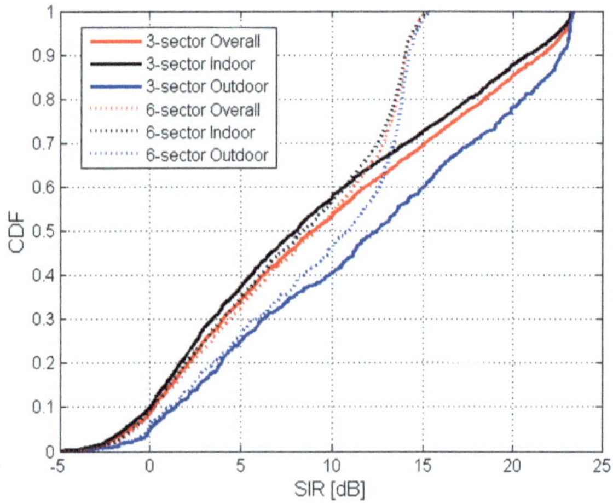

Fig. 11 CDF plot of SIR for 3-sector and 6-sector deployment

antenna of 32° HPBW, however it still degrades the mean SIR of cell. The positive point found from these results is that the degradation of SIR while shifting from 3-sector to 6-sector deployment is not significantly high. Therefore, in short the additional sectors with degraded SIR will still add more capacity to the system. Similar to 3-sector case, mainly the left centre part of the considered area experiences bad SIR samples due to lack of coverage and clear dominance. The gain of higher order sectorization is shown in Table 4.

Figure 11 shows the CDF plot of SIR in different environment for 3- and 6-sector site deployment. Considering the indoor users, it can be seen that there is no major impact of higher order sectorization on the users with low and moderate SIR values i.e. −5 to 10 dB. Similarly, the cell edge outdoor users with low SIR values i.e. −5 to 6 dB are not affected by higher order sectorization. It is found that higher order sectorization has neither improved nor decreased the quality of cell edge users, whereas the quality of users with strong SIR values is degraded with higher order sectorization. In the case of higher order sectorization, there is more number of cells in an area which results in increased interference, and it limits the maximum achievable quality. Therefore, the maximum SIR of 15.3 and 23.4 dB is attained with 3-, and 6-sector site, respectively. Both the sectored configurations were not able to provide homogeneous received signal level and quality over the considered area. Cell edge users equally deserve a better quality of services as site nearby users. Hence, it is important to maximize the cell edge capacity, and improve the user experience at cell edge. The optimization of network done through antenna down tilting and by using higher order sectorization is not able to increase the network capacity by huge margin. Therefore, a novel solution is required for radically enhancing the system capacity. Statistical analysis of signal to interference ratio for 3-, and 6-sector site configuration is shown in Table 3.

Figure 12 shows the SIR heat map acquired with SPMA. Exceptional results are obtained with SPMA in terms of achieving high SIR almost homogeneously over the considered area. It is interesting to see that interference can be avoided up to

Table 3 Statistical analysis of signal to interference ratio

Case	SIR 10 percentile [dB]	SIR 90 percentile [dB]	SIR median [dB]	SIR mean [dB]
3-sector indoor	0.03	20.84	8.06	9.22
3-sector outdoor	1.40	22.64	12.49	12.08
3-sector overall	0.25	21.66	9.02	9.93
6-sector indoor	0.05	14.02	8.41	7.76
6-sector outdoor	1.12	14.23	10.96	9.07
6-sector overall	0.27	14.07	9.09	8.09

Fig. 12 SIR heat map for SPMA case

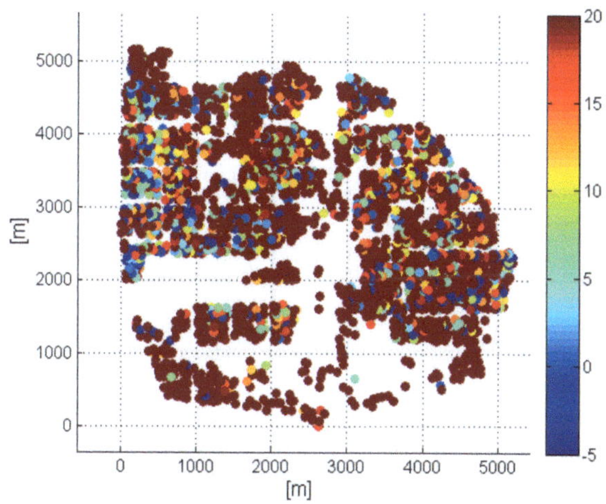

large extent by using the narrow needle beams of 0.5° × 0.2°. It is important to mention here that in Fig. 12, each user acts as interferer to the other user and the SIR values shown in Fig. 12 were computed assuming all (3450) users active at that particular instant. Not to forget that each user is not only interfered by the other SPMA nodes, rather other beams serving other users by the own serving SPMA node also cause interference. For the fairly close located users in the considered scenario, SPMA with 0.5° × 0.2° beam has been able to provide far better SIR compared to traditional cellular approach (3-sector and 6-sector site). SPMA overcomes the cell edge effect, and is able to provide equal services to all the users. However, the worst area (left central part) showed some sign of improvement but was not fully covered with the given SPMA nodes. Also few bad samples were found scattered over the whole area, possibly due to non-availability of distinct paths between the two nearby users, or due to lack of dominant (LOS or reflected) path.

Figure 13 shows the CDF plot of SIR for the case of SPMA. In simulations, the minimum and maximum supported values for SIR were set to −10 and 50 dB, respectively. Statistical analysis shows that 50% of the users enjoy around 30 dB of SIR, whereas 10 percentile is around −2 dB. In the considered scenario all the 3450 users were assumed active in time domain, however for the bad SIR locations if the closed by users are scheduled in different time instant then 10 percentile value can be improved. It is believed that by employing SPMA over TDMA, the problem of low SIR samples due to the presence of close by user (interferer) can be resolved.

Table 4 shows the cell spectral efficiency ($\bar{\eta}$) and area spectral efficiency ($\bar{\eta}_{area}$) for different cases. For post simulation spectral efficiency analysis only those cells (sectors) are considered which were providing coverage to the area under examination. Therefore, in Table 4 the cell density is 29 and 57 for 3- and 6-sector case, respectively. In case of SPMA, each user is assumed to have its own virtual cell. SPMA considers a global reuse factor of one, which means the same frequency

Fig. 13 CDF plot of SIR for SPMA

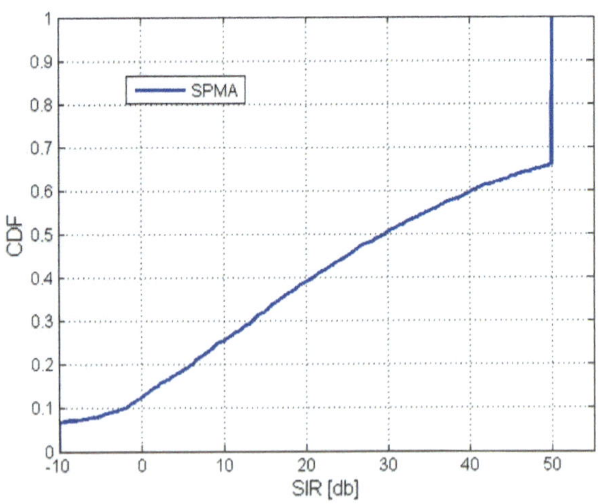

Table 4 Spectral efficiency results

Cases	Cell Density	Mean $\bar{\eta}$ [bps/Hz]	Mean $\bar{\eta}_{area}$ [bps/Hz/Area]	Mean Capacity gain [times]
3-sector	29	3.44	99.76	1.0×
6-sector	57	3.39	193.23	1.94×
SPMA	3450	9.19	31706	317.82×

Table 5 Overlapping zone analysis with 5 and 7 dB window

Case	Single server 5 dB [%]	Two servers 5 dB [%]	Three servers 5 dB [%]	≥ Four servers 5 dB [%]	Single server 7 dB [%]	Two servers 7 dB [%]	Three servers 7 dB [%]	≥ Four servers 7 dB [%]
3-sector	72.22	21.99	4.54	1.25	63.79	25.83	6.84	3.55
6-sector	76.09	19.98	3.17	0.76	67.71	24.9	5.59	1.80

resources can be reused by every user. Therefore, the SPMA has a cell density of 3450 virtual cells. Mean cell spectral efficiency is computed using mean cell SIR. Results presented in Table 4 shows that six sector site plan with optimized downtilting provides 1.94 times more capacity with respect to reference case of 3-sector site deployment. Whereas, SPMA shows exceptionally high area spectral efficiency, and in the considered scenario despite of some bad SIR samples SPMA was found to be 317.82 times more spectral efficient compared to traditional 3-sector site case.

Table 5 presents the cell overlapping results for 3-, and 6-sector site cases with 5 and 7 dB window. Signal strength and signal quality are the most commonly used metrics of radio network planning. However, for more in-depth detail, the cell overlapping area with multiple servers is also taken into account to approximate the handover region, or to learn about the possible pilot pollution. Single server (dominance) area is normally located near the site location, and cell border area is overlapped by multiple servers. Intuitively, increasing the number of sectors in a certain area decreases the single server dominance area, and increases the area occupied by multiple servers. However, the cell overlapping results presented in the Table 5 show that by adopting a narrow antenna pattern i.e. 32° HPBW in a horizontal plane for 6-sector site, a single server dominance area is not only maintained, rather it is further enhanced. With 5 dB window, a single serve dominance area is increased from 72.22 to 76.09%, and with 7 dB window it is raised from 63.79 to 67.71% by 6-sector site compared to 3-sector site. It means shifting from traditional 3-sector site to higher order sector site will not increase the handover area. Percentage of area covered with multiple servers is presented in Table 5.

5 Conclusions

In this article, we have studied the impact of higher order sectorization and different optimization techniques i.e. antenna downtilting and antenna beamwidth in traditional cellular networks, and then compared it with the performance of SPMA in macrocellular suburban environment. The system performance was evaluated from the quality (signal to interference ratio), cell spectral efficiency and area efficiency point of view. In the post simulation analysis for the considered scenario, the obtained results showed that an optimum antenna downtilting and selection of suitable antenna beamwidth for different sectored sites helps in optimizing the spectral efficiency of the system. In the considered simulation scenario, three sector sites were found performing better with 65° BW antenna and 6-sector sites with 32° BW antenna certain downtilt i.e. 8–9°. After adopting optimized antenna configuration the gain of using higher order sectorization (6-sector site) was only limited to 1.94 times with respect to the reference case of 3-sector site. However, SPMA showed an outstanding performance with the help of 0.5° × 0.2° needle beams. The obtained results indicate that narrow needle beams helps in avoiding the interference, and thus SPMA works in more spectral efficient way compared to traditional cellular networks. SPMA was found 317.82 times more spectrum efficient compared to the reference case of 3-sector site. In SPMA, it is recommended that if the two nearby users are experiencing bad signal quality then they should be scheduled at different time instant. SPMA over TDMA can help in overcoming this issue of bad SIR for the two nearby users. It was also observed that the site placement plays an important role in determining the radio channel conditions. No matter its traditional cellular approach or SPMA node is deployed, special attention should be given to site placement to cover the desired area.

Acknowledgements Authors would like to thank European Communications Engineering (ECE) Ltd and European Celtic-Plus project SHARING for supporting this research work.

References

1. Cisco Visual Networking Index: Global Mobile Data Traffic Forecast Up-date, 2013–2018, Cisco, 2014
2. Laiho J, Wacker A, Novosad T (2006) Radio network planning and optimization for UMTS, 2nd edn. Wiley, Chichester, England
3. Wacker A, Sipila K, Kuurne A (2002) Automated and remotely optimization of antenna subsystem based on radio network performance. In: IEEE 5th Symposium on Wireless Personal Multimedia Communications, pp 752–756
4. Sheikh MU, Lempiäinen J (2013) A flower tessellation for simulation purpose of cellular network with 12-sector sites. IEEE Wirel Commun Lett 2(3):279–282
5. 3GPP, E-UTRA and E-UTRAN Overall description: stage 2 (Release 10), 3GPP TS 36.300 V10.12.0, Jan. 215
6. Sheikh MU, Lempiäinen J (2014) Will new antenna material enable Single Path Multiple Access (SPMA)? Springer J Wirel Pers Commun 78(2):979–994

7. Sheikh MU, Säe J, Lempiäinen J (2015) Evaluation of SPMA and higher order sectorization for homogeneous SIR through macro sites. Springer J Wirel Netw
8. Lempiäinen J, Manninen M (2001) Radio interface system planning for GSM/GPRS/UMTS. Kluwer Academic Publishers
9. Sheikh MU, Ahnlund H, Lempiäinen J (2014) Advanced antenna techniques and higher order sectorization with novel network tessellation for enhancing macro cell capacity in DC-HSDPA network. Int J Wirel Mob Netw 5(5):65–84
10. Gomez-Diaz JS, Perruisseau-Carrier J (2012) Microwave to THz properties of graphene and potential antenna applications. In: International Symposium on Antennas and Propagation (ISAP), Nov 2012, pp 239–242
11. Schurig D et al (2006) Metamaterial electromagnetic cloak at microwave frequencies. Science 314:977–980
12. Gschwendtner BE (1994) Practical investigation using ray optical prediction techniques in microcells, Euro-COST 231 TD(94) 127
13. Huschka T (1994) Ray tracing models for indoor environments and their computational complexity. In: IEEE international symposium on personal, indoor and mobile radio communications (PIMRC), pp 486–490
14. Gschwendtner BE, Wolfle G, Burk B, Landstorfer FM (1995) Ray tracing versus ray launching in 3-D microcell modelling. In: 1st European personal and mobile communications conference (EPMCC), Bologna, Apr 1995, pp 74–79

Towards an Intelligent Deployment of Wireless Sensor Networks

Hadeel Elayan and Raed M. Shubair

1 Introduction

Localization has received a considerable amount of attention over the past decade since it is an essential capability enabling other applications. The localization problem is the process of determining the position of a node relative to a given reference frame of coordinates in a Wireless Sensor Network (WSN). Undeniably, the sensor location must be known for its data to be meaningful [1]. The prominence of this process comprises the ability to identify and correlate gathered data in a WSN as well as manage nodes located in a determined location. In fact, localization is achieved through the utilization of a subset of sensor terminals [2]. Nodes which have known states at all times and are aware of their own global coordinates a priori are called anchor nodes or beacon nodes. The anchor nodes are equipped with a Global Positioning System (GPS) thereby assisting in computing other nodes' locations by using a number of techniques such as lateration, angulation or a combination of both [3]. Nevertheless, sensor nodes which have a priori unknown states and need to determine their positions using a localization algorithm are referred to as agents [4].

Several localization algorithms have been proposed in the literature [5]. One of the most important aspects of localization involves measuring the transmission range of the wireless signal. Range-based localization relies on the availability of point-to-point distance (or angle information). The obtained measurements of different ranging techniques such as Time of Arrival (TOA), Time Difference of

H. Elayan (✉) · R.M. Shubair
Electrical and Computer Engineering Department, Khalifa University, Abu Dhabi, UAE
e-mail: hadeel.mohammad@kustar.ac.ae

R.M. Shubair
e-mail: raed.shubair@kustar.ac.ae

L. Ismail and L. Zhang (eds.), *Information Innovation Technology in Smart Cities*,
https://doi.org/10.1007/978-981-10-1741-4_16

Arrival (TDOA), Angle of Arrival (AOA), and Received Signal Strength (RSS) are the keys for range-based schemes [6, 7].

Most of the prevailing localization algorithms used in WSNs assume static nodes which do not move after deployment. Hence, the mobility effect is not explicitly considered in the analysis approach. The mobility effect cannot be ignored in certain WSN platforms such as intelligent transportation, patient monitoring, and habitual tracking. The major technical challenge in the localization of moving nodes is the rapidly changing localization scenarios resulting in inaccurate prediction of node locations [3]. It must be noted as well that mobile networks convey different characteristics from static networks. Other important aspects that deteriorate the performance of mobile sensor networks are issues related to the changing topology of the network, the varying connectivity, and latency problems [8]. These effects stimulate the need for developing robust yet accurate localization algorithms for moving nodes in WSNs.

The Monte Carlo localization (MCL) [9] was the first practical method for localization of mobile WSNs. MCL applies the Sequential Monte Carlo (SMC) method [10] to achieve localization. The reason behind this is that the posterior distribution of a sensor node after movement can be naturally formalized using a nonlinear discrete time model and the SMC method provides simple simulation-based approaches to estimate the distribution [8]. The method assumes no functional form, instead, it uses a set of random samples (also called particles) to estimate the posteriors. When the particles are properly placed, weighted, and propagated, posteriors can be estimated sequentially over time. This technique is more popularly known as the Particle Filter (PF) [11]. Nevertheless, previous SMC-based localization algorithms either suffer from low sampling efficiency or require high beacon density to achieve high localization accuracy.

In this chapter, we discuss the deployment of the Particle Filter (PF) framework into the localization of moving nodes in an open environment using TDOA measurements. This incorporation accurately captures the mean and covariance which improves the localization accuracy and helps achieve a robust performance. The rest of the chapter is organized as follows. Section 2 revisits the fundamental localization techniques. Section 3 presents the system model and problem statement. Section 4 describes the proposed TDOA-PF technique. In Sect. 5, numerical validation on simulated scenarios can be found. Finally, the conclusions are summarized in Sect. 6.

2 Localization Techniques

TOA, TDOA, RSS and DOA of the emitted signal are commonly used measurement techniques in WSN localization. Basically, TOAs, TDOAs and RSSs provide the distance information between the source and sensors while DOAs are the source bearings relative to the receivers. Nevertheless, finding the source position is not a trivial task because these measurements have non-linear relationships with the source position [12]. The signal models and their basic positioning principles are generalized as follows:

$$r = f(x) + n \tag{1}$$

where r is the measurement vector, x is the source position to be determined, $f(x)$ is a known nonlinear function in x, and n is an additive zero-mean noise vector.

2.1 Time of Arrival

TOA is defined as the difference between the sending time of the signal at the transmitter and the receiving time of the signal at the receiver (time delay). The time delay can be computed by dividing the separation distance between the nodes by the propagation velocity. TOA technique uses multilateration, since it includes ranges from more than three reference points. Mathematically, the TOA measurement model is formulated as follows. Let $x = [xy]^T$ be the unknown source position and $x_l = [x_l y_l]^T$ be the known coordinates of the l_{th} sensor $l = 1, 2, \ldots, L$, where $L \geq 3$ is the number of receivers. The distance between the source and the l_{th} sensor, denoted by d_l is:

$$d_l = [\|x - x_l\|]_2 = \sqrt{(x - x_l)^2 + (y - y_l)^2} \tag{2}$$

Without loss of generality, we assume that the source emits a signal at time 0, and the l_{th} sensor receives it at time t_l. That is, $\{t_l\}$ are the TOAs; a simple relationship between t_l and d_l is given by:

$$t_l = \frac{d_l}{c} \tag{3}$$

where c is the propagation speed of the radio signal (speed of light). In practice, TOAs are subject to measurement errors. As a result, the range measurement based on multiplying t_l by c denoted by $r_{(TOA,l)}$ is modeled as:

$$r_{(TOA,l)} = \mathrm{d}_l + \eta_{(TOA,l)} = \sqrt{(x - x_l)^2 + (y - y_l)^2} + \eta_{(TOA,l)} \tag{4}$$

where $\eta_{(TOA,l)}$ is the range error in $r_{(TOA,l)}$ which results from the TOA disturbance. In TOA, the nodes have to be synchronized and the signal must include the time stamp information. This requirement adds to the cost of the signal by demanding a highly accurate clock and increasing the complexity of the network.

To overcome such restrictions, the RTOA (Round-trip Time of Arrival) and TDOA are introduced [12]. RTOA is the most practical scheme in decentralized settings since it does not require a common time reference between the nodes. Actually, it measures the difference between the time when a signal is sent by the sensor and the time when the signal returned by a second sensor is received at the original sensor. Because the same clock is used to compute the round trip propagation time, there is no synchronization problem [13].

2.2 Time Difference of Arrival

The basic idea of TDOA is to determine the relative position of the mobile transmitter by examining the difference in time at which the signal arrives at a pair of sensors. This implies that clock synchronization across all receivers is required. Nonetheless, the TDOA scheme is simpler than the TOA method because the latter needs the source to be synchronized as well. Similar to the TOA, multiplying the TDOA by the known propagation speed leads to the range difference between the source and two receivers [12].

The TDOA measurement model is mathematically formulated as follows. We assume that the source emits a signal at the unknown time t_0 and the l_{th} sensor receives it at time t_l, $l = 1, 2, \ldots, L$ with $L \geq 3$. There are $L(L-1)/2$ distinct TDOAs from all possible sensor pairs, denoted by $t_{k,l} = (t_k - t_0) - (t_l - t_0)$, $k, l = 1, 2, \ldots, L$. Taking $L = 3$ as an example, the distinct TDOAs are $t_{2,1}$,$t_{3,1}$, and $t_{3,2}$. We easily observe that $t_{3,2} = t_{3,1} - t_{2,1}$, which is redundant. In order to reduce complexity without sacrificing estimation performance, we should measure all $L(L-1)/2$ TDOAs and convert them to (L−1) non-redundant TDOAs for source localization [14]. We consider the first sensor as the reference and the non-redundant TDOAs are $t_{l,1}$. The range difference measurements deduced from the TDOAs are modeled as:

$$r_{(TDOA,l)} = \mathrm{d}_{l,1} + \eta_{(TDOA,l)} = \sqrt{(x - x_l)^2 + (y - y_l)^2} + \eta_{(TDOA,l)} \tag{5}$$

where

$$\mathrm{d}_{l,1} = \mathrm{d}_l + \mathrm{d}_1 \tag{6}$$

and $\eta_{(TDOA,l)}$ is the range error in $r_{(TDOA,l)}$ which is proportional to the disturbance in $t(l,1)$.

2.3 Received Signal Strength

The RSS approach is used to estimate the distance between two nodes based on the strength of the signal received by another node. A sender node sends a signal with a determined strength that fades as the signal propagates. It is known that the bigger the distance to the receiver node, the lower the signal strength when it arrives to the node. In fact, the node can calculate its distance from the transmitter using the power of the received signal, knowledge of the transmitted power, and the path-loss model. The operation starts when an anchor node broadcasts a signal that is received by the transceiver circuitry and passed to the Received Signal Strength Indicator (RSSI) to determine the power of the received signal [9].

By assuming that the source transmitted power is P_t and there is no disturbance, the average power received at the lth sensor, denoted by $P(r, l)$ is modeled as:

$$P_{r,l} = K_l P_t d^{-\alpha} \tag{7}$$

where K_l accounts for all other factors which affect the received power, including the antenna height and antenna gain, while α is the path loss constant. Actually, the value of α can vary between 2 and 5. Particularly, $\alpha = 2$ depicts free space.

2.4 Direction of Arrival

The DOA estimation method requires the base stations to have multiple antenna arrays for measuring the arrival angles of the transmitted signal from the mobile station at the base stations. This technique can be further divided into two subclasses, those making use of the receiver antennas amplitude response and those making use of the receiver antenna phase response. The accuracy of the DOA measurements is limited by the directivity of the antenna, shadowing and multipath reflections [15]. Although this scheme does not require clock synchronization as in RSS-based positioning, an antenna array is needed to be installed at each receiver for DOA estimation. Assuming ϕ_1 be the DOA between the source and l_{th} receiver, we have:

$$\tan(\phi_l) = \frac{y - y_l}{x - x_l}, l = 1, 2, \ldots, L \tag{8}$$

with $L \geq 2$ Actually, ϕ_1 is the angle between the line-of-bearing from the l_{th} receiver to the target and the x-axis. The DOA in the presence of angle errors, denoted by $\{r_{(DOA,l)}\}$, are modeled as:

$$r_{DOA,l} = \phi_l + \eta_{DOA,l} = \tan^{-1}\left(\frac{y - y_1}{x - x_1}\right) + \eta_{DOA,l}, l = 1, 2, \ldots, L \tag{9}$$

where $\left\{\eta_{(DOA,l)}\right\}$ are the noises in $\{r_{(DOA,l)}\}$ which are assumed zero-mean uncorrelated Gaussian processes.

3 System Model and Problem Statement

3.1 Problem Statement

The problem to be investigated can be viewed in Fig. 1 and is described as follows. Basically, it consists of a mobile robot carrying a sensor node and of multiple anchor nodes. We refer to the mobile robot as a mobile node. Anchors whose locations are known ping out signals which are differentiable from the others.

Signals emitted from multiple anchors are used by the mobile node to determine its location. Although the mobile node recognizes the time of arrival of the pings, it has no information on the time of ping emission. Hence, it can only use the difference of arrival times of the pings to achieve localization. The following are the notations used for the solution:

- d_i = The distance between the moving node and the ith anchor.
- $d_{i1} = d_i - d_l$: range difference to the ith anchor and reference anchor.
- R_i = distance between the ith anchor and the reference anchor.
- R_r = distance between the moving node and the reference anchor.

3.2 System Model

1. **Mobility Model**

Time is divided into equal segments, ΔT, in which the mobile node moves along the direction at a constant value as depicted in Fig. 2. The movement is divided into several sub-paths according to the time segment where the mobile node progresses

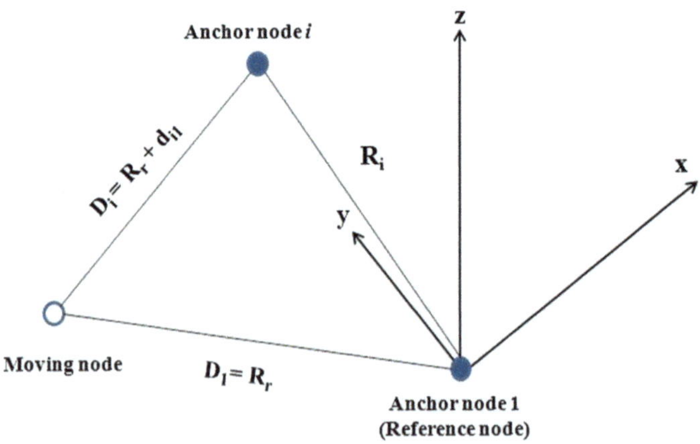

Fig. 1 The representation of the localization problem where the moving node depends on signals emitted from the anchors to estimate its location

at a constant velocity as presented in Fig. 3. After using a time segment, the mobile node can change the moving velocity thereby satisfying the uniform distribution in $[V_{min}, V_{max}]$, as shown in (10).

$$V \sim U[V_{min} V_{max}] \tag{10}$$

In Fig. 3 above, the triangle denotes the mobile node, Δs_i denotes the movement distance at time segment i by velocity v_i.

2. **Time Difference of Arrival Model**

TDOA is the difference in the time of the transmitted signal from the unknown sensor at a pair of anchors as explained in the section above. If the first anchor is assigned as a reference point, the range measurements based on the TDOAs are of the form [16].

$$\begin{aligned} r_{(TDOA,i)} &= (\mathrm{d}_i - \mathrm{d}_1) + \eta_{(TDOA,i)} \\ &= \left(\sqrt{(x - x_i)^2 + (y - y_i)^2} - \sqrt{(x - x_1)^2 + (y - y_1)^2}\right) + \eta_{(TDOA,i)} \end{aligned} \tag{11}$$

where $i = 2, 3, \ldots M$ and the range error $\eta_{TDOA,i}$ can be obtained from the difference of two TOA noise. So, $\eta_{TDOA,i}$ is $\eta_{TOA,i} - \eta_{TOA,i-1}$, $i = 2, 3, \ldots M$. Either linear or nonlinear algorithms may be deployed to achieve source localization. These algorithms basically involve minimizing the Least Squares (LS) or Weighted Least Squares (WLS) cost function. Another alternative involves analytically solving the problem through a closed form equation [7].

3. **System State Model**

The system state model for the mobile wireless sensor is represented by the following formula:

Fig. 2 Equal time segment

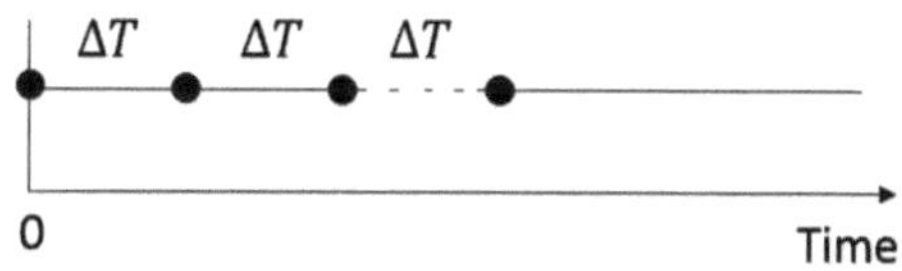

Fig. 3 Mobility model of a robot

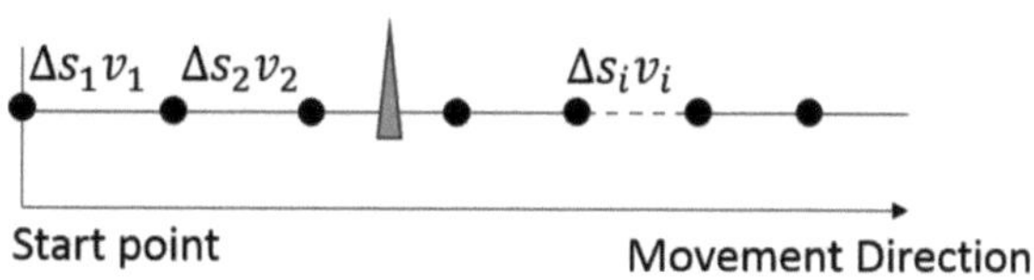

$$x_k = x_{k-1} + v_k \cdot \Delta T + \varepsilon_k \tag{12}$$

where x_k is the position of a mobile node from the anchor, ΔT is the time segment, v_k is the current velocity, and ε_k denotes the system state noise which obeys Gaussian distribution with zero mean and covariance Q_k, R_k. The TDOA measurement is provided in (11).

3.3 System Assumption

To simplify the problem, certain assumptions need to be made.

- All the sensor nodes own equal physical parameters.
- The movement velocity of mobile node stays the same at time ΔT, and the time of turning is ignored.
- The random velocity obeys normal uniform distribution as (10).
- The anchors are deployed at determined pattern and the position cannot be changed.

4 Localization of Moving Nodes Based on PF-TDOA Approach

Localizing a moving node is an interesting challenge due to the fact that it demands extensive computational effort at each iteration. To localize a moving sensor node, filtering methods are often applied. Kalman filter is a celebrated technique for recursive state estimation, and has been widely used in many applications. Due to its unbiased minimum variance estimation with white noise assumption, Kalman filter is an optimal solution in linear systems. However, it is limited in non-linear and non-Gaussian systems, which exist ubiquitously in the real world [16]. To address these limitations, a number of methodologies have been proposed, e.g., Extended Kalman Filter (EKF) [17] and Unscented Kalman Filter (UKF) [18]. Another alternative to the methods mentioned above is the Particle Filter (PF), which is a Bayesian inference based scheme.

In recent years, PF has become a research hotspot in the field of nonlinear filtering and estimation. The key idea [19] is to represent the required posterior probability density function (PDF) by a set of random samples (called particles) with associated weights and to compute estimates based on these samples and weights. As the number of particles increases, the Monte Carlo characterization becomes an equivalent representation to the usual functional description of the posterior PDF, and the PF approaches the optimal Bayesian estimator.

In order to formulate the PF scheme, assume the prior conditional probability density of a dynamic system is $p(x_0)$ and let $\{x^k_{i=0}, w^k_{i=0}\}^N_{i=1}$ represent both the measurement value and its weight of random sample during the time k. The PDF is $p(x_{0:k}|z_{1:k})$. The weights of x^k_i are normalized such that $\sum_{i=1}^{N_x} w^i_k = 1$. Then, the posterior density at time k can be approximated as

$$p(x_{0:k}|z_{1:k}) \approx \sum_{i=1}^{N_x} w^i_k \delta\left(x_{0:k} - x^i_{0:k}\right) \tag{13}$$

Following the derivation presented in [19], we get

$$p(x_k|z_{1:k}) \approx \sum_{i=1}^{N} w^i_k \delta\left(x_k - x^i_k\right) \tag{14}$$

The approximation of x yields

$$\hat{x}_k \approx \sum_{i=1}^{N} w^i_k x^i_k \tag{15}$$

Next, we compute an estimate of the effective number of particles as

$$\hat{N}_{eff} = \frac{1}{\sum_{i=1}^{N_x} \left(w^i_k\right)^2} \tag{16}$$

Actually, if the effective number of particles is less than a given threshold, $\hat{N}_{eff} < N_{thr}$, resampling must be performed.

The PF algorithm can be used to address mobile node localization by using TDOA. This approach was first presented in [20] for underwater localization. In this paper, we revisit the approach and extend it for the localization of moving nodes in the case of an open environment. In fact, the incorporation between the PF and TDOA makes use of both the internal motion information of the moving node and the interaction between the distance estimates of the anchor sensors which results in accurate localization of the moving node. Actually, localization based TDOA has turned out to be a promising approach when neither receiver positions nor the positions of signal origins are known a priori. Thus, the PF-TDOA technique is suitable for TDOA localization as the state can be computed online and it is robust to motion and measurement uncertainty [21].

This PF approach predicts the state of interest and then corrects the prediction based on the observed sensor data. The prediction and correction steps are repeated as the state progresses with time [23]. Each particle in the PF represents a state which translates into the moving node location in the case of TDOA localization.

The pseudo code of the PF can be viewed in Fig. 4. The code takes the following as inputs: the particle distribution at the previous sampling time X_{t-1}, the observed

---

$PF\ approach\ (X_{t-1}, \mathbf{z}_t, f_t, V)$

---

1. $X_t = Particle\ distribution = \emptyset$
2. $for\ i = 1$ to N do
3. $\quad x_t^{[i]} = Motion\ model\ (\mathbf{z}_t, x_{t-1}^{[i]})$
4. $\quad g_t^{[i]} = Sensor\ model\ (\boldsymbol{f}_t, x_t^{[i]}, \boldsymbol{V})$
5. $end\ for$
6. $for\ i = 1$ to N do
7. $\quad x_t^{[i]} = Resampling\ \left\{(\boldsymbol{x}_t, g_t^{[i]}) \mid \boldsymbol{j} = \mathbf{1}, \ldots\ldots, \boldsymbol{N}\right\}$
8. $end\ for$
9. $Return\ X_t$

Fig. 4 PF procedure with number of particles N

sensor data f_t, the internal moving node information z_t, the information on the surrounding local environment V that affects the sensor data, and the state x_t which includes the angle of roll, pitch, and yaw, respectively. The output of the code is the new set of particles X_t.

According to the PF approach, the prediction considers only the motion information of the sensor. Basically, the pose of the moving node x_t can be predicted using internal robot motion information such as motion command or odometer. Specifically, the internal motion information of a moving node, u_t, refers to the velocity command for surge, sway, heave, roll, pitch, and yaw motion, respectively. Moreover, the motion model of the moving node formulates the pose transition mathematically. The correction procedure calculates the degree of certainty signifying whether the predicted location indicates true location. Next, it resamples the probable location from the predicted location based on the assumption made. The certainty of each predicted location is calculated by comparing the real sensor data with the expected sensor data. Resampling aims at eliminating the degeneracy which might result when significant weight is concentrated on only one particle after some time steps. It solves this issue by discarding particles with negligible weights and enhances ones with larger weights.

5 Simulation Results

In this section, we apply particle filter method using the TDOA algorithm to enhance the localization accuracy. First, we set some basic parameters to the simulation scenario which are presented in Table 1.

The localization performance is evaluated by the estimation error in which the root mean square error (RMSE) is used to study the estimator's accuracy. The RMSE is defined as follows

$$RMSE = \sqrt{\frac{1}{T}\sum_{i}^{T}(x_k - \hat{x}_k)^2} \tag{17}$$

where T is the number of measurements (time steps), x_k is the real position of a mobile node and $\hat{x}_k$ is the estimation of the position. The resampling scheme used is Residual resampling since it provides lower conditional variance for all configurations of the weights [22]. At one round, we use the RMSE to evaluate the localization accuracy. We compute the localization accuracy by calculating the mean and variance of RMSE after many rounds. As a comparison, we present the result of the other filter algorithms, including the EKF, the UKF, and the PF based on TDOA measurement. The values of the mean and variance of the different filtering algorithms can be found in Table 2.

It can be noticed from Table 2 that the PF-TDOA approach has the least mean, which indicates that the algorithm always has the best localization accuracy in comparison to the others.

Table 1 Simulation parameters

Symbol	Meaning	Value
N	Number of particles	50
Q	Measurement variance	1
R	Process variance	3
V_{max}	Maximum velocity	5 m/s
V_{min}	Minimum velocity	1 m/s
N_x	Length of x augmentation	10
N_a	Number of anchors	6
N_{th}	Effective particles	10

Table 2 Mean and variance of different algorithms

Filter	Mean	Variance
EKF	4.411	7.1947
UKF	4.0583	1.6041
PF	3.433	0.69623
PF-TDOA	3.1624	0.67719

Figures 5 through 8 below demonstrate the accuracy of the proposed PF-TDOA approach in comparison to the other methods. In Fig. 5, the capability of the PF approach to predict and correct the state of interest based on the observed sensor data can be inferred. Each particle in the PF represents a state which translates into the moving node location in the case of TDOA localization. The PF-TDOA approach uses both the motion information as well as the distance estimates from the anchor nodes to be capable of providing a higher accuracy and enhanced robustness.

In Fig. 6, the TDOA-PF approach is proven to have the lowest mean of estimation in comparison to the other approaches. In Fig. 7, the moving node follows a randomly generated path which is shown in blue. It can be easily noticed that the TDOA-PF approach closely depicts the actual trajectory of the moving node as it

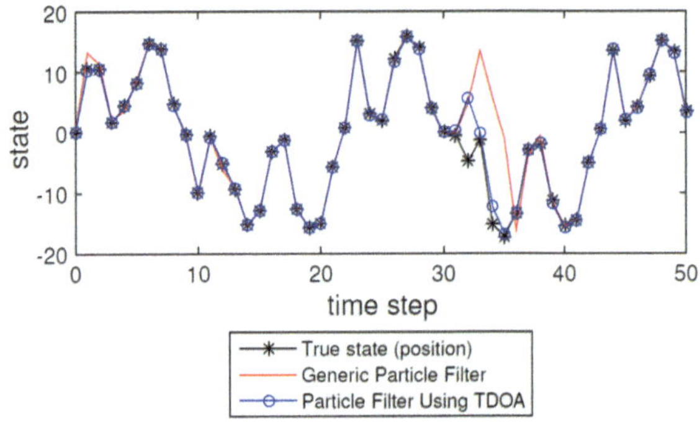

Fig. 5 Filter estimates versus true state

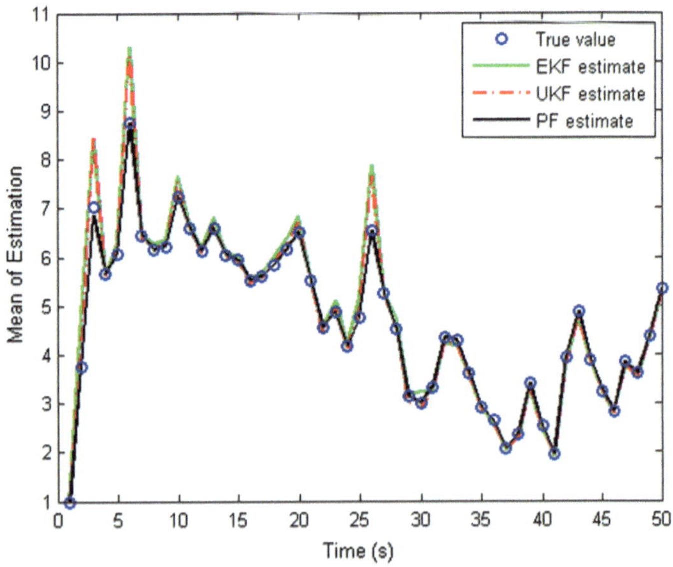

Fig. 6 Comparison between the performance of the PF, EKF and UKF filters

moves along the x and y coordinates in an open environment; hence, it is the most optimum. The fusion between the PF method and TDOA localization results in a reduced uncertainty region as illustrated in Fig. 8.

Next, we have evaluated the localization accuracy of both the general particle filter and that which utilizes the TDOA approach by assigning the simulation parameters with different values. Our goal is to find which parameters have a more significant influence on the localization accuracy and robustness.

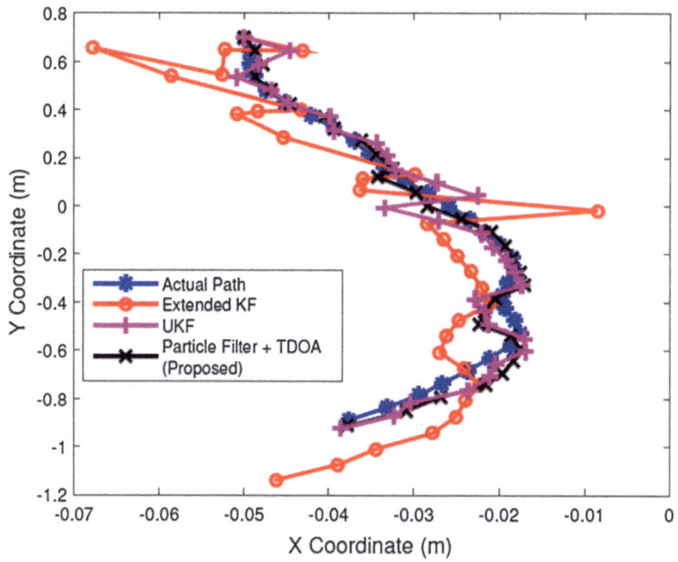

Fig. 7 Comparison between the performance of the PF, EKF and UKF filters in TDOA localization

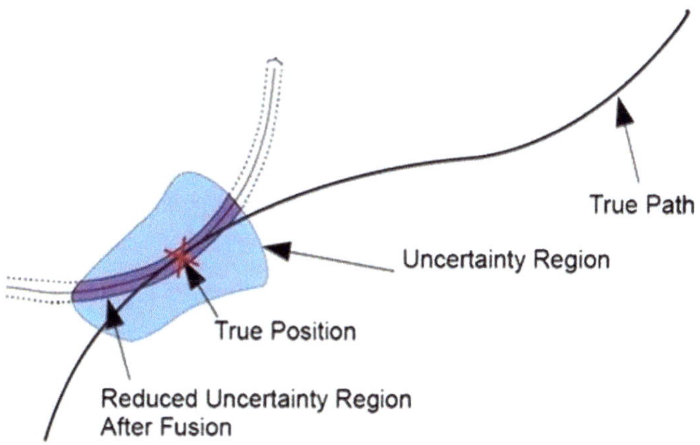

Fig. 8 Illustration of the TDOA-PF performance

5.1 Number of Particles

The localization accuracy increases when the amount of particles increase at most conditions. This result actually conforms to the feature of the particle filter.

5.2 Number of Anchors

The localization accuracy improves when increasing the number of anchors at all algorithms.

5.3 Number of Iterations

The particle filter uses the iterative process to eliminate the noise in system and measurement. As a matter of fact, as the number of iterations increases, the higher the accuracy that can be achieved in the generic system state. However, in our system model, the mobile node has to employ a random velocity when it moves forward in a time segment. Therefore, this might result in the accumulation of error which leads to some localization errors.

6 Conclusion

In this chapter, a technique for enhanced localization of moving nodes in WSNs has been presented. The proposed technique is based on the use of TDOA along with PF method in order to attain accurate detection and enhanced localization. Each particle in the PF represents a state which translates into the moving node location in the case of TDOA localization. The combined TDOA-PF technique utilizes the internal motion information of the moving node as well as the distance estimates that result from the interaction between the anchor nodes. The performance of the TDOA-PF is compared to the other existing methods used for mobile node localization. Simulation results proved that the proposed approach outperforms the other mentioned techniques in terms of the RMSE.

References

1. Boukerche A, Oliveira H, Nakamura E, Loureiro A (2007) Localization systems for wireless sensor networks. Wirel Commun, IEEE 14(6):6–12
2. Ilyas M, Mahgoub I, Kelly L (2004) Handbook of sensor networks: compact wireless and wired sensing systems. CRC Press Inc, Boca Raton, FL, USA
3. Mirebrahim H, Dehghan M (2009) Monte carlo localization of mobile sensor networks using the position information of neighbor nodes, in ad-hoc, mobile and wireless networks. Springer, pp 270–283
4. Sathyan T, Hedley M (2009) Evaluation of algorithms for cooperative localization in wireless sensor networks. In: Personal, indoor and mobile radio communications, 2009 IEEE 20th international symposium on, Sept 2009, pp 1898–1902
5. Hightower J, Borriello G (2001) Location systems for ubiquitous computing. Computer 8:57–66
6. Wang J, Ghosh RK, Das SK (2010) A survey on sensor localization. J Control Theory Appl 8 (1):2–11
7. Najibi M (2013) Localization algorithms in a wireless sensor network using distance and angular data, Ph.D. dissertation, Applied Sciences: School of Engineering Science
8. Zhang S, Cao J, Li-Jun C, Chen D (2010) Accurate and energy-efficient range-free localization for mobile sensor networks. Mobile Comput, IEEE Trans 9(6):897–910
9. Dellaert F, Fox D, Burgard W, Thrun S (1999) Monte carlo localization for mobile robots. In: Proceedings 1999 IEEE international conference on robotics and automation, vol. 2. IEEE, pp 1322–1328
10. Handschin J (1970) Monte carlo techniques for prediction and filtering of non-linear stochastic processes. Automatica 6(4):555–563
11. Rui Y, Chen Y (2001) Better proposal distributions: object tracking using unscented particle filter. In: Proceedings of the 2001 IEEE computer society conference on computer vision and pattern recognition 2001. CVPR 2001, vol 2. IEEE, pp II–786
12. So HC (2011) Source localization: algorithms and analysis. Handb Position Location: Theory, Pract, Adv pp 25–66
13. Laaraiedh M (2010) Contributions on hybrid localization techniques for heterogeneous wire-less networks, Ph.D. dissertation, Universit′ Rennes 1
14. So HC, Chan YT, Chan FKW (2008) Closed-form formulae for time-difference-of-arrival estimation. IEEE Trans Signal Process 56(6):2614–2620
15. Boudhir AA, Mohamed B, Mohamed BA (2010) New technique of wireless sensor networks localization based on energy consumption. Int J Comput Appl 9(12):25–28
16. Cheung KW, So HC, Ma WK, Chan YT (2006) A constrained least squares approach to mobile positioning: algorithms and optimality. EURASIP J Appl Sig Process, vol 2006, pp 150–150
17. Zhou F, Qin Z, Xiao C, Li S, Jiang W, Wu Y (2011) Tracking moving object via unscented particle filter in sensor network. Int J Digit Content Technol Appl 5(12)
18. Athans M, Wishner R, Bertolini A (1968) Suboptimal state estimation for continuous-time nonlinear systems from discrete noisy measurements. Autom Control, IEEE Trans 13(5):504–514
19. Wan E, Van Der Merwe R (2000) The unscented kalman filter for nonlinear estimation. In: Adaptive systems for signal processing, communications, and control symposium 2000. AS-SPCC. The IEEE 2000, pp 153–158
20. Arulampalam M, Maskell S, Gordon N, Clapp T (2002) A tutorial on particle filters for online nonlinear/non-gaussian bayesian tracking. Sig Process, IEEE Trans 50(2):174–188

21. Ko NY, Kim TG, Moon YS (2012) Particle filter approach for localization of an underwater robot using time difference of arrival. In: OCEANS, 2012—Yeosu, May 2012, pp 1–7
22. Bordoy J, Hornecker P, Hoflinger F, Wendeberg J, Zhang R, Schindelhauer C, Reindl L (2013) Robust tracking of a mobile receiver using unsynchronized time differences of arrival. In: (IPIN) 2013 International Conference on indoor positioning and indoor navigation, Oct 2013, pp 1–10
23. Djuric P, Kotecha JH, Zhang J, Huang Y, Ghirmai T, Bugallo M, Miguez J (2003) Particle filtering. Sig Process Mag IEEE 20(5):19–38

Secure Mobile Agent Protocol for Vehicular Communication Systems in Smart Cities

Dina Shehada, Chan Yeob Yeun, M. Jamal Zemerly, Mahmoud Al-Qutayri and Yousof Al Hammadi

1 Introduction

Nowadays the number of vehicles registered a remarkable increase compared to earlier years. Vehicles are a necessity to people and it is expected that the number will keep increasing in a non-steady way. Researchers also noticed that this remarkable growth is not exclusive to the number of vehicles, as the number of accidents and congestion rates also increased. New problem acquires a new solution, therefore, as part of the smart city vision, new solutions are introduced to meet the needs of the smart city framework. The vehicular communication system is a promising solution to the emerging problems. In these systems, vehicles exchange information about congestion information, state of the weather, state of the road, possible accidents, the existence of pedestrian crossing, divert routes, emergencies, and others. Sharing congestion and weather information with drivers saves them time and money through assisting them with the route choice. More importantly, sharing information about an accident or emergency events, prepares drivers to be able to make the appropriate reaction. Information can be exchanged between cars (Inter-Vehicle Communication) or between cars and roadside units (RSUs) (Car to RSU Communication) which are units mounted on the side of the road [1–3]. Figure 1 shows an example of a basic structure for vehicular communication systems.

In vehicular communication systems data is stored in distributed places and therefore, providing application users with a fast and efficient service is not an easy task [4, 5]. It is agreed that such smart systems are a must-have for any modern and comfortable city. As a solution, we propose the use of MAs in vehicular systems. Mobile agents (MAs) are intelligent pieces of software that have the ability to move between different nodes. Mobile agents are dispatched with specific tasks. They require low network bandwidth and are small in size [6].

D. Shehada (✉) · C.Y. Yeun · M. Jamal Zemerly · M. Al-Qutayri · Y.A. Hammadi
Khalifa University of Science, Technology and Research, Abu Dhabi, UAE
e-mail: dinashehada009@gmail.com

L. Ismail and L. Zhang (eds.), *Information Innovation Technology in Smart Cities*,
https://doi.org/10.1007/978-981-10-1741-4_17

Fig. 1 Basic structure of a vehicular communication system [7]

In addition to these features, MAs mobility, flexibility and fast migration, make them popular in distributed applications. Because of the distributed nature of MAs, their small size, and many other features that ensure fast and efficient retrieval process of data [7–9]. As vehicular communication systems rely on the wireless network established between vehicles and RSUs and due to MAs openness and flexibility, the system becomes vulnerable to many security attacks such as unauthorized access, replay, modification, repudiation, eavesdropping, masquerade and Man In the Middle (MITM) attacks [10, 11]. Being vulnerable to these attacks affects application security and privacy in addition to the overall efficiency and productivity of the application. Therefore, providing proper security and protection is an important part of any smart communication system from the very beginning. To ensure having a secure and trusted system, the following requirements should be provided [1, 12, 13]:

- Confidentiality: sensitive information should not be exposed to any unauthorized party.
- Integrity: sensitive information should be protected from unauthorized change by a malicious party and malicious changes should be detected.
- Anonymity: the identity of the system users should not be exposed.
- Authorization: entities should be verified, and accordingly are either granted or denied the access.
- Mutual authentication: any two parties communicating with each other should verify each other's identities.

- Accountability: traceability of malicious actions to attackers holding them responsible for their actions.
- Non-repudiation: a communication party cannot deny its actions, all actions should be traceable to their actors.

Many researchers have investigated MASs security. For example, in [14], the issue of user anonymity has been addressed. Moreover, confidentiality of path traversed by the MA is also ensured. Mixers are median nodes the MA visits during its journey. Data confidentiality and authorization of data access are provided. However, no proper authentication techniques are provided. It assumes that all nodes are legitimate and trusted. Moreover, the proposal is vulnerable to MITM and replay attacks and the loss of the MA indicates the total loss of all the collected data. On top of all that, the random choice of mixers introduces another important issue of the random time delay.

An anonymous authentication process is proposed in [15]. Information confidentiality and mutual authentication are provided. Moreover, protection from MITM, masquerade and replay attacks. However, data integrity, accountability, and non-repudiation are not provided. Another anonymity scheme is provided in [16]. The proposed protocol provides anonymity to owner platform. The protocol is vulnerable to replay, MITM, and some other attacks. Also, the loss of MA results in the loss of all the data collected [14]. In [17] a cryptographic and trusted platforms based protocol that ensures the confidentiality of a dynamic itinerary generated at runtime. Trusted platforms collect information about other platforms in a secure way and make the decision on the agent's next itinerary. The itinerary information is ensured to remain secret through encryption and can only be accessed by the intended platform. In addition, hash functions and asymmetric encryption are used to provide integrity to agent code. Although the protocol provides mutual authentication and other security features, it introduces a delay and timing overhead that is not suitable for many applications [18–20].

To address the replay attack, a protocol is proposed in [18]. The protocol ensures the protection of agents from replay attacks, allowing legitimate re-execution at the same time. MAs have a unique number called trip marker. Platforms keep track of trip markers and counter values of all the agents that execute on them. Authorization and mutual authentication are ensured, but again the same issue of loss of carried data in case of loss of agent is introduced [20]. A Visa Based Authentication Scheme (VBAS) for MASs is proposed in [21] to authorize and manage agents' information. Each agent is granted a passport that contains agent information, i.e. its identity, its owner information, time stamp, certificate and other data used to provide authentication. Checking each agent's Visa platforms can detect a malicious agent who is unauthorized and trying to access some resources. The protocol ensures the integrity of both passport and Visa. However, it is assumed that a secure communication channel exists to exchange passport and Visa. If at any time the channel got compromised then the passport and Visa sent in clear text are disclosed. Although the Visa and passport might not contain sensitive information, sending them in clear text, makes the system vulnerable to denial of service attack, where attackers might maliciously tamper with the passport and visa of an agent and send the tampered version to overwhelm legitimate platforms [22]. The work in [23] provides an extension to [24] protocol providing authorization. A malicious identification police (MIP) scans MAs for malicious code stored in a database of malicious codes previously detected. In [23] agents are also scanned to find out if they match the platform policy file. The scanning system introduces delays that jeopardized system security. Also, changes need to be reported to the Attack Identification Scanner (AIS) at once [25]. The authors also propose another protocol to address integrity of agent code. The protocol proposes an improvement to the classic Root Canal protocol [26] that was vulnerable to repudiation attacks. In addition to integrity, the proposed eXtended Root Canal (XRC) protocol provides protection from repudiation attacks as well [27].

Encryption keys generated from data produced at runtime are used in [27]. As a result, key guessing or prediction is quite hard. Authentication of agent owner, confidentiality of data, and integrity are provided. But vulnerability to replay attacks and the issue of agent loss affects the system functionality. The framework in [28] proposes the idea of multiple levels of access control of platforms resources in MASs. Platforms check the history of visited platforms of each agent through a signatures chain and then makes the decision of level of access this agent can have. One-way authentication and authorization are provided. However, as the path of the agent increases, it produces an overhead to verify the signature chain. Also, it assumes that platforms are able to identify malicious platforms in the system.

A secure information exchange system is proposed in [29]. Upon the verification of the platform, information required for decryption is sent. Data confidentiality and mutual authentication between sender and receiver are ensured and data are only be accessed by authorized parties [30, 31]. Proposed protocol in [32] is based on a trusted platform that verifies data integrity and reports any alteration. It provides confidentiality, integrity of MA data and non-repudiation of data are all provided. However, agent loss results in the loss of all the data. A knowledge-based system for collecting information about hosts' behaviors and actions is proposed in [33]. Collected information help in the decision making of the next destination platform. The protocol provides data confidentiality, non-repudiation, anonymity, and integrity. But, suffers from the same issue related to the loss of agent. Moreover, due to the expensive integrity checks occurring at trust hosts, it introduces a high computational cost [34]. In [35] agent code is split into two parts of the code, a critical and normal code. Critical codes are considered of high importance and therefore, only executed at trusted platform while, normal codes are carried by the MA. Trusted platforms have two main roles, some of them are responsible for executing the critical code while others monitor the system and provide a recovery policy in case of a system breakdown. Data is collected by MAs and sent to trusted platform to get the final results by executing the critical code. Trusted platforms detect replay attacks through issuing of timers. Confidentiality of data, integrity of data and code and the possibility to recover the agent code in case of modification or breakdown are some of the good features that are provided. As can be seen, although some of the reviewed related work provide some good security feathers. They suffer from many other drawbacks and bounded by many limitations i.e. random behaviors and centralized structure that is not fit with vehicular communication systems. Moreover, limited level of protection is provided in addition to the fact that they suffer from vulnerability to many attacks.

To address these issues, we propose a novel Secure MA Protocol (SMAP) for intelligent vehicular communication systems. SMAP overcomes most of the issues that the reviewed related protocols failed to address. Mutual authentication between vehicles, integrity, confidentiality, accountability, authorization, and non-repudiation are provided by SMAP. In addition to that, SMAP proposes a solution to the issue of loss of data in case of loss or killing of MA. Formal verification is conducted with Scyther [36], to validate the ability of SMAP to withstand many attacks and provide the claimed security features. The remaining part of this paper is structured as follows. In Sect. 2, the details of the novel SMAP are discussed. Followed by a security analysis and formal verification of SMAP in Sect. 3. In Sect. 4, the complexity of SMAP is studied. In Sect. 5, a scenario of a vehicular communication system is simulated with JADE [37, 38]. Finally, the paper findings and contributions are concluded in Sect. 6.

2 SMAP—A Novel Secure Mobile Agent Protocol

This section is dedicated to provide a detailed overview of our novel protocol SMAP. As will be seen later, SMAP provides the fundamental security requirements and protection from different security attacks. Moreover, its new hop-based architecture keeps the system from losing all its data in case of malicious loss or killing of agent provided that the first platform will not kill the agent.

Figure 2 shows an illustrative scenario of a vehicular communication system. In the proposed protocol the Public Key Infrastructure (PKI) is embedded into the communication system to ensure secure communication [39, 40]. Prior to the start

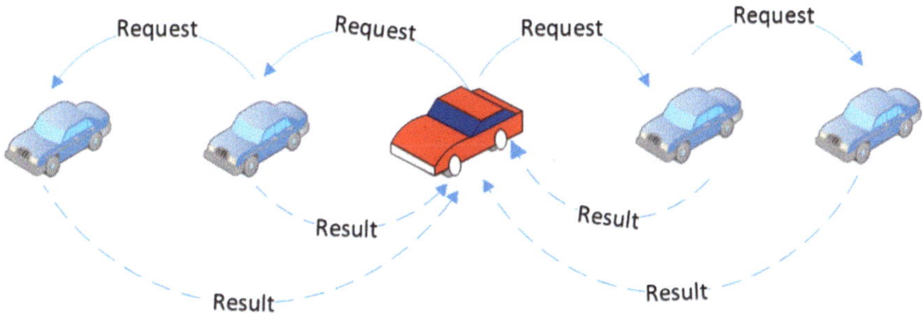

Fig. 2 Scenario of a vehicular communication system

Table 1 Acronyms and definitions

Acronym	Definition
Pla_i	Vehicle platform i
MAi	Mobile agent i
Pk_i	Public key of vehicle platform i
Sk_i	Private key of vehicle platform i
Req	Request
Res_i	Results generated by vehicle platform i
h()	Hash value
Add_i	Address of vehicle platform i
T_i	Time stamp generated by vehicle platform i
$\rightarrow$	Send

of the communication, all vehicles are verified by the Certificate Authority (CA) and granted public keys that are shared and distributed to other vehicles. In our system, the CA is the Traffic Management Center of the city. The communication starts when the user in the middle car requests information, therefore, copies of the MA are created carrying the request to close-by vehicles that send results back to the user and forward the request to other vehicles. Table 1 introduces some acronyms and their definitions that are needed for further explanation of SMAP.

Pla_1 is the owner (user vehicle) platform. Starting at Pla_1 the MA is created and launched carrying the request Req to the close-by vehicles. The owner platform Pla_1 generates MA1. SMAP provides a secure exchange of information between vehicles in the Vehicular Communication System. As shown in Fig. 3, the communication starts as the user in the owner platform Pla_1 requests information from nearby platforms, so if the Pla_1 wants to send the request to Pla_2, the owner platform Pla_1 does the following:

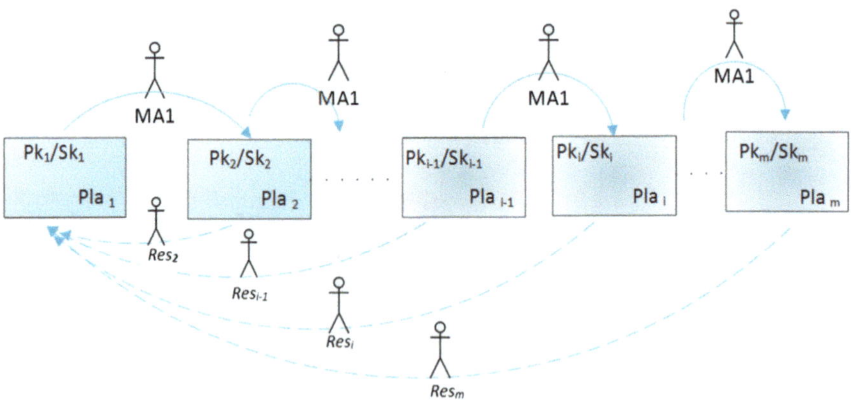

Fig. 3 Exchange of request and results with SMAP

- At Pla_1:

  1. Gets request Req from user.
  2. Generates a time stamp T_1.
  3. Generates $h(Req\|\ T_1\|\ add_2)$
  4. Signs $h(Req\|\ T_1\|\ add_2)$ with Sk_1 to generate $Sk_1(h(Req\|\ T_1\|\ add_2))$.
  5. Encrypts Req, add_2 and T_1 with Pk_2 to generate $Pk_2(Req\|\ add_2\|\ T_1)$.
  6. $Pla_1 \rightarrow Pla_2$:
     $Pk_2(Req\|\ add2\|\ T_1)\|\ Sk_1(h(Req\|\ add2\|\ T_1))$

After that Pla_2 receives the request and processes it as follows:

- At Pla_2:

  1. Decrypts $Pk_2(Req\|\ add_2\|\ T_1)$ with Sk_2.
  2. Checks freshness of T_1.
  3. Decrypts $Sk_1(h(Req\|\ add_2\|\ T_1))$ with Pk_1.
  4. Checks if $h(T_1\|\ Req\|\ add_2)$ in $Sk_1(h(Req\|\ add_2\|\ T_1))$ matches hash of sent T_1, add_2 and Req in $Pk_2\ (Req\|\ add_2\|\ T_1)$.
  5. If all checks are ok, then executes agent and generates result Res_2 and passes request to next platform Pla_3.
  6. Generates a time stamp T_2.
  7. Generates $h(Req\|\ T_1\|\ T_2\|\ add_2\|\ add_3)$.
  8. Signs $h(Req\|\ T_1\|\ T_2\|\ add_2\|\ add_3)$ with Sk_2 to generate $Sk_2\ (h(Req\|\ T_1\|\ T_2\|\ add_2\|\ add_3))$.
  9. Encrypts Req, add_2, add_3, T_2 and T_1 with Pk_3 to generate $Pk_3(Req\|\ T_1\|\ T_2\|\ add_2\|\ add_3)$.
  10. $Pla_2 \rightarrow Pla_3$:
      $Pk_3(Req\|\ T_1\|\ T_2\|\ add_2\|\ add_3)\|\ Sk_2(h(Req\|\ T_1\|\ T_2\|\ add_2\|\ add_3))\|\ Sk_1(h(Req\|\ add_2\|\ T_1))$.

After that, every vehicle platform will forward the request to the next platform. If Pla_i where, $2 \leq i < m$, has already received the request from Pla_{i-1} and wants to forward the request to Pla_{i+1}, Pla_i does the following:

- At Pla_i:

  1. Generates a time stamp T_i.
  2. Generates $h(Req\|\ T_1\|\ T_2\|\ \ldots\|\ T_{i-1}\|\ T_i\|\ add_2\|\ add_{i+1})$.
  3. Sign $h(Req\|\ T_1\|\ T_2\|\ \ldots\|\ T_{i-1}\|\ Ti\|\ add_2\|\ add_{i+1})$ with Sk_i to generate $Sk_i(h(Req\|\ T_1\|\ T_2\|\ \ldots\|T_{i-1}\|\ T_i\|\ add_2\|\ add_{i+1}))$.
  4. Encrypts Req, add_2, add_{i+1} and time stamps with Pk_{i+1} to generate $Pk_{i+1}(Req\|\ add_2\|\ add_{i+1}\|\ T_1\|\ T_2\|\ \ldots\|\ T_{i-1}\|\ T_i)$.
  5. $Pla_i \rightarrow Pla_{i+1}$:
     $Pk_{i+1}(Req\|\ add_2\|\ add_{i+1}\|\ T_1\|\ \ldots\|\ T_{i-1}\|\ T_i)\|\ Sk_i(h(Req\|\ T_1\|\ \ldots\|\ T_{i-1}\|\ T_i\|\ add_2\|\ add_{i+1}))\|\ Sk_1(h(Req\|\ add_2\|\ T_1))$.

All vehicle platforms will send back their results to the owner vehicle Pla_1 as soon as they are generated.

- At Pla_n, where, $2 \leq n \leq m$.

  1. Generates time stamp T_n.
  2. Generates hash of the result Req, Res_n and time stamps $h(Res_n\|\ Req\|\ T_1\|\ \ldots\|\ T_n)$.
  3. Encrypts result Res_n, Req and time stamps with Pk_1 to get $Pk_1(Res_n\|\ Req\|\ T_1\|\ \ldots\|T_n)$.
  4. Signs $(h(Req\|\ Res_n\|\ T_1\|\ \ldots\|T_n))$ with Sk_n to get $Sk_n(h(Req\|\ Res_n\|\ T_1\|\ \ldots\|\ T_n))$.
  5. $Pla_n \rightarrow Pla_1$
     $Pk_1(Res_n\|\ Req\|\ T_1\|\ \ldots\|T_n)\|\ Sk_n(h(Req\|\ Res_n\|\ T_1\|\ \ldots\|T_n))$

Finally at the owner platform Pla_1 receives the result from the vehicle platform Pla_n and does the following:

- At Pla_1:

  1. Decrypts $Pk_1(Res_n\|\ Req\|\|\ T_1\|\ \ldots\|\ T_n)$ with Sk_1.
  2. Checks freshness of T_1 through T_n.
  3. Decrypts $Sk_n(h(Req\|\ T_1\|\ \ldots\|\ T_n\|\ Res_n))$ with Pk_n.
  4. Checks if $h(Req\|\ Res_n\|\ T_1\|\ \ldots\|\ T_n)$ in $Sk_n(h(Req\|\ T_1\|\ T_2\|\ \ldots\|\ T_n\|\ Res_n))$ matches $h(Req\|\ T_1\|\ \ldots\|\ T_n\|\ Res_n)$ in $Pk_n(Res_n\|\ Req\|\|T_1\|\ \ldots\|T_n)$.
  5. If all checks are correct, displays results to the user.

SMAP secures the vehicle-to-vehicle communication. It ensures that requests and results are encrypted and secured. Moreover, because the vehicle sends its result as soon as it receives the request. The system provides a fast and efficient response to its users (Owners). After describing the proposed protocol in details, in the next section, security analysis and detailed results of formal verification test with Scyther are discussed.

3 Security Analysis and Formal Verification of SMAP

In this section, security analysis and details of formal verification of SMAP are discussed. After that, comparison of SMAP with related work is presented.

3.1 Security Analysis of SMAP

SMAP provides the following security requirements:

- Mutual authentication is the assurance that communicating entities can verify each other's identity. Take for example the interaction between owner platform and vehicle platform. Both platforms are sure of who sent the message because of the use of digital signatures. Messages sent are fresh due to the use of time stamps and both platforms can verify that the messages exchanged are actually intended for them because of the concatenation of the address of the destination in the message. Moreover, messages are received in order, and owner replies to owner platform after it receives the request. The interaction provides all the requirements for mutual authentication, therefore, SMAP provides mutual authentication.
- Confidentiality: the exchanged information between platforms remains secret throughout the communication due to the use of public key encryption in SMAP. At any time only the owner platform can access the output of the request because of the encryption. In addition, the user request and time stamps are also encrypted and remain confidential. The required information for decryption requires the knowledge of the private keys.
- Integrity: in SMAP the use of one-way hash function and signature support a standard integrity test. The integrity of request, the output of the request and all the time stamps are guaranteed at all times. Any change or manipulation is detected.
- Accountability: accountability is provided by SMAP. In SMAP, if any vehicle platform decides to act maliciously and change the request, then the owner platform, or other vehicle platforms can detect the change by comparing the hashes. The digital signatures provide proof of any malicious actions.
- Authorization: when the user accesses the application and provides his/her credentials, this information is verified to decide whether to grant the user access to the application services or not. In addition, the certificate authority checks the validity of the certificate and the identity of the platforms, therefore, only authorized platforms provide the user with the service. Moreover, by providing mutual authentication, platforms are able to verify each other and any communication request or message sent by a non-trusted platform is ignored. The request, time stamps and the output of the request are confidential and can only be decrypted by authorized platforms.
- Non-repudiation: the signature of platforms provides proof of the originator of the message sent. Platforms sign messages with their private keys or encrypted with session key, therefore, providing the possibility for action traceability.
- Next formal verification is used to prove the soundness of SMAP in providing the claimed security properties.

4 Formal Verification with Scyther

Formal model verification is done with the formal verification tool Scyther [36, 41]. Scyther provides analysis of an ordered set of events between two partners. Scyther is based on Security Protocol Description Language (SPDL) [42]. Scyther provides a set of claims to test many security features e.g. secrecy of information, synchronization, and authentication between communication partners. Providing synchronization indicates that messages are sent and received in order and by the intended parties. Moreover, it also indicates that the content of messages cannot be modified. Scyther match functions acting as logical comparison are used to verify integrity of message content [36, 42–44].

SMAP was tested for secrecy of request, results, secret keys, and time stamps. It was also tested for synchronization between vehicle platforms. A simple scenario of a vehicle communication system was implemented in which three vehicle platforms are simulated, representing the Owner vehicle platform (Owner-Vehicle) requesting information and two other vehicle platforms (Vehicle, Vehicle2) that receive the request and share their information with the Owner-Vehicle platform. Claims at Owner-Vehicle, Vehicle, and Vehicle2 platforms are defined as shown in Fig. 4, which was obtained from the output of the verification proof of Scyther for the defined claims. As can be seen, no attacks were detected for all tested claims at Owner-Vehicle, Vehicle and Vehicle2 platforms. Figure 5 shows the traced exchanged messages generated in Scyther. Passing the tests proves the strength of the proposed protocol in protecting the exchanged data. Next, a discussion of the results is shown and analyzed in details. Based on results of Scyther test claims, the following security requirements are provided:

- Mutual Authentication: synchronization among vehicle platforms was tested and proven to exist through the Non-injection synchronization (Nisynch) claims shown in claim; i1, claim; i8 and claim; i14. Synchronization provides the requirement of mutual authentication i.e. messages are sent and received by intended partners, are in order and are unmodified, therefore, it can be stated that SMAP provides mutual authentication between parties.
- Confidentiality: The test of secrecy claims presented in claim; i2 to claim; i7, claim; i9 to claim; i13 and claim; i15 to claim; i20 check the confidentiality of request, time stamps, output of request, private keys, and results exchanged between parties. Passing these claims proves that.
- Integrity: the match logical compare and hash functions supported by Scyther, are used to test the integrity of carried data i.e. the request, time stamps, and results. The received hash is compared to hash newly generated from message content and are checked for a match ensuring the integrity of the message. Communication is only in case of a match, otherwise, the communication is cut.
- Authorization: exchanged data remains secret and access of data requires proper decryption by a legitimate authorized partners. It was shown through claim; i7, claim; i13 and claim; i20 that secret keys remain confidential. As a result, only authorized parties can access message contents and hence providing protection from unauthorized accesses.

Claim				Status		Comn
Calim	Owner	Claim,i1	Nisynch	Ok		No attacks
		Claim,i2	Secret OR	Ok		No attacks
		Claim,i3	Secret T1	Ok		No attacks
		Claim,i4	Secret T2	Ok		No attacks
		Claim,i5	Secret T3	Ok		No attacks
		Claim,i6	Secret Req	Ok		No attacks
		Claim,i7	Secret sk(Owner)	Ok	Verified	No attacks.
	ServiceProvider	Claim,i8	Nisynch	Ok	Verified	No attacks.
		Claim,i9	Secret OR	Ok		No attacks
		Claim,i10	Secret T1	Ok		No attacks
		Claim,i11	Secret T2	Ok		No attacks
		Claim,i12	Secret Req	Ok		No attacks
		Claim,i13	Secret sk(ServiceProvider)	Ok	Verified	No attacks.
	ServiceProvider2	Claim,i14	Nisynch	Ok		No attacks
		Claim,i15	Secret OR2	Ok		No attacks
		Claim,i16	Secret T1	Ok		No attacks
		Claim,i17	Secret T2	Ok		No attacks
		Claim,i18	Secret T3	Ok		No attacks
		Claim,i19	Secret Req	Ok		No attacks
		Claim,i20	Secret sk(ServiceProvider2)	Ok	Verified	No attacks.

Fig. 4 Set of tested security claims in Scyther

- Accountability and non-repudiation: signature of a party is used as an indication of the message generator, therefore, any detected malicious action, and the party can be traced and accounted for its malicious behavior. The secrecy of the private keys was verified through claim; i7, claim; i13 and claim; i2. Ensuring their secrecy no party can deny data sent by it, therefore, non-repudiation and accountability are provided.

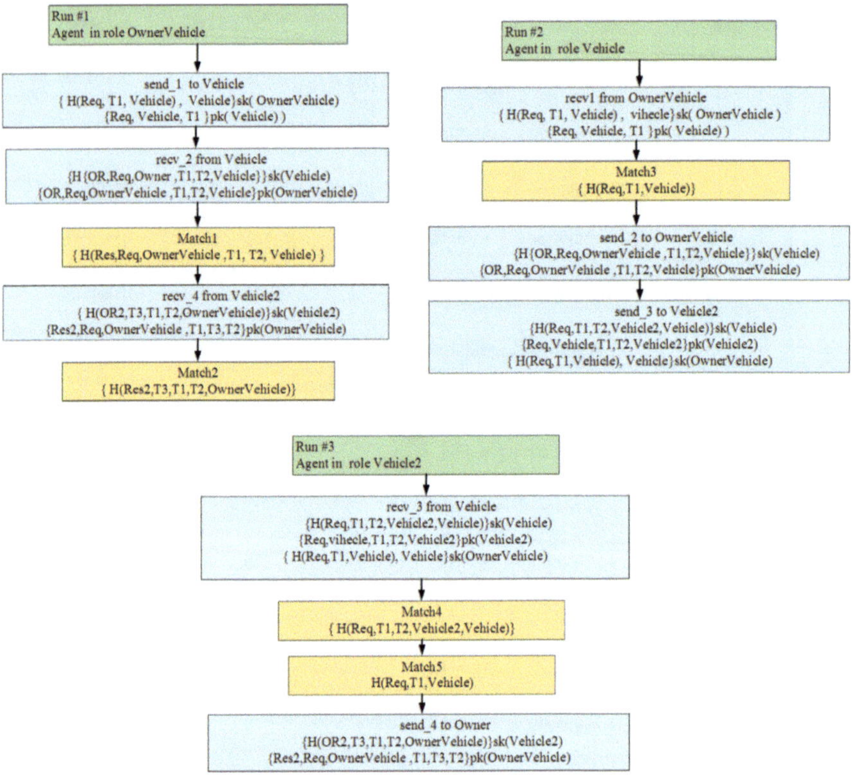

Fig. 5 Trace figures generated in Scyther

SMAP also provides protection from many security attacks. Protection from replay attacks is covered due to the usage of time stamps that are ensured to be secret and unchanged. Protection from MITM attacks is also provided; providing synchronization and information confidentiality ensures that communication between parties is secure and MITM attacks are detected. Moreover, parties cannot masquerade as another due to the signatures use and ensured secrecy of secret keys, hence, protection from masquerade attacks is provided as well. Modification attacks are also covered. Any malicious modification of the data is detected due to the use of hash integrity checks and assurance of synchronization.

In addition to all of that SMAP architecture, enables the fast retrieval of information due to the fact that platforms send their results back as soon as they receive the request and generate the results. Therefore, the user does not have to wait for other vehicles to receive an initial response.

This feature is very important in vehicle communication systems because sometimes the first car to receive the owner request about the condition of the road might as well be aware of a close-by accident therefore, the owner vehicle is informed a soon as possible and actions can be made faster. Moreover, unlike many other works proposed in the literature losing the MA does not necessarily indicate the loss of all the collected data. It can be said that the system remains functional as it receives results from platforms as soon as they are collected. Next, a comparison of the security features provided by SMAP and other security protocols is presented.

5 Comparison with Related Work

A comparison between the proposed SMAP and other similar protocols in the literature is carried out in this section. The criteria used for comparison is whether the protocol provides the following security requirements: anonymity, authentication, authorization, accountability, confidentiality, integrity, and non-repudiation. None of the proposed protocols in the literature provides availability, therefore, it is not included the comparison figure. Figure 6 shows the detailed comparison. In the figure Auth. represents authentication, Author. is authorization, Acc. is accountability, Conf. is confidentiality, Integ. is integrity and Non-rep. is non-repudiation. As shown, most of the compared works passed two or three of the seven security requirements, while [32] is found to provide five.

SMAP however, is the only protocol that provides six of the seven requirements. It provides mutual authentication, authorization, accountability, confidentiality, integrity, and non-repudiation. Another two important comparison factors are, whether the protocol takes into consideration the effect of potential loss or killing of an agent or the proposed protocol is verified by any other security verification tool. These two factors are important because the first affects the system functionality

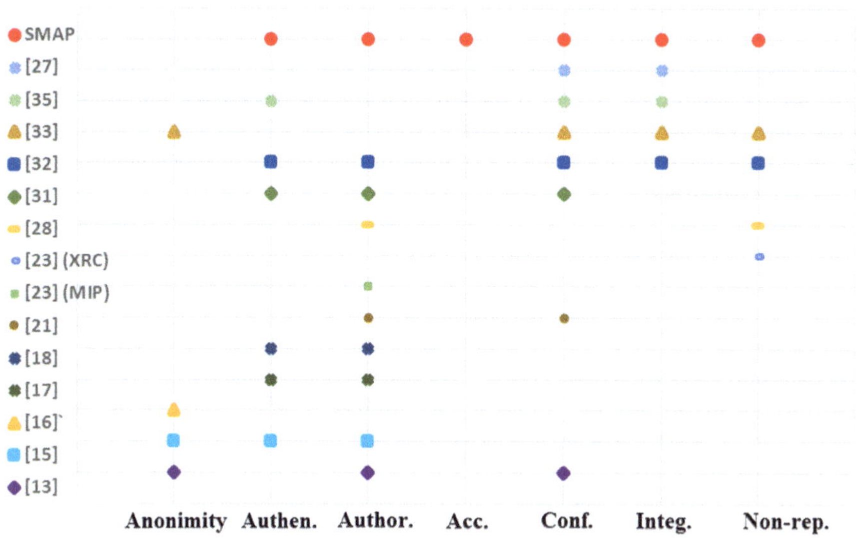

Fig. 6 Comparison of related protocols and SMAP

and the other verifies the correctness of the proposed protocol. In the related work, only [32, 35] consider agent loss by proposing the matrix hop and recovery method respectively. Moreover, the only work that verified its proposed protocol is [18] that was proven to withstand all simulated replay attacks in JADE, however, none of the works used formal verification methods.

Due to SMAP's enhanced multi-hop architecture, platforms send their results to owner as soon as they are generated therefore, SMAP has limited consideration of agent loss, in other words, loss of an agent at one stage does not necessarily indicate the loss of the whole data however, and no further results will be received at the owner platform. To further study the proposed protocol, a complexity analysis of SMAP is carried out. SMAP performance is also compared to some related protocols. Details are in the next section.

6 Complexity Analysis of SMAP

To evaluate the performance of SMAP, the total number of operations is approximated and compared to the other protocols. To give a fair comparison between protocols, the widely accepted evaluation method used by Kuo et al. [15] is applied. According to [15, 45], the computational cost of an asymmetric operation (A) is equivalent to one point operation which is equivalent to 1000 symmetric operations (S) and 10,000 hash operations (H). Therefore, every asymmetric, symmetric and hash operation is evaluated as 1, 0.001 and 0.0001 point operations, respectively.

Besides SMAP, the total number of operations is approximated for the following scheme [32, 27, 35, 23, 33]. The different schemes were evaluated for identical type of operations. All schemes are assumed to use identical operations, 256 AES symmetrical operations, 2048 RSA asymmetrical operations and 128-bit MD5 hash function. These schemes are chosen because they are similar to the proposed protocols in the sense that they are also based on cryptographic techniques. The total number of operations depends on N, which is the number of platforms the MA visits to collect results in a MAS. In Table 2, the total number of operations for each scheme is calculated.

Table 2 Comparison of proposed protocols and related protocols based on the number of operations

Scheme	Total operations for system with N platforms
(Guan et al., 2007) [32]	$11.0004 \times N + 6.002$
(Srivastava and Nandi, 2014) [27]	$6.0205 \times N + 4.0012$
(Ouardani et al., 2007) [35]	$12 \times N$
(Venkatesan et al., 2010) (XRC) [23]	$3.0001 \times N + 1.0001$
(Geetha and Jayakumar, 2011) [33]	$8.0003 \times N + 2.0001$
SMAP	$9.0005 \times N - 1.00001$

To evaluate the performance of protocols, the increase in the number of operations as the number of service provider's N increases is studied. Figure 7 shows the change in the total number of approximated operations as N increases. The key point that affects performance is the number of asymmetrical operations as they have the highest computational cost. The protocol in [35] is based on asymmetrical operations only and protocol in [32] is based on asymmetrical and hash operations. SMAP is ranked as the third protocols with the highest number of operations after [35, 32]. The protocol in [23] is also, an asymmetrical based scheme, however, it uses digital signature to provide integrity to agent's code and protection from non-repudiation attacks only. The protocol in [33] uses asymmetrical and hash operations to provide four security requirements anonymity, confidentiality, integrity, and non-repudiation. Moreover, the protocol performance is not only dependent on N; it also depends on a number of trusted hosts that the MA visits during its journey. To approximate the number of operations for this scheme we assume that only one trust host is visited, therefore, the approximation represents the minimum number of operations the system can have. The other scheme [27] relies on the hybrid approach, as Fig. 7 shows, are more efficient. It has a lower overall computational cost compared to the other schemes. However, the protocol in [27] provides only two security requirements, confidentiality, and integrity only. It is also vulnerable to replay attacks and does not consider the case of loss or malicious killing of MA causing the system to lose all the data collected by the MA.

Despite the fact that SMAP is not the most efficient protocol in comparison with other protocols in the literature, however, SMAP is the only one that provides six security requirements, mutual authentication, authorization, integrity, confidentiality, accountability, and non-repudiation. In addition, it also provides protection from different attacks such as replay, MITM, masquerade and repudiation attacks.

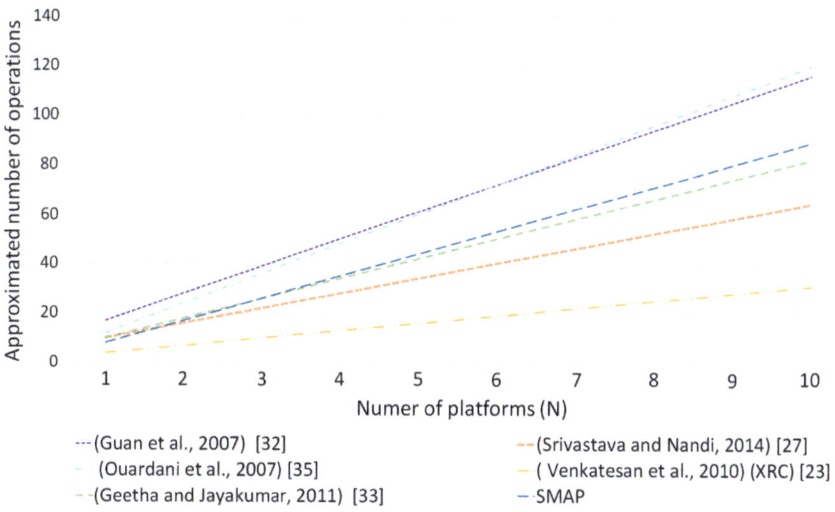

Fig. 7 Comparison in performance of SMAP and related work

In SMAP, malicious killing of a MA in the system does not mean the loss of the whole data. Taking all that into consideration, the number of operation overhead of SMAP compared to the provided security features is considered acceptable. Next, a simulation of a SMAP based vehicular communication system application is presented.

7 Simulation

To verify the viability of SMAP, a scenario of a vehicular communication system application that incorporates SMAP and MASs is simulated. In the application, vehicles exchange important information about the state of the road such as congestion information, possible accidents, weather information, the existence of pedestrian crossing, and other useful information. Implementation was conducted on the well-known JADE platform [37, 38]. To simulate the application, three containers were created as the Main-Container, Container-1 and Container-2, as shown in Fig. 8. Main-Container represents the owner vehicle platform while the other containers are other vehicle platforms in the road. Both mobile and stationary

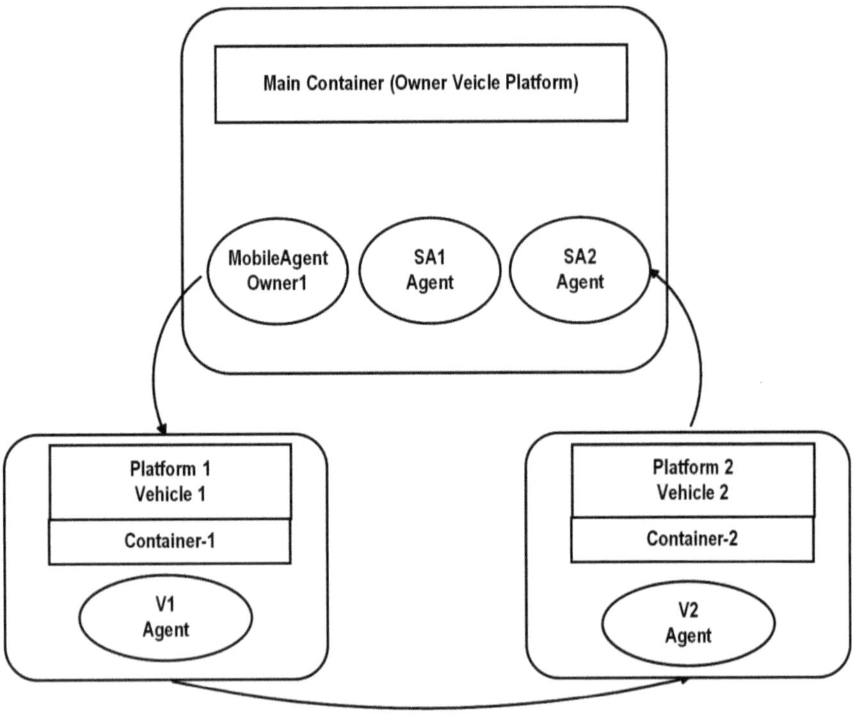

Fig. 8 Basic architecture of JADE setup

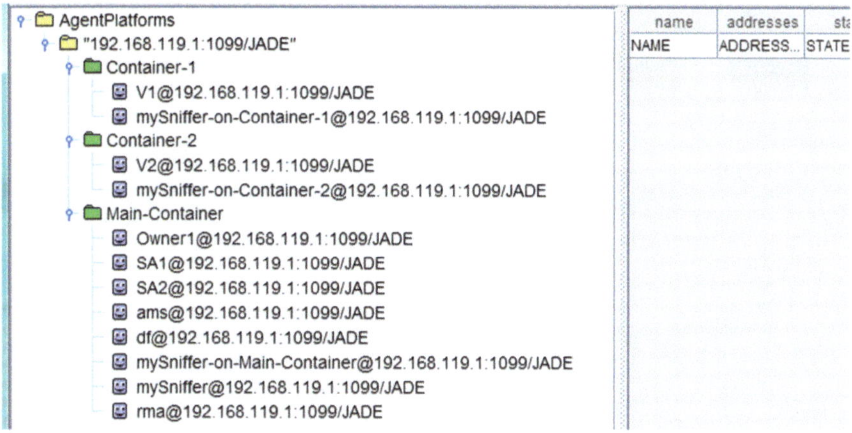

Fig. 9 Simulated containers and agents in JADE

agents are used. The Owner vehicle MA carries the request to the other nearby vehicles. Other stationary agents residing in the owner vehicle are responsible for receiving results from vehicles. Moreover, stationary agents reside at each vehicle exist to communicate with Owner MAs, process the request, and provide results. At the Main-Container, a copy of MA is created, Owner1.

The user request is carried to the different vehicles. Two stationary agents (V1 and V2) are created at each vehicle. Also, another two stationary agents (SA1 and SA2) are created at Owner vehicle. Figure 9 shows the setup of platforms and agents in JADE prior to the interaction process. JADE provides the user with the option of sniffing messages exchanged between two parties. A sniffer agent, mySniffer, is created at each container to trace interactions between the owner agents and vehicle agents. Figures 10 and 11 show sniffed messages that are exchanged between the owner MA and other vehicle agents during the process of the user request and sending of results.

Figure 12 shows the results of a successful interaction between the owner and other vehicles. The final results collected by MAs are printed to the user. The implemented application incorporates SMAP to provide confidentiality, integrity, mutual authentication, accountability, non-repudiation, and authorization. Also, the implementation verified the feasibility of SMAP.

8 Conclusion

Development of vehicular communication systems has become the interest of many researchers nowadays. To make use of such systems both, efficiency and security should be ensured. In this paper, a novel secure MAS protocol SMAP for vehicular communication systems is proposed. SMAP provides the basic fundamental

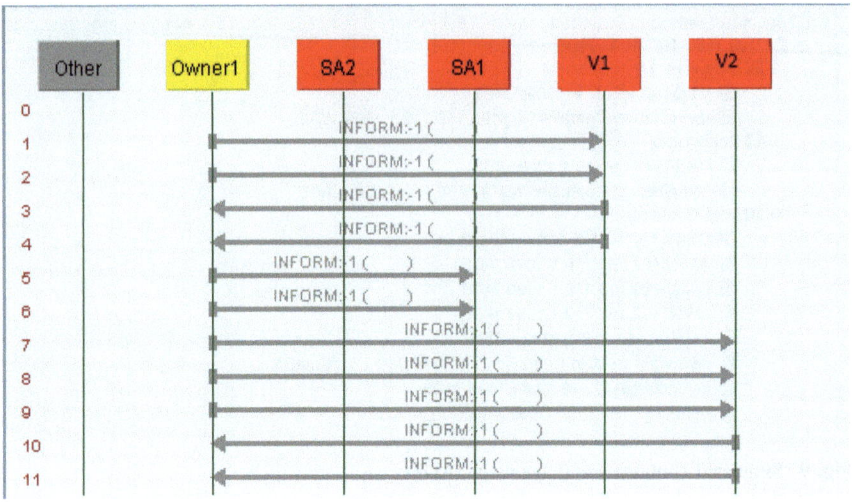

Fig. 10 Traced exchanged messages in JADE part 1

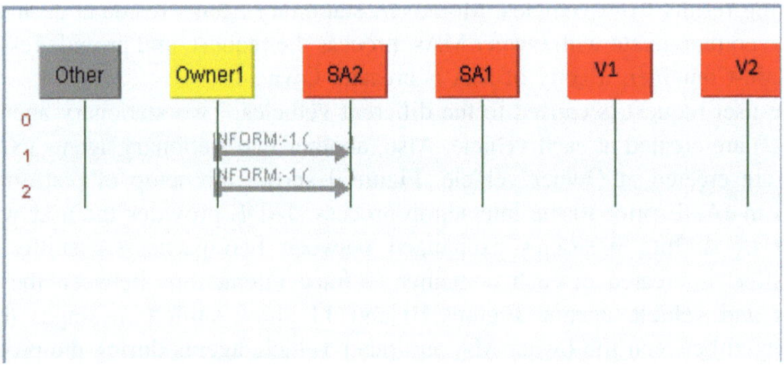

Fig. 11 Traced exchanged messages in JADE part 2

security requirements e.g. mutual authentication, authorization, integrity, confidentiality, accountability, and non-repudiation. SMAP also provides protection from many security attacks e.g. MITM, replay, masquerade, modification and unauthorized access attacks. In SMAP, owner vehicles receive results as soon as they are generated and therefore, providing fast information retrieval process. Moreover, another important feature of SMAP is that loss of the MA does not necessarily mean the loss of all the collected data, therefore the application functionality is maintained in this way.

Security analysis and formal verification proved the viability of SMAP in providing the claimed security features. Furthermore, the complexity analysis of SMAP showed that, despite the fact that SMAP is not the most efficient protocol compared to others, however, the computational overhead of SMAP compared to the provided

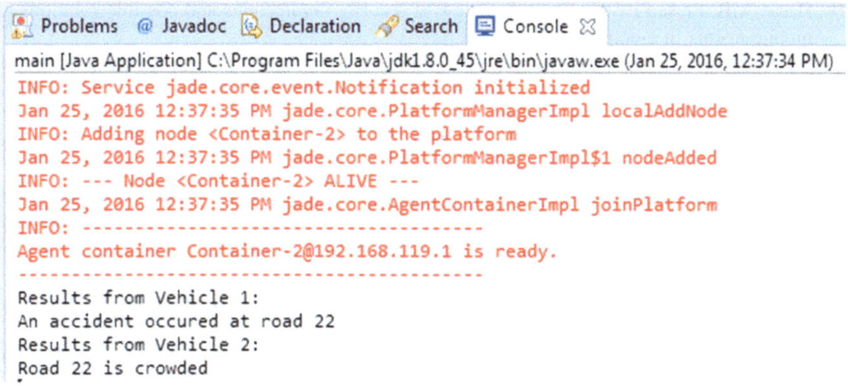

Fig. 12 Results shown to user in JADE complexity

security features is considered acceptable. Simulation of a SMAP based vehicular communication systems was implemented to prove the viability of SMAP. For future work, we plan to enhance SMAP to provide anonymity to system users to cover all the seven security requirements to provide a generic security protection.

Acknowledgements The authors wish to acknowledge Information and Communication Technology (ICT) fund for the continued support for the educational development and research.

References

1. Al-Qutayri M, Yeun C, Al-Hawi F (2009) Security and privacy of intelligent VANETs, Computational intelligence & modern heuristics, IN-TECH Publisher, November 2009, pp 191–219
2. Bariah L, Shehada D, Salahat E, Yeun CY (2015) Recent advances in VANET security. In: IEEE 82nd vehicular technology, Boston, September 2016, pp 1–7
3. Gajparia AS, Mitchell CJ, Yeun CY (2005) Supporting user privacy in location based services. In: IEICE transactions on communications
4. Gavalas D, Konstantopoulos C, Mastakas K, Pantziou G (2014) Mobile recommender systems in tourism. J Netw Comput Appl 39:319–333
5. Yeun C, Al-Qutayri M, Al-Hawi F (2009) Efficient security implementation for emerging VANETs. Ubiquit Comput Commun J 4(4):58–66
6. Burkle AHAMWWM (2009) Evaluating the security of mobile agent platforms. Auton Agent 18(2):295–311
7. Shehada D, Yeun CY, Zemerly MJ, Al-Qutayri M, Al Hammadi Y (2015) A secure mobile agent protocol for vehicular communication systems. In: International conference on innovations in information technology (IIT), November 2015, pp 92–97
8. Baig Z (2012) Multi-agent systems for protecting critical infrastructures: a survey. J Netw Comput 35(3):1151–1161
9. Rashvand HF, Salah K, Calero JMA, Harn L (2010) Distributed security for multi-agent systems—review and applications. IET Inf Secur 4(4):188–201

10. Hasan MB, Prasad PWC (2009) A review of security implications and possible solutions for mobile agents in e-commerce
11. Shemaili MB, Yeun CY, Mubarak K, Zemerly MJ (2012) A new lightweight hybrid cryptographic algorithm for the internet of things. In: international conference for internet technology and secured transactions, pp 87–92.
12. Jung Y, Kim M, Masoumzadeh A, Joshi JBD (2012) A survey of security issue in multi-agent systems. Artif Intell Rev 37(3):239–260
13. Cavalcante RC, d'Silva IIBAP, Silva M, Costa E, Santos R (2012) A survey of security in multi-agent systems. A survey of security in multi-agent systems, vol 39, no 5, pp 4835–4846
14. Raji F, Ladani BT (2010) Anonymity and security for autonomous mobile agents. IET Inf Secur 4(4):397–410
15. Kuo W-C, Wei H-J, Cheng J-C (2014) An efficient and secure anonymous mobility network authentication scheme. J Inf Secur Appl 19(1):18–24
16. Leszczyna R (2007) Anonymity architecture for mobile agent systems. Holonic Multi-Agent Syst Manuf 4659:93–103, Springer
17. Garrigues C, Robles S, Borrell J (2008) Securing dynamic itineraries for mobile agent applications. J Net Comp App 31(4):487–508
18. Garrigues C, Migas N, Buchanan W, Robles S, Borrell J (2009) Protecting mobile agents from external replay attacks. J Syst Softw 82(2):197–206
19. Garrigues C, Robles S, Borrell J, Navarro-Arribas G (2010) Promoting the development of secure mobile agent applications. J Syst Softw 83(6):959–971
20. Menacer DE, Drias H, Sibertin-Blanc C (2013) Towards a security solution for mobile agents. Adv Inf Syst Technol 206:969–979, Springer
21. Fong C-H, Parr G, Morrow P (2011) Security schemes for a mobile agent based network and system management framework. J Netw Syst Manage 19(2):230–256
22. Ganapathy S, Kulothungan K, Muthurajkumar S, Vijayalakshmi M, Yogesh P, Kannan A (2013) Intelligent feature selection and classification techniques for intrusion detection in networks: a survey. EURASIP J Wirel Commun Network pp 1–16
23. Venkatesan S, Chellappan C, Vengattaraman T, Dhavachelvan P, Vaish A (2010) Advanced mobile agent security models for code integrity and malicious availability check. J Network Comput Appl 33(6):661–671
24. Venkatesan S, Chellappan C (2008) Protection of mobile agent platform through attack identification scanner (AIS) by malicious identification police (MIP). In: International conference on emerging trends in engineering and technology (ICETET'08)
25. Venkatesan S, Baskaran R, Chellappan C, Vaish A, Dhavachelvan P (2013) Artificial immune system based mobile agent platform protection 35(4):365–373
26. Venkatesan S, Chellappan C (2008) Identifying the split personality of the malicious host in the mobile agent environment. In: International IEEE conference on intelligent systems (IS'08)
27. Srivastava S, Nandi G (2014) Self-reliant mobile code: a new direction of agent security. J Network Comput Appl 37:62–75
28. Kapnoullas T, Chang E, Dillon T, Damiani E (2003) Security framework for mobile agent platforms (SFMAP). In: On the move to meaningful internet systems 2003: OTM 2003 workshops, 2889, Springer, pp 845–858
29. Sulaiman R, Sharma D (2011) Enhancing security in e-health services using agent. In: Electrical engineering and informatics, *International conference electrical engineering and informatics* (ICEEI)
30. Martins RA, Correia ME, Augusto AB (2012) A literature review of security mechanisms employed by mobile agents. In: 2012 7th Iberian conference on information systems and technologies (CISTI)
31. Sulaiman R, Huang X, Sharma D (2009) E-health services with secure mobile agent. Communication Networks and Services Research Conference (CNSR'09)
32. Guan H, Zhang H, Chen P, Zhou Y (2007) Mobile agents integrity research. IFIP—the international federation for information processing, vol 251, pp 194–201

33. Geetha G, Jayakumar C (2011) Data security in free roaming mobile agents. In: Communications in computer and information science, vol 196, Springer, pp 472–482
34. Geetha G, Jayakumar C (2015) Implementation of trust and reputation management for free-roaming mobile agent security. IEEE Syst J 9(2):556–566
35. Ouardani A, Pierre S, Bouchene H (2008) A security protocol for mobile agents based upon the cooperation of sedentary. J Network Comput Appl 30(3):1228–1243
36. Cremers CJF (2008) The Scyther tool: verification, falsification, and analysis of security protocols. Lect Notes Comput Sci 5123:414–418
37. Bellifemine F, Bergenti F, Caire G, Boggi A (2005) Jade—a java agent development framework. In: Multi-agent programming, vol 15, Springer, pp 125–147
38. Bellifemine F, Caire G, Poggi A, Rimassa G (2003) JADE a white paper, 9 2003. [Online]. Available: http://jade.tilab.com/papers/2003/WhitePaperJADEEXP.pdf
39. Han K, Mun H, Shon T, Yeun CY, Park J (2012) Secure and efficient public key management in next generation mobile networks. Pers Ubiquit Comput 16(6):677–685
40. Han K, Yeun CY, Shon T, Park J, Kim K (2011) A scalable and efficient key escrow model for lawful interception of IDBC-based secure communication. Inter J Com Sys 24(4):461–472
41. Cremers C, Mauw S (2012) Operational semantics and verification of security protocols, Springer
42. Cremers C (2014) Scyther user manual, 2 2014. [Online]
43. Kahya N, Ghoualmi N, Lafourcade P (2012) Key management protocol in WIMAX revisited. Adv Comput Sci, Eng Appl 167, Springer, pp 853–862
44. Al-Hammadi HA, Yeun CY, Zemerly MJ, Al-Qutayri M, Ghawanmeh A (2011) Formal modeling and verification of DLK protocol. Internet Technology and Secured Transactions (ICITST), Abu Dhabi, pp 578-583.
45. Yanrong L, Xiaobo W, Xiaodong Y (2015) A secure anonymous authentication scheme for wireless communications using smart cards. Int J Network Secur 17(3):237–245

In Vivo Communication in Wireless Body Area Networks

Hadeel Elayan, Raed M. Shubair and Nawaf Almoosa

1 Introduction

Wireless Body Area Networks (WBANs) are a new generation of Wireless Sensor Networks (WSNs) dedicated for healthcare monitoring applications. The aim of these applications is to ensure continuous monitoring of the patients' vital parameters, while giving them the freedom of moving which results in an enhanced quality of healthcare [1]. In fact, a WBAN is a network of wearable computing devices operating on, in, or around the body. It consists of a group of tiny nodes that are equipped with biomedical sensors, motion detectors, and wireless communication devices [2]. Actually, advanced healthcare delivery relies on both body surface and internal sensors since they reduce the invasiveness of a number of medical procedures [3]. Electrocardiogram (ECG), electroencephalography (EEG), body temperature, pulse oximetry (SpO_2), and blood pressure are evolving as long-term monitoring sensors for emergency and risk patients [4].

One attractive feature of the emerging Internet of Things (IoT) is to consider in vivo networking for WBANs as an important application platform that facilitates continuous wirelessly-enabled healthcare [5]. Internal health monitoring [6], internal drug administration [7], and minimally invasive surgery [8] are examples of the pool of applications that require communication from in vivo sensors to body surface nodes. However, the study of in vivo wireless transmission, from inside the body to external transceivers is still at its early stages. Figure 1 shows a modified

H. Elayan (✉) · R.M. Shubair · N. Almoosa
Electrical and Computer Engineering Department, Khalifa University, Abu Dhabi, UAE
e-mail: hadeel.mohammad@kustar.ac.ae

R.M. Shubair
e-mail: raed.shubair@kustar.ac.ae

N. Almoosa
e-mail: nawaf.almoosa@kustar.ac.ae

L. Ismail and L. Zhang (eds.), *Information Innovation Technology in Smart Cities*,
https://doi.org/10.1007/978-981-10-1741-4_18

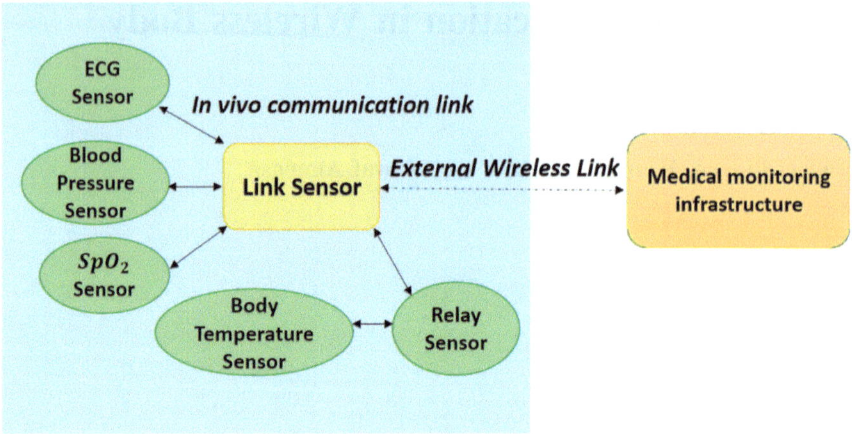

Fig. 1 Simplified overview of the in vivo communication network

network organization for interconnecting the biomedical sensors. The data is basically not directly transferred from the biomedical sensors to the hospital infrastructure. Indeed, sensors send their data via a suitable low-power and low-rate in vivo communication link to the central link sensor (located on the body like all other sensors). Any of the sensors may act as a relay between the desired and the central link sensor if a direct connection is limited. An external wireless link enables the data exchange between the central link sensor and the external hospital infrastructure [4].

Wireless in vivo communication creates a wirelessly-networked cyber-physical system of embedded devices. Such systems utilize real-time data to enable rapid, correct as well as cost-conscious responses for surgical, diagnostic, and emergency circumstances [3]. The crucial element that should be carefully regarded when referring to in vivo communications is modeling the in vivo wireless channel. The ability to understand the characteristics of the in vivo channel is fundamental to achieve optimum processing and design effective protocols that enable the arrangement of WBANs inside the human body [3].

This chapter surveys the existing research which investigates the state-of-art of the in vivo communication. It also focuses on characterizing and modeling the in vivo wireless channel and contrasting this channel with other familiar ones. MIMO in vivo is also of interest since it significantly enhances the performance gain and data rates. Finally, this chapter introduces in vivo nano-communication as a novel communication paradigm. The rest of the chapter is organized as follows. In Sect. 2, we present the state-of-art of in vivo communication. Conducted research on in vivo channel characterization is provided in Sect. 3. The MIMO in vivo system is described in Sect. 4. In vivo nano-communication is addressed in Sect. 5. Finally, we draw our conclusions and summarize the chapter in Sect. 6.

2 State-of-Art of In Vivo Communication

In vivo communication is a genuine signal transmission field which utilizes the human body as a transmission medium for electrical signals [9]. The body becomes a vital component of the transmission system. Electrical current induction into the human tissue is enabled through sophisticated transceivers while smart data transmission is provided by advanced encoding and compression. Figure 2 shows the main components of an in vivo communication link.

A transmitter unit permits sensor data to be compressed and encoded. It then conveys the data by a current-controlled coupler unit. The human body acts as the transmission channel. Electrical signals are coupled into the human tissue and distributed over multiple body regions. On the other hand, the receiver unit is composed of an analog detector unit that amplifies the induced signal and digital entities for data demodulation, decoding, and extraction [4].

Developing body transmission systems have shown the viability of transmitting electrical signals through the human body. Nonetheless, detailed characteristics of the human body are lacking so far. Not a lot is known about the impact of human tissue on electrical signal transmission. Actually, for advanced transceiver designs, the effects and limits of the tissue have to be cautiously taken into consideration [4, 10]. The main requirements of an in vivo system include low power, low latency, less complexity, robustness to jamming, reliability, and size compactness [11]. In vivo communication is involved in a wide array of practical medical usages. For instance, in vivo sensors are utilized in health monitoring applications in order to keep track of glucose and blood pressure levels. In vivo actuators are also important

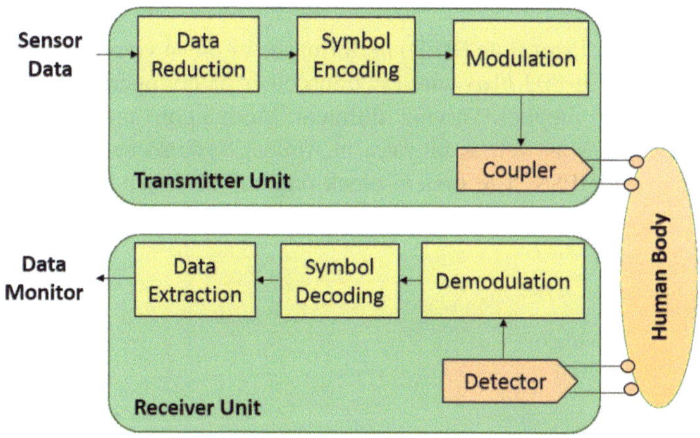

Fig. 2 In vivo communication for data transmission between sensors enabled by transmitter and receiver units

for implanted insulin pump as well as bladder controllers. Moreover, in vivo technology is involved in both medical nanorobotic device communication and in therapeutic nanoparticles employed in malignant tumor elimination processes. Such distinctive communication can add an effective contribution in the development of Prosthetics including artificial retina, cochlear implants and brain pacemakers for patients with Parkinson's disease.

2.1 Human Body Model

Research into in vivo communications primarily used the ANSYS HFSS [12] Human Body Model software to conduct the simulations. This software is a high-performance full-wave electromagnetic field simulator which enables the complete electromagnetic fields prediction and visualization. Hence, important parameters such as S-Parameters, resonant frequency, and radiation characteristics of antennas can be computed and plotted. The human body is modeled as an adult male body with more than 300 parts of muscles, bones and organs, having a geometric accuracy of 1 mm and realistic frequency dependent material parameters. The original body model only has the parameters from 10 Hz to 10 GHz. However, the maximum operating frequency is increased to 100 GHz by manually adding the values of the parameters to the datasets [13].

2.2 System Level Setup

To evaluate the Bit Error Rate (BER) performance of the in vivo communication, an OFDM-based (IEEE 802.11n) wireless transceiver model operating at 2.4 GHz is setup. This model implies varying different Modulation and Coding Schemes (MCS) index values as well as bit rates in Agilent SystemVue for various in vivo channel setups in HFSS. The system block diagram is shown in Fig. 3 [14].

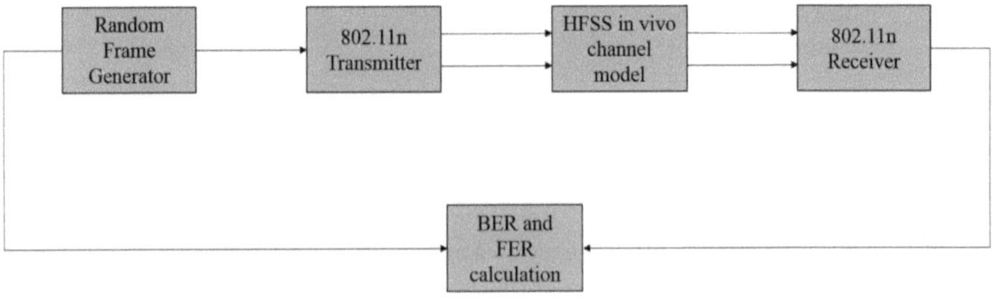

Fig. 3 Block diagram of system level simulation with HFSS in vivo channel model

3 In Vivo Channel Modeling and Characterization

The in vivo channel is a novel paradigm in the field of wireless propagation; thus, it is very different when compared to other frequently analyzed wireless environments such as cellular, Wireless Local Area Network (WLAN), and deep space [2]. Figure 4 illustrates the classic multi path channel and the in vivo multipath channel.

Basically, in an in vivo channel, the electromagnetic wave passes through various dissimilar media that have different electrical properties, as depicted in Fig. 4. This leads to the reduction in the wave propagation speed in some organs and the stimulation of significant time dispersion that differs with each organ and body tissue [14]. This effect coupled with attenuation due absorption by the different layers result in the degradation of the quality of the transmitted signal in the in vivo channel.

The authors in [4] compare the characteristics of wireless technologies including the WLAN, Bluetooth, Zig-bee, and active Radio Frequency Identification (RFID) as shown in Table 1. Their aim is to seek a novel transmission technique for in vivo communication which focuses on transmission power below 1 mW, data rates of 64 kbit/s, and the possibility for miniaturization to integrate the transceiver modules into band-aids and implantable pills.

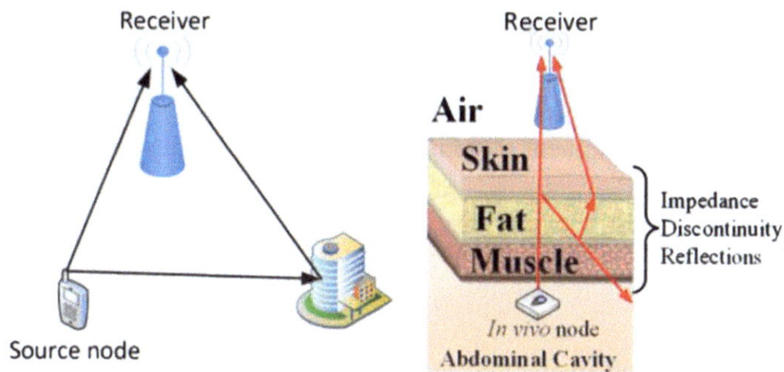

Fig. 4 Classic multi-path channel versus in vivo multi-path channel [3]

Table 1 Characteristics of wireless technologies [4]

Technology	Frequency	Data rate	Transmission power (mW)	Size
WLAN	2.4/5.1 GHz	54 Mbit/s	100	PC card
Bluetooth	2.4 GHz	723.1 kbit/s	1	PCB module
Zigbee	868 MHz	20 kbit/s	10	PCB module
Active RFID	134 kHz	128 bit/s	<1	Pill
In vivo communication	<1 MHz	>64 kbits/s	<1	Band-aid/pill

In addition, since the in vivo antennas are radiating into a complex lossy medium, the radiating near fields will strongly couple to the lossy environment. This signifies that the radiated power relies on both the radial and angular positions; hence, the near field effect has to be always taken into account when functioning in an in vivo environment [15]. The electric and magnetic fields behave differently in the radiating near field compared to the far field. Therefore, the wireless channel inside the body necessitates different link equations [16]. It must be noted as well that both the delay spread and multi-path scattering of a cellular network are not directly applicable to near-field channels inside the body. The reason behind this is the fact that the wavelength of the signal is much longer than the propagation environment in the near field [17].

The authors in [3] used an accurate human body to investigate the variation in signal loss at different radio frequencies as a function of position around the body. They noticed significant variations in the Received Signal Strength (RSS) which occur with changing positions of the external receive antenna at a fixed position from the internal antenna [3]. Nevertheless, their research did not take into account the basic characterization of the in vivo channel. In [11], the authors used an immersive visualization environment to characterize RF propagation from medical implants. Based on 3-D electromagnetic simulations, an empirical path loss (PL) model is developed in [18] to identify losses in homogeneous human tissues. In [19], the authors carried out numerical and experimental investigations of biotelemetry radio channels and wave attenuation in human subjects with ingested wireless implants.

Modeling the in vivo wireless channel including building a phenomenological path loss model is one of the major research goals in this field. A profound understanding of the channel characteristics is required for defining the channel constraints and the subsequent systems' constraints of a transceiver design [4].

3.1 Path Loss

Path loss in in vivo channels can be investigated using either a Hertzian dipole antenna or a monopole antenna. The authors in [17] carried out their study based on Hertzian dipole in which path loss is examined with minimal antenna effects. The length of the Hertzian dipole is so small resulting in little interaction with its surrounding environment. The path loss can be calculated as

$$Path\ Loss = 20 * \log_{10}\left(\frac{|E|_{r=0}}{|E|_{r,\theta,\phi}}\right) \tag{1}$$

where r represents the distance from the origin, i.e. the radius in spherical coordinates, θ is the azimuth angle and φ is the polar angle. $|E|_{r,\theta,\phi}$ is the magnitude of the electric field at the measuring point and $|E|_{r=0}$ is the magnitude of electric field at the origin.

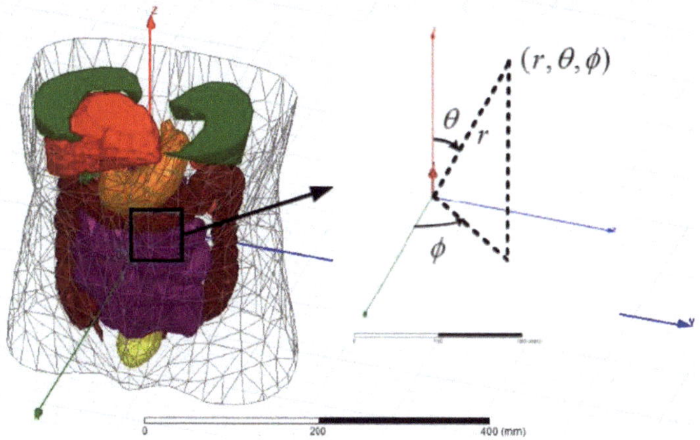

Fig. 5 Truncated human body with a Hertzian dipole at the origin in spherical coordinate system [17]

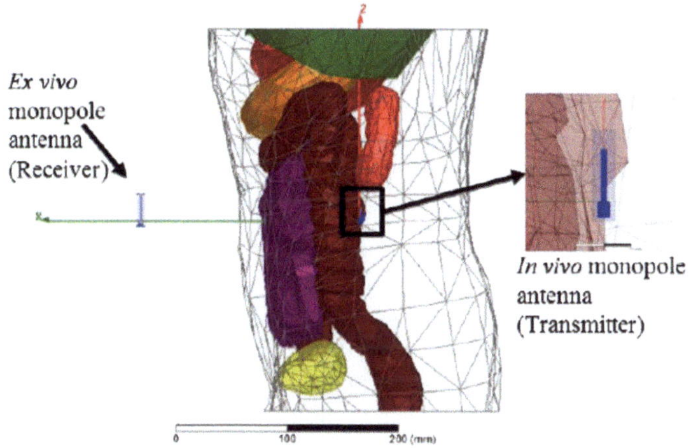

Fig. 6 Simulation setup by using monopoles to measure the path loss [17]

Due to the fact that the in vivo environment is an inhomogeneous medium, it is mandatory to measure the path loss in the spherical coordinate system [17]. The setup of this approach is depicted in Fig. 5 which includes the truncated human body, the Hertzian dipole, and the spherical coordinate system.

The authors in [3] carried out their study based on monopole antenna. Actually, monopoles are good choice of practical antennas since they are small in size, simple and omnidirectional.

The path loss can be measured by scattering parameters (S parameters) that describe the input-output relationship between ports (or terminals) in an electrical system [3]. According to Fig. 6, if we set Port 1 on transmit antenna and Port 2 on receive antenna, then S_{21} represents the power gain of Port 1 to Port 2, that is

$$|S_{21}|^2 = \frac{P_r}{P_t} \tag{2}$$

where P_r is the received power and P_t is the transmitted power. Therefore, we calculate the path loss by the formula below

$$\mathit{Path\ Loss}\ (\mathrm{dB}) = 20\log_{10}|S_{21}| \tag{3}$$

Based on the simulations presented in [17], it can be observed that there is a substantial difference in the behaviors of the path loss between the in vivo and free space environment. In fact, significant attenuation occurs inside the body resulting in an in vivo path loss that can be up to 45 dB greater than the free space path loss. Fluctuations in the out-of-body region is experienced by the in vivo path loss. On the other hand, free space path loss increases smoothly. The inhomogeneous medium results as well in angular dependent path loss [17].

3.2 *Comparison of Ex Vivo and In Vivo Channels*

The different characteristics between ex vivo and in vivo channels are summarized in [17] as shown in Table 2.

4 MIMO In Vivo

Due to the lossy nature of the in vivo medium, attaining high data rates with reliable performance is considered a challenge [5]. The reason behind this is that the in vivo antenna performance may be affected by near-field coupling as mentioned earlier and the signals level will be limited by a specified Specific Absorption Rate (SAR) levels. The SAR is a measurement of how much power is absorbed per unit mass of conductive material, in our case, the human organs [20]. This measurement is limited by the Federal Communications Commission (FCC) which in turns limits the transmission power [20].

4.1 *Capacity of MIMO In Vivo*

The MIMO in vivo system capacity is the upper theoretical performance limit that can be achieved in practical systems, and can provide insight into how well the system can perform theoretically and give guidance on how to optimize the MIMO in vivo system [14].

Table 2 Comparison of ex vivo and in vivo channel [14]

Features	Ex vivo	In vivo
Physical wave propagation	Constant speed Multipath (reflection, scattering and diffraction)	Variable speed Multipath and penetration
Attenuation and path loss	Lossless medium Decreases inversely with distance	Very lossy medium Angular (directional) dependent
Directionality	Propagation essentially uniform	Propagation varies with direction Directionality of antennas changes with position.
Near field communications	Deterministic near-field region around the antennas	Inhomogeneous medium Near field region changes with angles and position inside the body
Antenna gains	Constant	Gains highly attenuated
Power limitations	Average and Peak	Average, peak as well as SAR (specific absorption ratio)
Shadowing	Follows a log-normal distribution	To be determined
Multipath fading	Flat fading and frequency-selective fading	To be determined
Wavelength	The speed of light divided by wavelength	$\frac{\lambda}{\sqrt{\varepsilon_r f}}$ at 2.4 GHz, average dielectric constant $\varepsilon_r = 35$, which is roughly 6 times smaller than the wavelength in free space

The achievable transmission rates in the in vivo environment have been simulated using a model based on the IEEE 802.11n standard [21] because this OFDM-based standard supports up to 4 spatial streams (4 × 4 MIMO). Owing to the form factor constraint inside the human body, current studies are restricted to 2 × 2 MIMO.

The OFDM system can be modeled as:

$$Y_k = H_k X_k + W_k, \ k = 1, 2, \ldots N_{data} \tag{4}$$

where $Y_k, X_k, W_k \in C^2$ denote the received signal, transmitted signal, and white Gaussian noise with power density of N_0 respectively at OFDM subcarrier k. The symbol N_{data} is the total number of subcarriers configured in the system to carry data. The complex frequency channel response matrix at subcarrier k is denoted by $H_k \in C^{2*2}$.

The SVD (Singular Value Decomposition) of H_k is given as:

$$H_k = U_k \sum_k V_k^H \tag{5}$$

where U_k, $V_k^H \in C^{2*2}$ are unitary matrices, and $\sum_k$ is the nonnegative diagonal matrix whose diagonal elements are singular values of $\sqrt{\lambda_{k1}}$, $\sqrt{\lambda_{k2}}$, respectively.

The system capacity for subcarrier k is [22]:

$$C_k = E\left[\sum_{i=1}^{2} \log_2\left(1 + \frac{\lambda k_i P}{2N_0 BW}\right)\right] \quad (6)$$

where P is the total transmit signal power of the two transmitter antennas, BW is the configured system bandwidth in H_k, and E denotes expectation. In this chapter, only time-invariant Gaussian channels will be considered. Hence, the expectation in the capacity calculation will be ignored.

The total system capacity is calculated as:

$$\frac{1}{T_{sym}} \sum_{k=1}^{N_{data}} C_k = (\frac{BW}{N_{total}} + TGI) \sum_{k=1}^{N_{data}} C_k \quad (7)$$

where T_{sym} is the duration of each OFDM symbol, N_{total} is the total number of subcarriers available in bandwidth BW, and TGI is the guard interval.

4.2 Results of MIMO In Vivo

The authors in [14] analyzed the Bit Error Rate for a MIMO in vivo system. By comparing their results to a 2 × 2 SISO in vivo, it was evident that significant performance gains can be achieved when using a 2 × 2 MIMO in vivo. This setup allows maximum SAR levels to be met which results in the possibility of achieving target data rates up to 100 Mbps if the distance between the transmit (Tx) and receive (Rx) antennas is within 9.5 cm [20]. Figure 7 demonstrate the simulation setup that shows the locations of the MIMO antennas. Two Tx antennas are placed inside the abdomen while two Rx antennas are placed at different locations inside the body at the same planar height [14].

The antennas used in Fig. 7 are monopole antennas designed to operate at the 2.4 GHz ISM band in their respective medium which is either free space for the ex vivo antennas or the internal body for the in vivo antennas [14]. For the in vivo case, the monopole's performance and radiation pattern varies with position and orientation inside the body; therefore, the performance of the in vivo antenna is strongly dependent on the antenna type [3, 15].

Further, in [5], it was proved that not only MIMO in vivo can achieve better performance in comparison to SISO systems but also considerably better system capacity can be observed when Rx antennas are placed at the side of the body. Figure 8 compares the in vivo system capacity for front, right side, left side, and back of the body. In addition, it was noticed that in order to meet high data rate requirements of up to 100 Mbps with a distance between the Tx and Rx antennas greater than

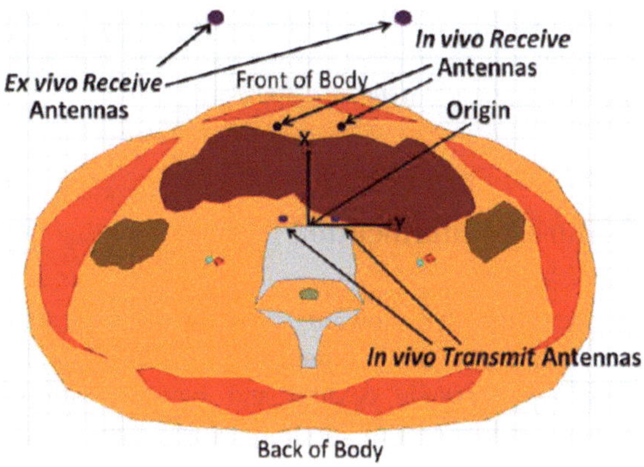

Fig. 7 Simulation setup showing locations of MIMO antenna [14]

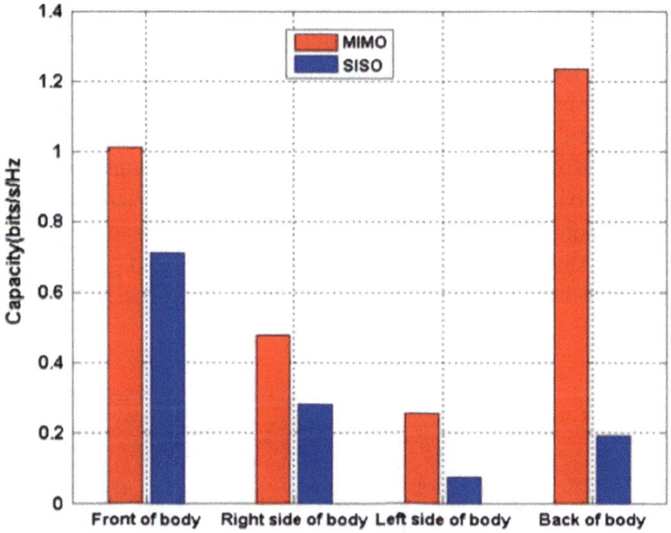

Fig. 8 2 × 2 MIMO and SISO in vivo system capacity comparison [5]

12 cm for a 20 MHz channel, relay or other similar cooperative networked communications are necessary to be introduced into the WBAN network [5].

4.3 Applications of MIMO In Vivo

One prospective application for MIMO in vivo communications is the MARVEL (Miniature Anchored Remote Videoscope for Expedited Laparoscopy) [23].

MARVEL is a wireless research platform for advancing MIS (Minimally Invasive Surgery) that necessitates high bit rates ($\sim$ 80–100 Mbps) for high-definition video transmission with low latency during surgery [24].

5 In Vivo Nano-Communication

Nanotechnology opens the door towards a new communication paradigm that introduces a variety of novel tools. This technology enables engineers to design and manufacture nanoscale electronic devices and systems with substantially new properties [25]. These devices cover radio frequencies in the Terahertz (THz) range and beyond, up to optical frequencies. The interconnections of nanodevices build up into nanonetworks enabling a plethora of potential applications in the biomedical, industrial, environmental and military fields.

In vivo nanosensing systems [26], which can operate inside the human body in real time, have been recently proposed as a way to provide faster and more accurate disease diagnosis and treatment than traditional technologies based on in vitro medical devices. However, the sensing range of each nanosensor is limited to its close nano-environment; thus, many nanosensors are needed to cover significant regions or volumes. Moreover, an external device and user interaction are necessary to read the actual measurement. By means of such communication, nanosensors will be able to overcome their limitations and expand their applications [27]. Indeed, nanosensors will be able to transmit their information in a multi-hop fashion to a gateway or sink, react to instructions from a command center, or coordinate between them in case that a joint response to an event or remote command is

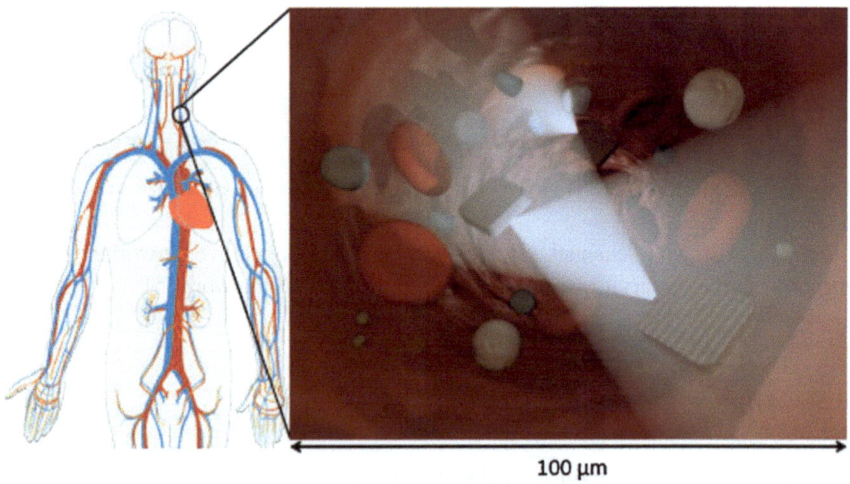

Fig. 9 In vivo nanosensor network inside a blood vessel [28]

needed. For instance, Fig. 9 shows several in vivo nanosensors that communicate as they travel through a blood vessel.

A number of challenges exist in the creation of in vivo nanosensor networks, which range from the development of nano-antennas for in vivo operation to the characterization of the intra-body channel environment from the nanosensor perspective [28].

In order to develop in vivo wireless nanosensor networks (iWNSNs), plasmonic nano-antennas for intra-body communication must be utilized [29]. In addition, a new view on intra-body channel modeling must be presented. In traditional channel models, the human body is modeled as a layered material with different permeabilities and permittivities. However, from the nanosensor perspective, and when operating at very high frequencies, the body is a collection of different elements (cells, organelles and proteins, among others), with different geometry and arrangement, as well as different electrical and optical properties. Further, coupling and interference effects among multiple nanosensors must be investigated and utilized at the basis of novel protocols for iWNSNs. The very high density of nanosensors in the envisioned applications results in non-negligible interference effects as well as electromagnetic coupling among nano-devices [28].

The future vision of in vivo networks entails the distribution of nano-machines that will patrol in the body, take measurements wherever necessary, and send collected data to the outside [30]. As a result, the development process and the operation of these devices invoke careful measures and high requirements. Moreover, it is important to understand in-body propagation at THz, since it is regarded as the most promising band for electromagnetic paradigm of nano-communication. Actually, the THz Band (0.1–10 THz) is envisioned as a key technology that satisfies the increasing demand for higher speed wireless communication. THz communication alleviates the spectrum scarcity and capacity limitations of current wireless systems, and hence enables new applications both in classical networking domains as well as in novel nanoscale communication paradigms. Nevertheless, a number communication challenges exist when operating at the THz frequency such as propagation modeling, capacity analysis, modulation schemes, and other physical and link layer design metrics [30].

6 Conclusion

This chapter provided an overview of the in vivo communication and networking. The overview focuses on the state of art of the in vivo communication, the in vivo channel modeling and characterization, and the concept of MIMO in vivo. The chapter also addresses in vivo nano-communication which is considered a novel communication paradigm that is going to revolutionize the concept of wireless body area networks. However, several challenges exist which open the door towards further research in this genuine field.

References

1. Honeine P, Mourad F, Kallas M, Snoussi H, Amoud H, Francis C (2011) Wireless sensor networks in biomedical: body area networks. In: 2011 7th international workshop on systems, signal processing and their applications (WOSSPA), May 2011, pp 388–391
2. Cypher D, Chevrollier N, Montavont N, Golmie N (2006, April) Prevailing over wires in healthcare environments: benefits and challenges. IEEE Commun Mag 44(4):56–63
3. Ketterl T, Arrobo G, Sahin A, Tillman T, Arslan H, Gitlin R (2012) In vivo wireless communication channels. In: 2012 IEEE 13th annual wireless and microwave technology conference (WAMI-CON), April 2012, pp 1–3
4. Wegmüller MS et al (2007) Intra-body communication for biomedical sensor networks. Ph.D. dissertation, Diss., Eidgenössische Technische Hochschule ETH Zürich, Nr. 17323
5. He C, Liu Y, Ketterl T, Arrobo G, Gitlin R (2014) Performance evaluation for mimo in vivo wban systems. In 2014 IEEE MTT-S International microwave workshop series on RF and wireless technologies for biomedical and health-care applications (IMWS-Bio), Dec 2014, pp 1–3
6. Piel E, Gonzalez-Sanchez, Gross H-G, van Gemund A (2011) Spectrum-based health monitoring for self-adaptive systems. In: 2011 fifth IEEE international conference on self-adaptive and self-organizing systems (SASO), Oct 2011, pp 99–108
7. Chow E, Beier B, Ouyang Y, Chappell W, Irazoqui P (2009) High frequency transcutaneous transmission using stents configured as a dipole radiator for cardiovascular implantable devices. In: IEEE MTT-S international microwave symposium digest, 2009. MTT '09, June 2009, pp 1317–1320
8. Sun Y, Anderson A, Castro C, Lin B, Gitlin R, Ross S, Rosemurgy A (2011) Virtually transparent epidermal imagery for laparo-endoscopic single-site surgery. In: 2011 annual international conference of the IEEE engineering in medicine and biology society, EMBC, Aug 2011, pp 2107–2110
9. Zimmerman TG (1995) Personal area networks (pan): near-field intra-body communication. Master of Science in Media Arts and Sciences, Massachusetts Institute of Technology
10. Lindsey D, McKee E, Hull M, Howell S (1998) A new technique for transmission of signals from implantable transducers. IEEE Trans Biomed Eng 45(5):614–619
11. Sayrafian-Pou K, Yang W-B, Hagedorn J, Terrill J, Yekeh Yazdandoost K, Hamaguchi K (2010) Channel models for medical implant communication. Int J Wirel Inf Netw 17(3–4):105–112. [Online]. Available: http://dx.doi.org/10.1007/s10776-010-0124-y
12. Ansoft, "ANSYS HFSS, 3D Full-wave Electromagnetic Field Simulation". [Online]. Available: http://www.ansoft.com/products/hf/hfss/
13. Gabriel C, Gabriel S (1996) Compilation of the dielectric properties of body tissues at rf and microwave frequencies. King's Coll London (United Kingdom) Dept of, Tech. Rep.
14. He C, Liu Y, Ketterl T, Arrobo G, Gitlin R (2014) Mimo in vivo. In: 2014 IEEE 15th annual wireless and microwave technology conference (WAMICON), June 2014, pp 1–4
15. Skrivervik A (2013) Implantable antennas: the challenge of efficiency. In: 2013 7th European conference on antennas and propagation (EuCAP), April 2013, pp 3627–3631
16. Schantz H (2005) Near field phase behavior. In: 2005 IEEE antennas and propagation society international symposium, July 2005, vol 3B, pp 134–137
17. Liu Y, Ketterl T, Arrobo G, Gitlin R (2014) Modeling the wireless in vivo path loss. In: 2014 IEEE MTT-S international microwave workshop series on RF and wireless technologies for biomedical and healthcare applications (IMWS-Bio), Dec 2014, pp 1–3
18. Kurup D, Joseph W, Vermeeren G, Martens L (2012, June) In-body path loss model for homogeneous human tissues. IEEE Trans Electromagn Compat 54(3):556–564
19. Alomainy A, Hao Y (2009) Modeling and characterization of biotelemetric radio channel from ingested implants considering organ contents. IEEE Trans Antennas Propag 57(4):999–1005

20. Ketterl T, Arrobo G, Gitlin R (2013) Sar and ber evaluation using a simulation test bench for in vivo communication at 2.4 GHz. In: 2013 IEEE 14th annual wireless and microwave technology conference (WAMICON), April 2013, pp 1–4
21. “IEEE standard for information technology–local and metropolitan area networks–specific requirements–part 11: wireless lan medium access control (mac)and physical layer (phy) specifications amendment 5: enhancements for higher throughput,” IEEE Std 802.11n-2009 (Amendment to IEEE Std 802.11-2007 as amended by IEEE Std 802.11 k-2008, IEEE Std 802.11r-2008, IEEE Std 802.11y-2008, and IEEE Std 802.11w-2009), pp 1–565, Oct 2009
22. Tse D, Viswanath P (2005) Fundamentals of wireless communication. Cambridge University Press, Cambridge
23. Castro C, Smith S, Alqassis A, Ketterl T, Sun Y, Ross S, Rosemurgy A, Savage P, Gitlin R (2012) Marvel: a wireless miniature anchored robotic videoscope for expedited laparoscopy. In: 2012 IEEE international conference on robotics and automation (ICRA), May 2012, pp 2926–2931
24. Castro C, Alqassis A, Smith S, Ketterl T, Sun Y, Ross S, Rosemurgy A, Savage P, Gitlin R (2013) A wireless robot for networked laparoscopy. IEEE Trans Biomed Eng 60(4):930–936
25. Russer P, Fichtner N (2010) Nanoelectronics in radio-frequency technology. IEEE Microw Mag 11(3):119–135
26. Eckert MA, Zhao W (2013) Opening windows on new biology and disease mechanisms: development of real-time in vivo sensors. Interface Focus 3(3):20130014
27. Akyildiz IF, Jornet JM (2010) Electromagnetic wireless nanosensor networks. Nano Commun Netw 1(1):3–19
28. Jornet JM (2014) Fundamentals of plasmonic communication for in vivo wireless nanosensor networks. In: 36th annual international conference on IEEE engineering in medicine and biology society (EMBC), Chicago, IL, USA
29. Park Q-H (2009) Optical antennas and plasmonics. Contemp Phys 50(2):407–423
30. Akyildiz IF, Jornet JM, Han C (2014) Terahertz band: next frontier for wireless communications. Phys Commun 12:16–32

Part VI
Physical Energy Systems, Energy Efficiency

Achieving Spectral and Energy Efficiencies in Smart Cities Multi-cell Networks

Bilal Maaz, Kinda Khawam, Samer Lahoud, Jad Nasreddine and Samir Tohme

1 Introduction

Mobile communication is considered as one of the building blocks of smart cities, where citizens should be able to enjoy telecommunications services wherever they are and whenever they want in a secure and non-costly way. This can be done by dense deployment of broadband mobile systems such as Long Term Evolution (LTE) and its successors. This dense deployment will lead to higher energy consumption, and thus more gas emission and pollution. Therefore, it is crucial from environmental point of view to reduce the energy consumption. In this context, the focus of this chapter is to introduce radio resource management methods that increase energy efficient, and thus reduce pollution and power wastage. Most of the work that tackles the problem of energy efficiency in cellular networks considers the case of one single cell. In this chapter, we propose a game theoretical approach for the problem of energy efficiency in multicell LTE networks. We address the problem of ICIC in the downlink of LTE OFDMA-based systems, where the power level selection for frequency subcarriers is portrayed as a non-cooperative game in the context of self-organizing networks. The existence of Nash equilibriums (NEs) for the modeled game shows that stable power allocations can be reached by selfish eNBs. To attain these NEs, we propose a decentralized algorithm based on Best Response dynamics. In order to evaluate our proposal, we compare the obtained results to an optimal global Coordinated Multi-Point (CoMP) solution

B. Maaz (✉) · K. Khawam · S. Tohme
University of Versailles, Versailles, France
e-mail: bilal.maaz@ens.uvsq.fr

S. Lahoud
University of Rennes I, Rennes, France

J. Nasreddine
Rafik Hariri University, Beirut, Lebanon

L. Ismail and L. Zhang (eds.), *Information Innovation Technology in Smart Cities*,
https://doi.org/10.1007/978-981-10-1741-4_19

where a central controller is the decision maker. Numerical simulations assessed the good performance, in terms of throughput and energy efficiency, of the proposed distributed approach in comparison with the centralized approach.

2 The Network Model

We consider an LTE network comprising a set of eNBs denoted by J. We focus on the downlink in this chapter. The time and frequency radio resources are grouped into time-frequency Resource Blocks (RBs). An RB is the smallest radio resource unit that can be scheduled to a mobile user. Each RB consists of N_s OFDM symbols in the time dimension and N_f sub-carriers in the frequency dimension (in LTE $N_s = 7$ and $N_f = 12$ in the most common configuration). The set of RBs is denoted by K, and the set of users is denoted by I. We consider the Single Input Single Output (SISO) technology in this chapter and we will consider Multi Input Multi Output (MIMO) technology in a future work. In the following, we make the following assumptions:

- We consider a fixed cell assignment and we don't consider mobility in our network model, each user typically compares the received signal power from each eNB and chooses to connect to the eNB with the strongest signal.
- In order to evaluate the maximum system performance, we consider permanent downlink traffic where each eNB has persistent traffic towards its users. We also assume that all RBs are assigned on the downlink at each scheduling epoch.
- We adopt the widely used Proportional Fair (PF) scheduler for serving active users. Symbols and variables used within this chapter are defined in Table 1.

The power consumption of eNB $j \in J$ is modeled as a linear function [1] of the average transmit power per site as: $\mathrm{p_j} = \mathrm{p_j^1}\pi_\mathrm{j} + \mathrm{p_j^0}$. where $\mathrm{p_j}$ and π_j denote the average consumed power by eNB j and its transmit power, respectively. The coefficient p_j^1 accounts for the power consumption, that scales with the transmit power due to radio frequency amplifier and feeder losses.

Table 1 Sets, parameters and variables in the chapter

Variables	Signification	Variables	Signification
J	Set of eNBs	π_{jk}	Transmit power of eNB j on RB k
I	Total set of users	θ_{ik}	Percentage of time user i is associated with RB k
K	Set of Resource blocks	G_{ijk}	Channel power gain (user i on RB k on eNB j)
N_0	Noise power	ρ_{ijk}	SINR of user i associated eNB j served on RB k
$I(j)$	Set of users associated to eNB j	α_{jk}	Interference impact on RB k of eNB j among other eNBs

The coefficient p_j^0 models the power consumed independently of the transmit power due to signal processing and site cooling. The transmit power of each eNB is allocated to resource blocks serving the users in the network. The total transmit power of eNB j is the sum of the transmit power on each RB $k \in K : \pi_j = \sum_{k \in K} \pi_{jk}$. The total power consumed by any eNB j is given by:

$$P_j = p_j^1 \sum_{k \in K} \pi_{jk} + p_j^0. \tag{1}$$

Given user i associated with eNB j (i.e. $i \in I(j)$), the signal-to-interference-plus-noise-ratio (SINR) of this user when served on RB k is given by:

$$\rho_{ijk} = \frac{\pi_{jk} G_{ijk}}{N_0 + \sum_{j' \neq j} \pi_{j'k} G_{ij'k}} \tag{2}$$

where G_{ijk} is the path gain of user i on resource block k on eNB j (i.e. the average path gain over the sub-carriers in the resource block), and N_0 is the noise power, which is, without loss of generality, assumed to be the same for the all users on all resource blocks.

Assuming a proportional fairness service by each eNB on each resource block, the system utility function is given by what follows:

$$\begin{aligned} U(\theta) &= \sum_{j \in J} \sum_{i \in I(j)} \frac{g(|I(j)|)}{|I(j)|} \sum_{k \in K} \log(\rho_{ijk}) \\ &= \sum_{j \in J} \sum_{i \in I(j)} \sum_{k \in K} \frac{g(|I(j)|)}{|I(j)|} \log\left(\frac{\pi_{jk} G_{ijk}}{N_0 + \sum_{j' \neq j} \pi_{j'k} G_{ij'k}} \right). \end{aligned} \tag{3}$$

where $g(|I(j)|) = \sum_{s=1}^{|I(j)|} 1/s$, as we consider the PF scheduler with a fast varying fading channel (Rayleigh fading). In the following sections we will provide solution for the problem of maximizing the above mentioned utility function.

3 Centralized Power Control Approach

We cast hereafter the centralized power control problem:

$$P(\pi) : \underset{\pi}{\text{maximize}} \sum_{j \in J} \sum_{i \in I(j)} \sum_{k \in K} \frac{g(|I(j)|)}{|I(j)|} \log\left(\frac{\pi_{jk} G_{ijk}}{N_0 + \sum_{j' \neq j} \pi_{j'k} G_{ij'k}} \right) \tag{4a}$$

$$\text{Subject to} \sum_{k\in K} \pi_{jk} \leq p_j^{max}, p_j^{min} \leq \pi_{jk}, \forall\, k \in K. \forall\, j \in J \tag{4b}$$

Problem (4a, b) is non-linear and apparently difficult, non-convex optimization problem. However, it can be transformed into a convex optimization problem in the form of geometric programming by performing a variable change $\hat{\pi}_{jk} = \log(\pi_{jk})$ and defining $\hat{N}_0 = \log(N_0)$ and $\hat{G}_{ijk} = \log(G_{ijk})$. The resulting optimization problem deemed $P(\hat{\pi})$ is given by the following:

$$P(\hat{\pi}) : \underset{\pi}{\text{maximize}}\, \mathrm{U_j}(\hat{\pi}), \text{with} \quad \mathrm{U_j}(\hat{\pi})$$

$$= \sum_{j\in J}\sum_{i\in I(j)}\sum_{k\in K}\frac{g(|I(j)|)}{|I(j)|}\left(\log(\pi_{jk}) + \log(G_{ijk}) - \log\left(N_0 + \sum_{j'\neq j}\pi_{j'k}G_{ij'k}\right)\right)$$

$$= \sum_{j\in J}\sum_{i\in I(j)}\sum_{k\in K}\frac{g(|I(j)|)}{|I(j)|}\left(\hat{\pi}_{jk} + \widehat{G}_{ijk} - \log\left(N_0 + \sum_{j'\neq j}\exp(\log(\pi_{j'k}G_{ij'k}))\right)\right)$$

$$= \sum_{j\in J}\sum_{i\in I(j)}\sum_{k\in K}\frac{g(|I(j)|)}{|I(j)|}\left(\hat{\pi}_{jk} + \widehat{G}_{ijk} - \log\left(\exp\left(\widehat{N}_0\right) + \sum_{j'\neq j}\exp\left(\hat{\pi}_{j'k} + \widehat{G}_{ij'k}\right)\right)\right). \tag{5a}$$

$$\text{Subject to} \log\left(\sum_{k\in K}\exp(\hat{\pi}_{jk})\right) - \log\left(p_j^{max}\right) \leq 0, \forall j \in J, k \in K. \tag{5b}$$

$$-\hat{\pi}_{jk} + \log\left(p_j^{min}\right) \leq 0, \forall j \in J, k \in K. \tag{5c}$$

Proposition 3.1 *The resulting optimization problem* $\boldsymbol{P}(\hat{\boldsymbol{\pi}})$ *is convex and hence can be efficiently solved for global optimality even with a large number of users.*

Proof The first term of the objective is a linear function, thus concave (and convex). The second term contains log-sum-exp expression which is convex. The opposite of the sum of convex functions being concave, this completes the proof of the concavity of the objective function. As for the new constraints: constraints (5b) are convex by virtue of the properties of the log-sum-exp functions and (5c) are linear function and hence convex.

4 Distributed Power Control Approach

Although optimal, a central power allocation is complex and necessitates having recourse to a central control that harvest signaling information from eNBs to allocate power optimally. We turn here to distributed schemes to diminish complexity at the cost of slow convergence time and lower performance.

Here, we propose a cooperative algorithm that makes profit from the X2 interface between neighboring eNBs in LTE, Any local optimum π^* of the centralized convex problem (5a, b, c) must satisfy the KKT conditions, i.e. there exist unique Lagrange multipliers $\forall j \in J$ such that: Any local optimum π^* of the centralized convex problem (5a, b, c) must satisfy the Karush-Kuhn-Tucker (KKT) conditions, i.e. there exist unique Lagrange multipliers $\forall j \in J$ such that:

$$\frac{\partial U_j(\hat{\pi})}{\partial \hat{\pi}_{jk}} + \sum_{l \neq j} \frac{\partial U_l(\hat{\pi})}{\partial \hat{\pi}_{jk}} = \mu_j - \lambda_j^k; \quad \forall k \in K. \tag{6a}$$

$$\mu_j \cdot \left(\log(P_j^{max}) - \log\left(\sum_{k \in K} \exp(\hat{\pi}_{jk}) \right) \right) = 0. \tag{6b}$$

$$\lambda_j^k \cdot \left(\hat{\pi}_{jk} - \log(p_j^{min}) \right) = 0; \quad \forall k \in K. \tag{6c}$$

$$\mu_j \geq 0 \ and \ \lambda_j^k \geq 0; \quad \forall k \in K. \tag{6d}$$

We come back to the solution space in π instead of $\hat{\pi}$. In particular, we have what follows:

$$\frac{\partial U_j(\pi)}{\partial \pi_{jk}} = \frac{\partial \hat{\pi}_{jk}}{\partial \pi_{jk}} \frac{\partial U_j(\hat{\pi})}{\partial \hat{\pi}_{jk}} = \frac{1}{\pi_{jk}} \frac{\partial U_j(\hat{\pi})}{\partial \hat{\pi}_{jk}}.$$

Accordingly, we obtain the following set of equations:

$$\pi_{jk} \cdot \left(\frac{\partial U_j(\pi)}{\partial \pi_{jk}} + \sum_{l \neq j} \frac{\partial U_{lk}(\pi)}{\partial \pi_{jk}} \right) = \mu_j - \lambda_j^k; \forall k \in K. \tag{7a}$$

$$\mu_j \cdot \left(P_j^{max} - \sum_{k \in K} \pi_{jk} \right) = 0. \tag{7b}$$

$$\lambda_j^k \cdot \left(\pi_{jk} - p_j^{min} \right) = 0; \forall k \in K. \tag{7c}$$

$$\mu_j \geq 0 \ and \ \lambda_j^k \geq 0; \quad \forall k \in K. \tag{7d}$$

Using the KKT conditions, we give a decomposition of the original problem into $|J|$ subproblems. Following [2] we define the interference impact I_{ijk} for user i associated to BS j on RB k such as:

$$I_{ijk}(\pi_{-j}) = \sum_{l \neq j} \pi_{lk} G_{ilk} + N_0; \quad \forall i \in I(j). \tag{8}$$

Further, we define the derivative of $U_{ijk} = \frac{g(|I(j)|)}{|I(j)|} \log\left(\frac{\pi_{jk} G_{ijk}}{N_0 + \sum_{j' \neq j} \pi_{j'k} G_{ij'k}}\right)$ relative to the interference impact as follows:

$$\frac{\partial U_{ijk}}{\partial I_{ijk}} = \frac{g(|I(j)|)}{|I(j)|} \frac{-1}{I_{ijk}}.$$

Using (8), condition (7a) can be re-written as:

$$\pi_{jk}\left(\frac{\partial U_{jk}}{\partial \pi_{jk}} - \sum_{l \neq j} \sum_{k \in K} \sum_{i \in I(l)} \frac{g(|I(j)|)}{|I(j)|} \frac{G_{ijk}}{I_{ijk}}\right) = \mu_j - \lambda_j^k; \quad \forall k \in K, \forall j \in J. \tag{9}$$

Given fixed interference and fixing the power profile of any eNB except eNB j, it can be seen that (9) and conditions (7b–d) are the KKT conditions of the following optimization sub-problems $\forall j \in J$:

$$\underset{\pi_j}{\text{maximize}} \, V_j(\pi_j, \pi_{-j})$$

$$= \sum_{k \in K} \sum_{i \in I(j)} \frac{g(|I(j)|)}{|I(j)|} \log\left(\frac{\pi_{jk} G_{ijk}}{N_0 + \sum_{j' \neq j} \pi_{j'k} G_{ij'k}}\right) - \sum_{k \in K} \pi_{jk} \alpha_{jk}. \tag{10a}$$

$$\text{Subject to}: \sum_{k \in K} \pi_{jk} \leq p_j^{max}, p_j^{min} \leq \pi_{jk}; \forall k \in K. \tag{10b}$$

where α_{jk} is the interference impact on RB k of eNB j on other eNBs, and given by:

$$\alpha_{jk} = \sum_{\substack{l \in J \\ l \neq j}} \sum_{i \in I(l)} \frac{g(|I(j)|)}{|I(j)|} \frac{G_{ijk}}{\left(\sum_{\substack{j' \in J \\ j' \neq l}} \pi_{j'k} G_{ij'k} + N_0\right)}. \tag{11}$$

However, we choose to replace α_{jk} by $\bar{\alpha}_{jk} = \frac{\alpha_{jk}}{|J|}$, which is the mean interference impact on RB k inflicted by eNB j on other eNBs. Hence, we formulate a new non-cooperative game $G' = \langle J, S, \overline{V} \rangle$, where:

$$\overline{V}_j(\pi_j, \pi_{-j}) = \sum_{k \in K} \left(\sum_{i \in I(j)} U_{ijk} - \bar{\alpha}_{jk}\pi_{jk} \right); \quad \forall j \in J. \tag{12}$$

The first term of the new utility function $\sum_{i \in I(j)} U_{ijk}$ is a non-decreasing function in π_{jk} while the second term $-\bar{\alpha}_{jk}\pi_{jk}$ is decreasing in π_{jk}, which permits to strike a good balance between spectral efficiency and energy efficiency. Hence, the higher is the mean interference harm inflected on neighboring eNBs on a given RB k, the lower will be the chosen power amount π_{jk}.

For every j, $\overline{V}_j$ is concave w.r.t. π_j and continuous w.r.t. $\pi_l, l \neq j$. Hence, a Nash Equilibrium (NE) exists [3]. Furthermore, the game at hand is super-modular. In fact, the strategy space S_j is obviously a compact convex set of $\mathbb{R}^k$, while the objective function of any eNB j is super-modular [4]:

$$\frac{\partial \overline{V}_{jk}}{\partial \pi_{lk} \partial \pi_{jk}} = \sum_{\substack{s \in J \\ s \neq \{j,l\}}} \sum_{i \in I(s)} \frac{g(|I(s)|)}{|\mathrm{J}||I(s)|} \frac{G_{ijk}G_{ilk}}{\left(\sum_{j' \in J, j' \neq s} \pi_{j'k}G_{ij'k} + N_0\right)^2} \left(1 - \frac{G_{ijk}\pi_{jk}}{\left(\sum_{j' \in J, j' \neq s} \pi_{j'k}G_{ij'k} + N_0\right)}\right) \geq 0.$$

$\forall\, l \in J - \{j\}$ and $\forall\, k \in K$, as we can fairly assume with at least 6 neighboring eNBs for any eNB s that $\frac{G_{ijk}\pi_{jk}}{\left(\sum_{j' \in J, j' \neq s} \pi_{j'k}G_{ij'k} + N_0\right)} < 1$.

As we proved that we are in presence of a super-modular game, we know that a Best response algorithm enables attaining the NEs. The main idea behind this algorithm is for each eNB j to iteratively solve the optimization problem in (10a, b) given the current interference impact and power profile of the other eNBs and then to recalculate the corresponding interference impact until convergence. Formally, we summarize this as follows:

1. Each eNB j chooses an initial power profile π_j satisfying the power constraint.
2. Using (11), each eNB j calculates the mean interference price vector $\bar{\alpha}_j$ given the current power profile and announces it to other eNBs.
3. At each time t, one eNB j is randomly selected to maximize its payoff function $\overline{V}_j(\pi_j, \pi_{-j})$ and update its power profile, given the other eNBs power profiles π_{-j} and price vectors, i.e., $\pi_j(t+1) = arg\max_{\pi_j \in S_j} \overline{V}_j(\pi_j, \pi_{-j}(t))$.

4.1 The Power Expression at Equilibrium

We begin by solving the unconstrained convex optimization problem $\max_{\pi_j} \overline{V_j}(\pi_j, \pi_{-j})$. Then, to obey the bounding constraints on power levels, any eNBs j must do locally a projection step in order to get back to the feasible region defined by S_j. The optimal values of the unconstrained problem are either on the boundaries of the strategy space or resulting from the following derivation $\forall j \in J, \forall k \in K$:

$$\frac{\partial \overline{V}_j(\pi_j, \pi_{-j})}{\partial \pi_{jk}} = 0 \Rightarrow \tag{13a}$$

$$\pi_{jk}^2 \cdot \left(\sum_{l \neq j} \sum_{i \in I(l)} G_{ijk}^2 C_l B_{jl} \right) + \pi_{jk} \cdot \left(\sum_{l \neq j} \sum_{i \in I(l)} G_{ijk} A_{ik} C_l (2B_{jl} - 1) \right) + \sum_{j \neq j} \sum_{i \in I(l)} C_l A_{ik}^2 = 0 \tag{13b}$$

where

$$C_l = \frac{g(|I(j)|)}{|\mathrm{J}||I(j)|}, B_{jl} = \frac{g(|I(j)|)}{g(|I(j)|)} \; and \; A_{ik} = \left(\sum_{\substack{j' \in J \\ j' \neq \{j,l\}}} \pi_{j'k} G_{ij'k} + N_0 \right).$$

Consequently, π_{jk} is the solution of the second degree equation in (13b). After obtaining the various sπ_{jk}, the projection algorithm 1 is run by every eNB j at each iteration as follows:

Algorithm 1 Projection algorithm for eNB *j*

procedure POWERPROJECTION (π_j)
  S(K) ← SORTINDECREASINGORDER(K)
  for all $k \in s(K)$ **do**
    if $\pi_{jk} < p_j^{min}$ **then**
      $\pi_{jk}^{p} \leftarrow P_j^{min}$
    end if
  end for
  if $\pi_j \notin S_j$ **then**
    $\rho(k) \leftarrow \pi_{jk} + \frac{1}{k} \times \left(P_j^{maz} - \sum_{i \in s(K), i \leq k} \pi_{ji}\right)$
      $\forall k \in s(K) and\ \pi_{jk} > P_j^{min}$
    $\rho^* \leftarrow \text{argmax}_{k \in s(K)}\{v(k)\}$
    $\lambda \leftarrow \frac{1}{\rho^*} \times \left(P_j^{max} - \sum_{i \in s(K), i \leq k} \pi_{ji}\right)$
    for all $k \in$ S(K) **do**
      if $\pi_{jk} > p_j^{min}$ **then**
        $\boldsymbol{\pi}_{j,k}^{p} \leftarrow max\{\pi_{jk} - \lambda . p_j^{min}\}$
      end if
    end for
  end if
Return $\boldsymbol{\pi}_j^p = \left(\boldsymbol{\pi}_{j,k}^p, \forall k \in K\right)$
end procedure

5 Performance Evaluation

We consider a Bandwidth of 5 MHz with 25 RBs in a 9 hexagonal cells network, the number of UE ranging from 4 to 14 per eNB uniformly distributed in any cell. Further, we consider the following parameters listed in the 3GPP technical specifications TS 36.942 [5]: the mean antenna gain in urban zones is 12 dBi (900 MHz). Transmission power is 43 dBm (according to TS 36.814) which corresponds to 20 W (on the downlink). The eNBs have a frequency reuse of 1, with $W = 180$ kHz. As for noise, we consider the following parameters: user noise figure 7.0 dB, thermal noise −104.5 dBm which gives a receiver noise floor of $pN = -97.5$ dBm.

In this chapter, we conducted preliminary simulations in a Matlab simulator, where various scenarios were tested to assess the performances of the power control schemes.

For each approach, 25 simulations were run, where in each cell a predefined number of users is selected; users' positions were uniformly distributed in the cells. For each simulation instance, the same pool of RBs, users and pathloss matrix are given for both algorithms.

In Fig. 1, we can see the similarity of power economy efficiency between the centralized algorithm and the semi-distributed algorithm. Both power control schemes permit a considerable power economy in comparison with the Max Power policy, that uses full power P^j_{max} for each eNB, as we can see in Fig. 1 where the power economy percentage for all eNBs vary from 55 to 65% in comparison with the Max power policy, which is a sensible power economy.

In fact, the existence of the power cost $-\sum_{k\in K}\pi_{jk}\bar{\alpha}_{jk}$ in the utility function (12), diminishes the selfishness of eNBs that are tempted to transmit at full power on all RBs.

This power economy is obtained while maintaining good performances as we can see in Fig. 2 where the utility function in (3) is depicted as a function of the number of users for the centralized algorithm and the semi-distributed algorithms.

In Fig. 3, we report the mean convergence time per eNBs for the semi-distributed algorithm for various scenarios. We note that each eNBs attains in average the NE within 19–27 iterations. At each iteration, one eNB is randomly selected to maximize its payoff function given in (13). The iteration period coincides with one TTI (Transmit Time Interval), which equals 1 ms in LTE.

We noted during the extensive simulations conducted, that the power levels attain 90% of the values reached at convergence in less than 8 iterations. We represented in Fig. 4 the power distribution of 25 RBs for an eNB selected randomly and for which convergence time was equal to 22 iterations. Low convergence time in conjunction with high performances is an undeniable asset for our semi-distributed schemes.

The Decentralized algorithms can adapt to fast changes of network state though it is difficult to avoid converging to local optimum. It turns out that even though the distributed game results are sub-optimal, the low degree of system complexity and

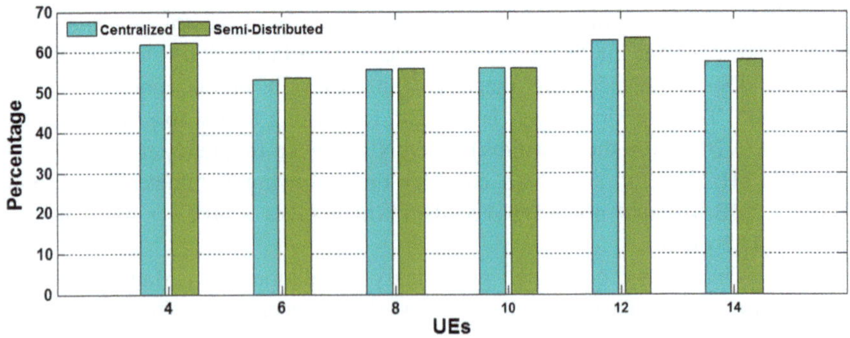

Fig. 1 Percentage of power economy as a function of the number of users for centralized and semi-distributed versus Max power algorithms

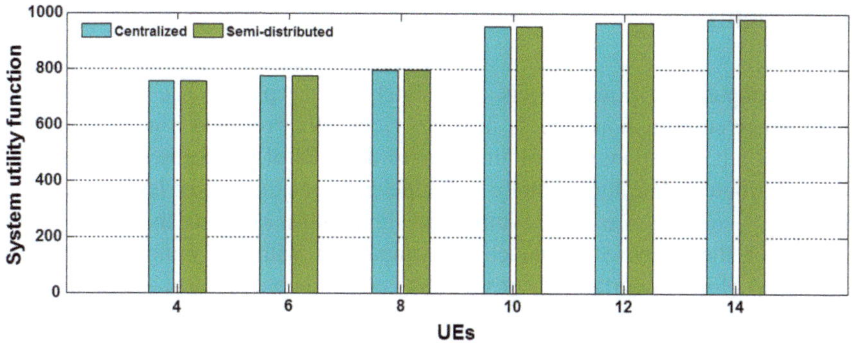

Fig. 2 Utility function as function of the number of users for centralized and semi-distributed

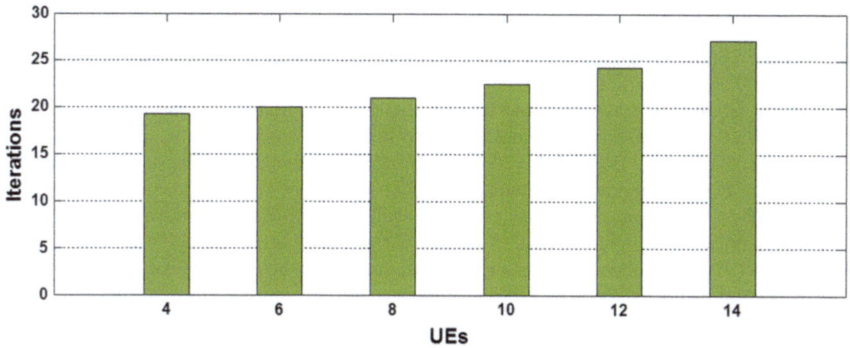

Fig. 3 Total convergence time by eNBs as function of the number of users for semi-distributed algorithm

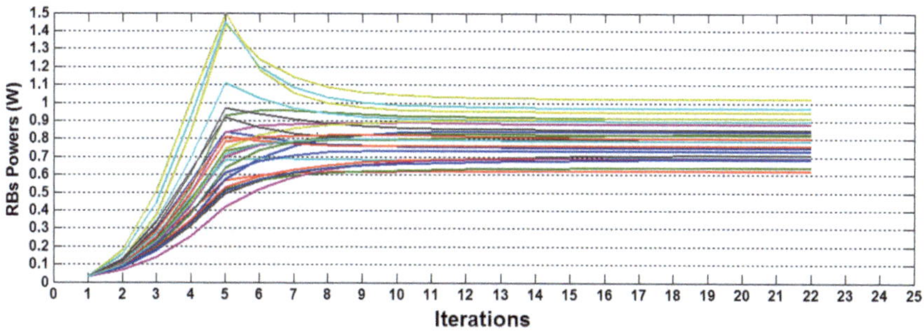

Fig. 4 Power distribution by RBs before reaching convergence for semi-distributed algorithm

the inherent adaptability make the decentralized approach promising especially for dynamic scenarios. The fast convergence time, the near optimal results and the lower complexity degree of the semi-distributed approach makes it a very attractive solution.

6 Conclusion

In this chapter, the power levels are astutely set as part of the LTE Inter-cell Interference coordination process in smart cities. We proposed a non-cooperative game and a best response algorithm to reach the NEs of the portrayed game. This semi-distributed algorithm astutely and efficiently set the power levels with relatively low convergence time. Numerical simulations assessed the good performances of the proposed approach in comparison with the optimal centralized approach. More importantly, considerable power economy and signaling optimization can be realized.

References

1. Auer G, Blume O, Giannini V, Godor I, Imran M, Jading Y, Katranaras E, Olsson M, Sabella D, Skillermark P et al (2010) D2. 3: Energy efficiency analysis of the reference systems, areas of improvements and target breakdown, INFSOICT-247733. EARTH (Energy Aware Radio and NeTwork TecHnologies), Tech. Rep.
2. Huang J, Berry RA, Hoing ML (2006) Distributed interference compensation for wireless networks. IEEE J Sel Areas Commun 24(5):1074–1084
3. Rosen JB (1965) Existence and uniqueness of equilibrium points for concave n-person games. Econometrica 33
4. Topkis (1979) Equilibrium points in non-zero sum n-person sub-modular games. SIAM J Control Optim 17(6):773–787
5. GPP TR 36.942 V12.0.0 Release 12. Evolved Universal Terrestrial Radio Access; (E-UTRA); Radio Frequency (RF) system scenarios, October 2014

Maximization of Social Welfare in Deregulated Electricity Markets with Intermediaries

Ryo Hase and Norihiko Shinomiya

1 Introduction

The structure of electricity markets has changed in the world recently. In many countries, regulated electricity markets with the vertically integrated business structure have been considered first to supply electricity safely. However, the regulated electricity markets have some issues on the transparency of electricity prices due to the centralized structure in which there are only a few suppliers [1]. To solve the issues, the competition between electricity suppliers has been integrated into the electricity markets. Development of electricity management techniques using ICT has also changed the structure of electricity markets [2]. Entering electricity markets becomes easier than ever by exploiting *smart meters*, which enable suppliers to measure and collect the electricity consumption of each consumer via information networks. Increasing participants make the market structure more complicated; hence, previous methods to examine centralized markets cannot be applied to deregulated markets [3]. Nevertheless, characteristics of deregulated markets should be examined carefully before the deregulation. In fact, insufficient examination caused California electricity crisis in 2001.

For those reasons, numerous kinds of research have been conducted to examine characteristics of the electricity markets [4]. There are especially various papers focusing on efficiency, which can be examined by social welfare that is the total of payoffs of all market participants. Swami considers social welfare maximization in electricity markets with considering congestion of transmission lines [5]. Dong et al. build an optimization model for demand response considering generation capacity and

R. Hase (✉) · N. Shinomiya
Graduate School of Engineering, Soka University, Tokyo, Japan
e-mail: ryo_hase.0428@icloud.com

N. Shinomiya
e-mail: shinomi@ieee.org

L. Ismail and L. Zhang (eds.), *Information Innovation Technology in Smart Cities*,
https://doi.org/10.1007/978-981-10-1741-4_20

electricity demands in smart grids [6]. Yamamoto et al. proposes a pricing mechanism for multi-regional electricity trades for smart grids [7]. In [8], we proposed an algorithm to find optimal matching in an electricity market model based on network market, which is proposed in [9]. However, these papers have not focused on electricity retailers, because the previous market models contain only suppliers and consumers, not a retailer. In [10], Babic notes advantages of an agent-based modeling technique for electricity retail markets; however, no characteristics regarding retailers are demonstrated in the paper. Kleinberg et al. consider optimal price setting on a tripartite graph by exploiting a game theoretical approach to model activities of retailers in network markets [11]. Nevertheless, the method cannot deal with a multi-unit commodity such as electricity because it is assumed that market participants trade only a single commodity. Besides, Nava introduces the competition model utilizing network flows in oligopolistic markets [12]. Even though the model can cope with retailers dealing with multi-unit commodities, the role of each participant is eventually determined by equilibrium in the model. Therefore, the model cannot be applied for electricity markets because the roles of participants in electricity markets are determined before equilibrium prices are discovered.

This chapter presents algorithms to determine prices and efficient trades in an electricity market model with agents including electricity retailers. In this paper, we formulated a determination problem for efficient electricity trades on a static market model coping with electricity as a multi-unit commodity. To solve that problem, we constructed algorithms to choice electricity trades in the market model by integrating a similar price setting mechanism proposed in [11] and unsplittable flows [13]. Simulation results show efficiency by examining the social welfare of determined electricity trades. This chapter is structured as follows. Section 2 introduces our electricity market model. Section 3 explains a price setting game for the electricity market model. Section 4 proposes objective functions to consider social welfare maximization. Section 5 defines an evaluation metric for electricity trades determined the objective functions. In Sect. 6, two algorithms to determine electricity trades are presented. Section 7 shows simulation results, and Sect. 8 concludes this chapter.

2 Market Model with Three Types of Agents

This section presents an agent model representing deregulated electricity markets.

2.1 Network Representing Agent Model

Our market model is denoted by tripartite network $G = (S \cup B \cup T, A)$. G is composed of three types of agents: *buyer* $b_j \in B(1 \leq j \leq |B|)$, *seller* $s_i \in S(1 \leq i \leq |S|)$, and *trader* $t_k \in T(1 \leq k \leq |T|)$. Arc set A contains arcs denoted by (s_i, t_k) or (t_k, b_j) due to the following three constraints. First, since G is a tripartite network, each arc in G must connect two nodes that are not belonging to the same types of nodes each other. Second, b_j must be provided electricity from t_k. Third, t_k must purchase electricity from s_i.

2.2 Supposed Participants Assigned to Agent Model

In this chapter, an agent model represents a day-ahead electricity market in which electricity prices are determined hourly or half hourly. Supposed market participants are the following four types: the public utility, independent power producers, retailers, and consumers. These participants are described by agents explained in Sect. 2.

- *Public utility PU* conducts electricity generation and supply. The PU is a large firm that has conducted electricity generation and supply to consumers since the market was regulated. In the deregulated market, the PU can purchase electricity from the other generators. The PU is denoted by a pair of seller and trader.
- *Retailer* $R_y \in R(1 \leq y \leq |R|)$ conducts electricity trades with generators and consumers. A retailer is described by one pair of a trader and a seller if the retailer has its own generator. Otherwise, one trader denotes a retailer.
- *Independent power producer* $IPP_z \in IPP(1 \leq z \leq |S| - |R| - 1)$ has its own generator to provide wholesale electricity. An IPP is assigned to one of the sellers, and it can supply electricity to any retailers and the PU.
- *Consumer* $C_j \in C$ is an end-user of electricity. A consumer purchases electricity from one of the best suppliers (the PU and retailers) connected to the consumer. The consumer is assigned to one of the buyers.

The market model has only one PU, which had previously existed in an electricity market before deregulation. Besides, the market model contains some IPP and retailers that have newly joined to the market after deregulation. The number of the newly joining participants is not restricted in our model. As noted above, the four types of participants are denoted by agent s_i, b_j, and t_k explained in Sect. 2.1. Figure 1 shows an example of an electricity market model consisting of the four types of participants. The model has one PU, two IPP, two retailers, and ten consumers. The model has arcs between the PU and all consumers. To describe deregulated markets, each consumer in our model must be connected with one or more retailers. In our model, retailer t_k is connected to each consumer with the probability represented by $prob(t_k, b_j) \in [0, 1]$. Thus, all arcs between retailers and a consumer are constructed if $prob\,(t_k, b_j) = 1$.

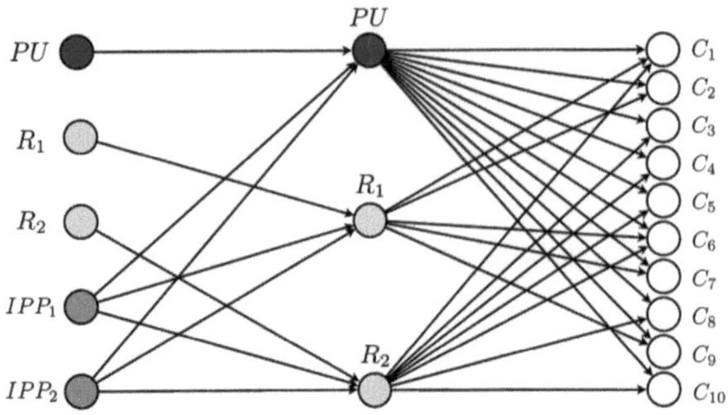

Fig. 1 An example of an agent model with four types of market participants

2.3 Supply Capacity and Demand of Electricity Flow

In the electricity market model, each seller has *supply capacity* c_{s_i}, and each buyer has *demand* d_{b_j}. In our model, network flow is utilized to describe electricity flow. This kind of problem is similar to generalized assignment problem explained in [14]. Integer x_{ba} denotes the quantity of electricity flow on arc (b, a). Lower bound and upper bound of x_{ba} is represented by lb_{ba} and ub_{ba} respectively. If $x_{ba} > 0$, electricity currents on (b, a); otherwise, there is no electricity flow on (b, a). s_i can supply electricity flow up to c_{s_i}, and there is no electricity flow if s_i does not trade any electricity. Hence, $lb_{s_i t_k} = 0$, and $ub_{s_i t_k} = c_{s_i}$. Besides, b_j requires d_j units of electricity, and this demand must be satisfied. Therefore, flow constraints on (t_k, b_j) are denoted by $lb_{t_k b_j} = d_{b_j}$ and $ub_{t_k b_j} = d_{b_j}$. Figure 2 indicates these capacity constraints on electricity flow on a market model.

Our model utilizes *unsplittable flow* to denote electricity trades. In Fig. 3, both $s_1 - t_1 - b_1$ and $s_1 - t_2 - b_1$ are $s_1 - b_1$ paths. The flow is called splittable flow if both of the paths supply b_1 with electricity. Only $s_1 - b_1$ path, however, must be selected because no buyer can purchase electricity from more than one trader in our model. The electricity flow is called *unsplittable flow* if only one $s_1 - b_1$ path provides b_1 with electricity. $s_1 - b_1$ path through t_k is represented by

$$x_{t_k s_i b_j} = \begin{cases} 1 & \left(x_{s_i t_k} > 0 \cap x_{t_k b_j} > 0\right), \\ 0 & \left(x_{s_i t_k} = 0 \cup x_{t_k b_j} = 0\right). \end{cases}$$

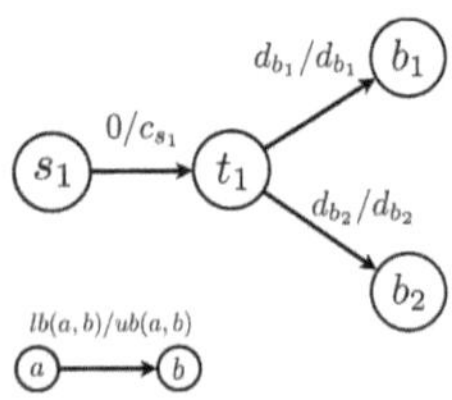

Fig. 2 Flow constraints on arks

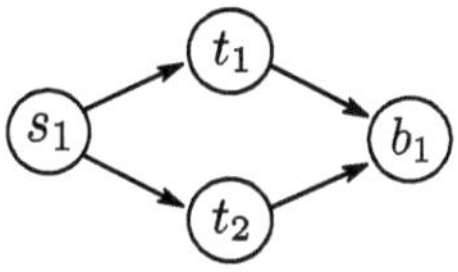

Fig. 3 Two $s_1 - b_1$ paths

3 Price Setting Game on the Market Model

As in [11], prices offered by traders in the model are determined by a price setting game. In the market model, each seller and buyer has *valuation*, which describes the utility for trading one unit of electricity. v_{b_j} indicates the valuation of b_j for purchasing one unit of electricity. v_{s_i} denotes the valuation of s_i for supplying one unit of electricity. The sets of each valuation are defined by

$$\mathbf{v}_s = \{v_{s_i} | s_i \in S, v_{s_i} > 0\}, \mathbf{v}_b = \{v_{b_j} | b_j \in B, v_{b_j} > 0\}.$$

Electricity trades are conducted between seller s_i and buyer b_j via trader t_k. About a trade dealing with one unit of electricity, *trade value* is distributed to s_i, b_j, and t_k as the payoff of each of them. Because we assume that costs for supplying electricity through arcs between s_i and b_j are zero regardless of t_k, the trade value is described by $w_{s_ib_j} = (v_{b_j} - v_{s_i})d_{b_j}$. If t_k conducts a trade between b_j and s_i, t_k has its own strategy denoted by $(\alpha_{t_kb_j}, \beta_{t_ks_i})$ consisting of two types of prices, which are called *ask price* $\alpha_{t_kb_j}$ and *bid price* $\beta_{t_ks_i}$. $\alpha_{t_kb_j}$ is offered to b_j by t_k which is one of the traders adjacent to b_j. Since there will be agents lost money if $\beta_{t_ks_i} > \alpha_{t_kb_j}$, a strategy of t_k must be a *no-crossing strategy* represented by $\beta_{t_ks_i} \leq \alpha_{t_kb_j}$ [15]. Hence, $w_{s_ib_j}$ should satisfy $w_{s_ib_j} \geq 0$.

In addition, b_j must purchase d_{b_j} units of electricity from one of the traders since demand of b_j is d_{b_j}. The payoff of b_j for purchasing electricity from t_k is represented by $p_{b_jt_k} = (v_{b_j} - \alpha_{t_kb_j})d_{b_j}$. b_j tries to get electricity from one of the traders maximizing $p_{b_jt_k}$. Hence, the total payoff of b_j is denoted by $P_{b_j} = p_{b_jt_k}$. Besides, $\beta_{t_ks_i}$ is offered to s_i from t_k. One of the traders offering the bid price maximizing payoff of s_i will be chosen to provide electricity to a buyer. Payoff of s_i for supplying d_{b_j} units of electricity to b_j through t_k is denoted by $p_{s_i(t_k,b_j)} = (\beta_{t_ks_i} - v_{s_i})d_{b_j}$. Each seller can provide one or more buyers with electricity if the total of demands does not exceed the supply capacity of the seller. The set of pairs including each pair of b_j and t_k provided electricity from s_i is denoted by $pair(s_i)$. Therefore, the total payoff of s_i supplying electricity is represented by $P_{s_i} = \sum_{(t_k,b_j)\in pair(s_i)} p_{s_i(t_k,b_j)}$.

Payoff of t_k for trading d_{b_j} units of electricity between s_i and b_j is denoted by $p_{t_k(s_i,b_j)} = (\alpha_{t_kb_j} - \beta_{t_ks_i})d_{b_j}$. Let $S(t_k) \in S$ be the set of s_i connected to t_k, and let $B(t_k) \in B$ be the set of b_j adjacent to t_k. The total payoff of t_k is $P_{t_k} = \sum_{s_i\in S(t_k), b_j\in B(t_k)} p_{t_k(s_i,b_j)} x_{t_ks_ib_j}$. In the market models, efficiency means the optimal allocation of the electricity with appropriate prices [16]. *Social welfare* is the total of payoffs of all participants. Let x and $W(x)$ represent trades and the social welfare of x respectively. If x are more efficient than x', $W(x) > W(x')$.

4 Problem Formulation

This section proposes objective functions to determine electricity trades on the model.

4.1 Optimization Problem for Each Trader

First, an objective function for an optimization problem of each trader is explained. In this problem, each trader greedily selects electricity trades maximizing its total payoff. The following integer program finds the maximum payoff of t_k.

$$\begin{aligned} \max P_{t_k} &= \sum_{s_i \in S(t_k), b_j \in B(t_k)} p_{t_k(s_i,b_j)}\, x_{t_k s_i b_j}. \\ & 0 \leq x_{t_k s_i b_j} \leq 1, p_{t_k(s_i,b_j)} \geq 0, \sum_{b_j \in B(t\_k)} x_{t_k s_i b_j}. \end{aligned} \tag{1}$$

4.2 Optimization Problem to Satisfy Overall Supply and Demand

However, if every trader determines all trades by (1), demands for some sellers might exceed their own supply capacity. Hence, safe electricity trades are independently determined by an institution called Independent System Operator (ISO) to avoid the excess of demands for sellers. The ISO does not participate in the market but observes the trades. For the determination conducted by the ISO, the maximum unsplittable flow problem is utilized. Bipartite network $G_{bi} = (S \cup B, A_{bi})$ is utilized to consider the problem. Arc set A_{bi} corresponds to the set of possible trades x_t. Thus, for all $t_k \in T$, A_{bi} contains—if $x_{t_k s_i b_j} = 1$ in (1). For all $a \in S \cup B$, let $adj(a)$ be the set of t_k connected to a. The capacity of flow on (s_i, b_j) is denoted by $0 \leq x_{s_i b_j} \leq 1$. s_i can supply flow up to c_{s_i}, and demand of flow of b_j is d_{b_j}. Finally, the following integer program gives $W(x_t)$ that is the maximum social welfare on G_{bi}.

$$\max W(x_t) = \sum_{(s_i,b_j)\in A_{bi}} x_{s_ib_j} w_{s_ib_j}.$$
$$x_{s_ib_j} \geq 0, \sum_{s_i\in adj(b_j)} x_{s_ib_j} \leq 1, \sum_{b_j\in adj(s_i)} x_{s_ib_j} d_{b_j} \leq c_{s_i}. \tag{2}$$

5 Evaluation Metric

To evaluate the objective functions described above, we utilize a metric on the efficiency of determined electricity trades. Although $W(x_t)$ is the maximum social welfare on G_{bi}, $W(x_t)$ does not necessarily correspond to $W(x)$ that is the maximum social welfare on G. Even if there is the difference between $W(x_t)$ and $W(x)$, the difference should be small to keep high efficiency. Hence, the performance of our algorithm in terms of social welfare can be evaluated by comparing $W(x_t)$ with $W(x)$. For the comparison, $W(x)$ can be obtained by constructing bipartite network $G'_{bi} = \left(\mathrm{S} \cup \mathrm{B}, A'_{\mathrm{bi}}\right)$ from G. If s_i can trade with b_j through at least one trader in G, arc set A'_{bi} contains an arc (s_i, b_j). Therefore, A'_{bi} represents the set of possible trades x on G'_{bi}. With the network G'_{bi}, the following integer program gives $W(x)$.

$$\max W(x) = \sum_{(s_i,b_j)\in A'_{bi}} x_{s_ib_j} w_{s_ib_j}.$$
$$0 \leq x_{t_ks_ib_j} \leq 1, w_{s_ib_j} \geq 0, \sum_{b_j\in B} x_{t_ks_ib_j} d_{b_j} \leq c_{s_i}. \tag{3}$$

Then, the comparison between $W(x_t)$ and $W(x)$ can be conducted by examining Efficiency Rate (ER), such that $ER(x_t, x) = \{W(x\_t)/W(x)\} \times 100\,[\%]$.

6 Algorithms to Determine Electricity Trades

This section provides algorithms to determine equilibrium prices and efficient trades on the electricity market model.

6.1 Price Setting Algorithm

First, Algorithm 1 shows a price setting algorithm that has a similar price setting mechanism explained in [11]. In this algorithm, each seller and buyer finds its maximum payoff for trading electricity by considering the valuation of the other agents. The payoff of s_i and b_j for trading one unit of electricity are denoted by $p(s_i)$

and $p(b_j)$ respectively. Then, ask and bid prices offered by each trader are set based on the maximum payoff of each seller and buyer. Ask and bid prices finally determined by Algorithm 1 are equilibrium on the market model. A real number $\mu (0 < \mu \leq 0.5)$ is a parameter used to adjust the payoff of each seller and buyer. The range of μ is set to realize the no-crossing strategy explained in Sect. 3. Therefore, μ adjusts the payoff of each participant and ensures that no participant obtains the payoff exclusively.

Algorithm 1 getPrices($G, \mathbf{v}_s, \mathbf{v}_b, \mu$).

for $j \leftarrow 1$ to M **do**
    if $\mid adj(b_j) \mid = 1$ **then**
        $p(b_j) = 0$.
    else
        Find $\hat{s_i}$ having the min $v_{\hat{s_i}}$.
        $p(b_j) = (v_{b_j} - \hat{v_{s_i}})\mu$.
        while $p(b_j) = (v_{b_j} - v_{\tilde{s_i}})\mu \ (\tilde{s_i} \neq \hat{s_i}, v_{\tilde{s_i}} \geq v_{\hat{s_i}})$ **do**
            Decrease $p(b_j)$.
        end while
    end if
end for
for $i \leftarrow 1$ to N **do**
    if $\mid adj(s_i) \mid = 1$ **then**
        $p(s_i) = 0$.
    else
        Find $\hat{b_j}$ having the max $v_{\hat{b_j}}$.
        $p(s_i) = (\hat{v_{b_j}} - v_{s_i})\mu$.
        while $p(s_i) = (v_{\tilde{b_j}} - v_{s_i})\mu \ (\tilde{b_j} \neq \hat{b_j}, v_{\tilde{b_j}} \leq v_{\hat{b_j}})$ **do**
            Increase $p(s_i)$.
        end while
    end if
end for
for $k \leftarrow 1$ to L **do**
    return $\alpha_{t_k b_j} = v_{b_j} - p(b_j)(b_j \in B(t_k))$.
    return $\beta_{t_k s_i} = v_{s_i} + p(s_i)(s_i \in S(t_k))$.
end for

6.2 *Algorithm for Trade Determination*

Second, the overall process to determine efficient trades is described in Algorithm 2. First, equilibrium prices are calculated by Algorithm 1. Then, every trader discovers trades maximizing payoff of the trader. After that, efficient trades x_t will be determined in all trades that the traders want to conduct. From x_t, social welfare $W(x_t)$ can be calculated. Finally, the algorithm computes the maximum social welfare $W(x)$ to compare it with $W(x_t)$.

Algorithm 2 getSocialWelfare($G, \mathbf{v}_s, \mathbf{v}_b, \mu$).

getPrices($G, \mathbf{v}_s, \mathbf{v}_b, \mu$).
for $t_k \in T$ **do**
  Determine $x_{t_k s_i b_j}(s_i \in S(t_k), b_j \in B(t_k))$ by (1).
end for
Construct G_{bi} by using $x_{t_k s_i b_j}(t_k \in T, s_i \in S, b_j \in B)$.
Obtain $W(x_t)$ and x_t by solving (2) with G_{bi}.
Construct G'_{bi} from G.
Obtain $W(x)$ by solving (3) with G'_{bi}.
return $W(x_t), x_t,$ and $W(x)$

7 Experimental Results

This section shows simulation results of our algorithms. The aim for the simulations is to investigate average and standard deviation of ER. For conducting simulations, a simulation software for our model was developed with Java and lp_solve, which is an integer programming solver.

7.1 Conditions

Table 1 shows parameters used in the simulations. With regard to valuation of sellers, the valuation of IPP and retailers was relatively lower than the valuation of PU. This condition means newly joining participants, such as IPP and retailers, can offer cheaper electricity than PU.

In the simulations, we assumed that supply capacity of newly joining participants depends on the period passing after the beginning of deregulation. Table 2 shows Capacity Patterns (CP), which are the conditions of supply capacity of each seller. CP 1 indicates newly joining participants do not have large supply capacity because not a long period has passed since the beginning of deregulation.

Table 1 Conditions of parameters in the simulations

Parameters	Value
# of agents	$\lvert S\rvert = 5, \lvert T\rvert = 3, \lvert B\rvert = 10$
# of participants	PU: 1 (s_1 and t_1), R: 2 (s_2 and t_2, s_3 and t_3), IPP: 2 (s_4, s_5), C: 10 ($\{b_j \vert j \in \mathbb{N}, 1 \leq j \leq 10\}$)
v_{b_j}	20 for all buyers
v_{s_i}	$v_{s_1} = 10, v_{s_2} = 4, v_{s_3} = 3, v_{s_4} = 9, v_{s_5} = 8$
d_{b_j}	Equal to ID of $b_j(\{d_{b_j} = j \vert j \in \mathbb{N}, 1 \leq j \leq 10\})$
$prob(t_k, b_j)$	0.1, 0.2, 0.3, 0.4, 0.5, 0.6, 0.7, 0.8, 0.9, 1.0

Table 2 Patterns of supply capacity of sellers in the simulations

CP	$c_{s_1}(PU)$	$c_{s_4}(IPP_1)$	$c_{s_5}(IPP_2)$	$c_{s_2}(R_1)$	$c_{s_3}(R_2)$
1	55	13	13	6	6
2	55	27	27	13	13
3	55	55	55	27	27

Furthermore, CP 2 implies the situation in which the difference of supply capacity between market participants became smaller than CP 1. Besides, CP 3 indicates the difference came to be smaller than CP 2. For all CP, supply capacity of PU is equal to the total of all demands to guarantee safe supply. Therefore, all consumers can purchase electricity at least from PU in these CP.

7.2 Results and Discussion

Since the model structure depends on $prob(t_k, b_j)$, 1000 times of simulations for every $prob(t_k, b_j)$ and CP were conducted. Then, the average and standard deviation of ER obtained in the simulation were examined.

Figure 4 shows the average of $ER(x_t, x)$ in the 1000 times of simulations. Our algorithm demonstrated high ER because the average was larger than 90% for any $prob(t_k, b_j)$ and CP. Regardless of CP, the average of ER decreased as $prob(t_k, b_j)$

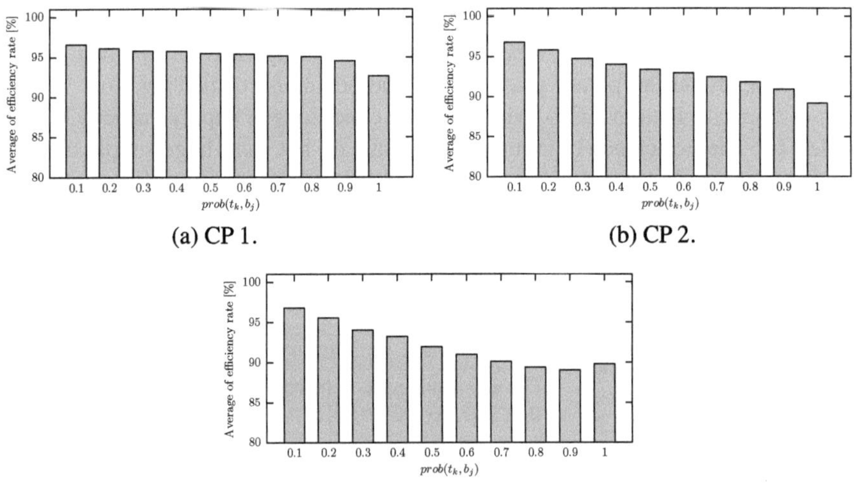

(a) CP 1. (b) CP 2. (c) CP 3.

Fig. 4 Average of $ER(x_t, x)$

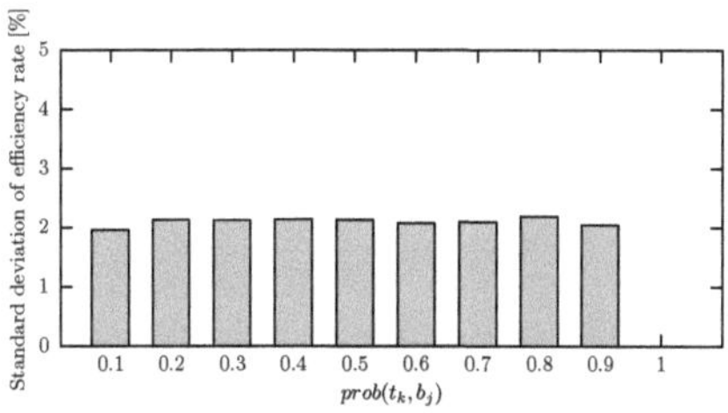

(a) CP 1.

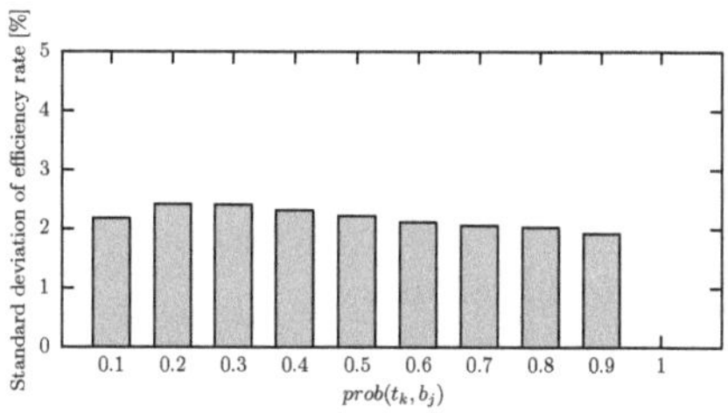

(b) CP 2.

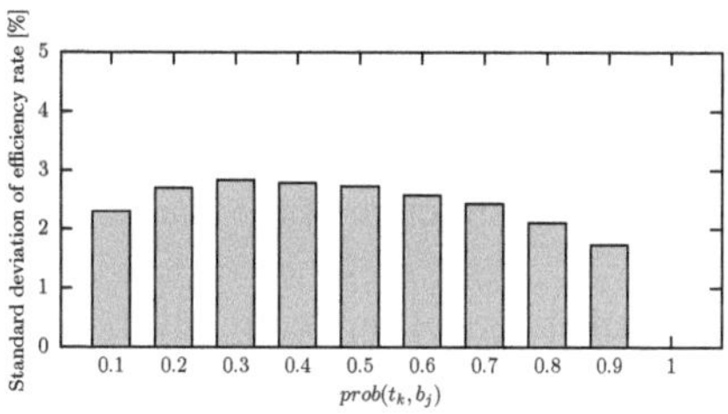

(c) CP 3.

Fig. 5 Standard deviation of $ER(x_t, x)$

increased. This decrease might relate to the number of possible trades on G. G has larger number of possible trades if $prob(t_k, b_j)$ is high, and the possibility to determine trades with the maximum social welfare decreased in that case.

Figure 5 shows the standard deviation of $ER(x_t, x)$. The standard deviation became 0 when $prob(t_k, b_j) = 1$ since G_{bi} was the same as G'_{bi} if $prob(t_k, b_j) = 1$. Regarding $prob(t_k, b_j)$, each figure shows similar characteristics; the shape of the plot in each figure is like a mountain. Regarding CP, peak of the plot in each figure became small when retailers and IPP have smaller capacity than PU (CP 1 or 2). Thus, our algorithms demonstrated better results when supply capacity of participants is similar to CP 1 since the standard deviation of ER is smaller than the other CP.

Our algorithms were successfully able to determine trades with electricity as a multi-unit commodity. This point is better than the algorithm presented in [11] dealing with a single commodity. However, although our algorithms found trades in which $ER(x_t, x) = 100\%$ in some simulations, our algorithms cannot keep $ER(x_t, x) = 100\%$ every simulation. In other words, electricity trades with the maximum social welfare were not discovered in all simulations. These results are worse than the algorithm proposed in [11] because their algorithm gives 100% of $ER(x_t, x)$ at all times. Thus, our algorithms should be modified to acquire higher ER.

In terms of CP, the results indicate that better ER was acquired when the supply capacity of IPP and retailers is relatively lower than PU. The results adversely became worse when the difference of supply capacity of each participant was small. These facts imply that our algorithms achieve better results if deregulation has started not long ago, and IPP and retailers have relatively smaller supply capacity than PU. On the other hand, performance of our algorithm might become worse if long period passes since the deregulation, and supply capacity of participants becomes similar to that of PU.

8 Conclusion

This chapter proposed methods to determine efficient trades in a deregulated electricity market models. By conducting simulation experiments, the efficiency of electricity trades determined by our algorithms was examined by social welfare, which is the total of payoffs of all participants. As a result, our algorithm discovered efficient electricity trades on the market model. The efficiency of determined electricity trades depended on the structure of the market model and supply capacity of market participants.

Since our electricity market model has currently been represented as a static model, our model cannot observe any characteristics about the dynamic behavior of market participants. To construct more realistic electricity market models, integrating dynamic interactions between market participants into the model is left as our future works.

References

1. Sioshansi FP (2008) Competitive electricity markets: design, implementation. Elsevier Science, Performance
2. Gunter T, McRae I, Leerdam GV (2014) Making smart grid a reality. The North Highland Company
3. Mazer A (2007) Electric power planning for regulated and deregulated markets. Wiley-IEEE Press, pp 180–191
4. Kwon RH, Frances D (2012) Handbook of networks in power systems I. Springer, Berlin, pp 41–60
5. Swami R (2012) Social welfare maximization in deregulated power system. Am Math Monthly 1(4):4–8
6. Dong Q, Yu L, Song W-Z, Tong L, Tang S (2012) Distributed demand and response algorithm for optimizing social-welfare in smart grid. In: IEEE 26th international parallel & distributed processing symposium (IPDPS), pp 1228–1239
7. Yamamoto H, Tsumura K (2012) Control of smart grids based on price mechanism and network structure. Mathematical engineering technical reports, pp 1–15
8. Hase R, Shinomiya N (2014) A matching problem in electricity markets using network flows. In: Ninth international conference on systems, pp 79–82, 2014
9. Kranton RE, Minehart DF (2001) A theory of buyer-seller networks. Am Econ Rev 91 (3):485–508
10. Babic J, Podobnik V (2014) Energy informatics in smart grids: agent-based modelling of electricity markets. In: Erasmus Energy Forum 2014 Science Day, pp 1–15
11. Blume LE et al (2009) Trading networks with price-setting agents. Games Econ Behav 67 (1):36–50
12. Nava F (2015) Efficiency in decentralized oligopolistic markets. J Econ Theory 157:315–348

13. Kleinberg JM (1996) Single source unsplittable flow. In: 37th annual symposium on foundations of computer science, pp 68–77
14. Shmoys DB, Tardos E (1993) An approximation algorithm for the generalized assignment problem. Math Program Ser A and B 62:461–474
15. Sadrieh A (1998) The alternating double auction market: a game theoretic and experimental investigation. Springer, Berlin
16. Jackson MO (2008) Social and economic networks. Princeton University Press, Princeton, p 206

Multi-objective Optimization Modeling for the Impacts of 2.4-GHz ISM Band Interference on IEEE 802.15.4 Health Sensors

Mohammad Hamdan, Muneer Bani-Yaseen and Hisham A. Shehadeh

1 Introduction

Wireless sensor network (WSN) has become an important factor in real life like monitoring of wildlife, fire detection, security monitoring, and importantly in health care monitoring [1, 2]. In last decade health monitoring system at home is common. Here we monitor the health status issues from home without the need to go to hospital or health care centers. This is very important in countries with large population. It can also reduce the incurred costs in hospitals and health centers [3]. In this era, many types of short range wireless technologies have been created and developed to achieve getting a flexible and portable connectivity. Actually, wireless technologies are gaining a decent position especially at home and in general in building automations [4, 5]. Guo et al. [6] have provide a laconic predicts the possibility of using IEEE 802.15.4 in networking, building automation, sensing and controlling new construction. ZigBee technology [7] along with IEEE 802.15.4 standard [8] opens the way for applying the wireless communication for field devices like zoon or room controllers, electrical meters, associated room temperature, variable frequency drives, and point modules at prominent industrial facilities and commercial. These devices are shifted from research stage to commercial deployment by home and building automation companies. Healthcare system depends mainly on a set of sensors that represent the backbone of healthcare system like bed sensor, pillow sensor, heart pulps sensor, and Electro-cardio-graph

M. Hamdan (✉)
Computer Science, Yarmouk University, Irbid, Jordan
e-mail: hamdan@yu.edu.jo; m.hamdan@hw.ac.uk

M. Hamdan
Computer Science, Heriot-Watt University, Dubai, UAE

M. Bani-Yaseen · H.A. Shehadeh
JUST, Irbid, Jordan

L. Ismail and L. Zhang (eds.), *Information Innovation Technology in Smart Cities*,
https://doi.org/10.1007/978-981-10-1741-4_21

(ECG) sensor [3]. They are composed of: (1) radio unit for communication, (2) Sensing unit for capturing data, (3) Processing unit for processing received data, (4) Energy source like battery [9].

This article is an extension of the previous study on the energy efficiency and throughput as optimization problems [10]. Both issues are important in such systems. The packet error rate (PER) affects both issues. PER refers to incorrectly received number of packets divided by all received packets. Interference, attenuation and noise by other appliances increases PER. This book chapter is organized as follows. Section 2 presents related work. Background on IEEE 802.15.4 (ZigBee) healthcare is given in Sect. 3. Section 4 describes potential interference sources in 2.4 GHz band. The objective models in Sect. 5. Section 6 presents experimentation and Result. Finally conclusion is in Sect. 7.

2 Related Works

Hamdan et al. [11] have proposed a multi objective optimization model for electrocardiogram health network in home. They have tried to minimize average end to end delay and to maximize energy efficiency and as both objectives depend on packet payload size by using a set of genetic algorithms like SPEA-II, NSGA-II and OMOPSO. Berre et al. [12] have tried to optimize a set of objective models for wireless sensor network like network life time model, financial cost model, and coverage model. They tested these models by using a set of algorithms such as MOACO, SPEA-II, and NSGA-II. Abidin et al. [13] have proposed a set of objective models for a wireless sensor network such as energy consumption and coverage ratio comparison. Also, they have used them to compare between multi objective algorithm Multi-objective TPSMA (MOTPSMA) and single objective algorithm Territorial Predator Scent Marking Algorithm (TPSMA).

3 Background on IEEE 802.15.4 (ZigBee) in Health Care

IEEE 802.15.4 standard supports MAC packet size up to 127 bytes in which the payloads size is about 114 bytes. Figure 1 presents the IEEE 802.15.4 data frame. Figure 2 shows the media access control (MAC) and PHY layer for IEEE 802.15.4 (ZigBee). The PHY layer provides three distinct frequency bands as mentioned in Table 1 [8].

In this work we take attention on ZigBee PHY layer that operates on 2.4 GHz band because it has a higher bandwidth rather than other bands and important one it available in worldwide. Furthermore, this band is probably vulnerable to interference and intrusion by other devices like Microwave oven and Wi-Fi.

Types of healthcare sensor for measuring patient's vital signs at home include [14]: (1) Force Sensitive Resistors (FSR) sensors like bed and pillow sensors

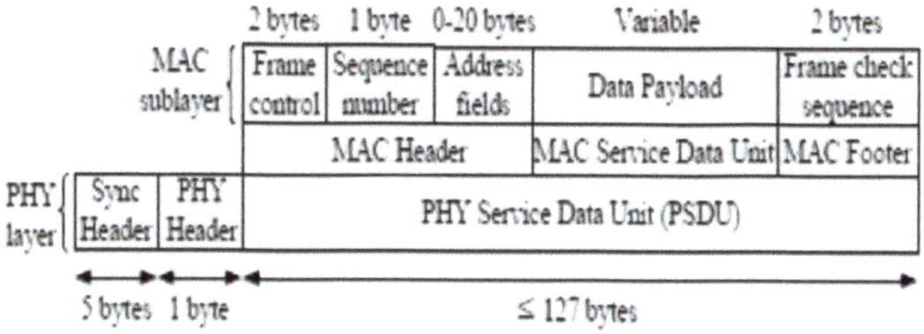

Fig. 1 The IEEE 802.15.4 data frame

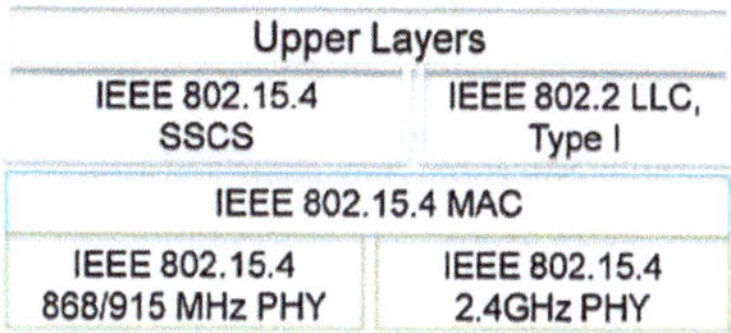

Fig. 2 IEEE 802.15.4 architecture

Table 1 The IEEE 802.15.4 standard frequency bands

Frequency bands	Area	Frequency range	Data rate	Channel number(s)
2.4 GHz	Asia, Worldwide	2405–2480	250	16 channels
868 MHz	Europe	868.3	20	1 channels
915 MHz	Australia, America	902–928	40	10 channels

(2) Pulse Rate Sensor, (3) Temperature Sensor, (4) ECG Sensor. But because of increasing the demand on comfort-ability and portability, the wearable sensor is created. This sensor can be used to measure human vital sign like (oxygen saturation, skin and body temperature, heart rate, respiration rate, blood pressure and electrocardiogram) when the patient wears it [15]. Figure 3 represents IEEE 804.15.4 health sensors [16].

4 Potential Interference Sources in 2.4 GHz Band

Home appliances that work on the same band (2.4 GHZ ISM band) can interfere with the operation of health (IEEE 802.15.4) sensors since they all operate on the same band. Next we outline few of them [17].

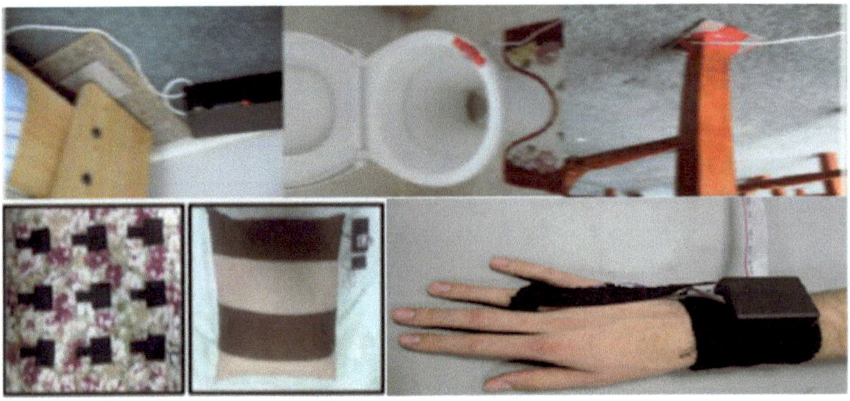

Fig. 3 Pillow, hand wearable, bed, bathroom and chair sensors

4.1 IEEE 802.11g

To access Internet from home, the majority of user use Wi-Fi. This is based on the IEEE 802.11 standard with Wi-Fi protocol. When IEEE 802.11g standard operates on 2.4-GHz frequency range, it provides high bandwidth (30 Mb/s for practice and 54-Mb/s for throughput). The IEEE 802.11 g standard can work on 11 channels available in the ISM 2.4-GHz band. The default Channels are numbers 1, 6 and 11 (which is the most widely used). Figure 4 presents IEEE 802.11 channel distribution [17].

The Wi-Fi interference model (Packet Error Rate) on ZigBee communication can be given as [17]:

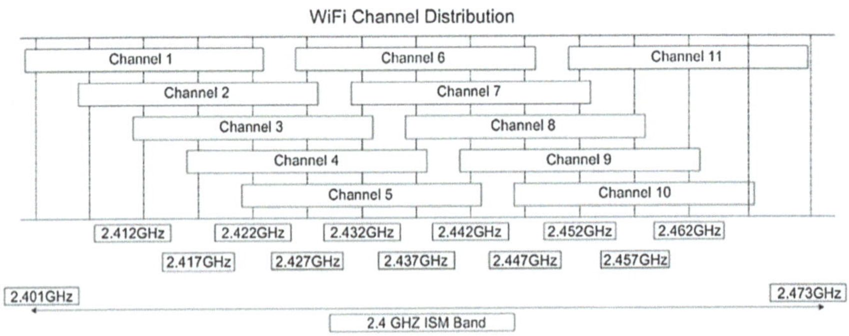

Fig. 4 IEEE 802.11 channel distribution

$$PER(x,y) = \left\{ \begin{array}{ll} \left|\ln\left(\frac{y}{2}\right) \times 10^{-2}\right| + D_N, & y = x \\ x \cdot \ln\left(\frac{x}{y}\right) \times 10^{-2}\Big| + D_N, & Others \end{array} \right\} \quad (1)$$

where x: the distance between transceiver (Tx) and receiver (Rx), and y: the distance from interference source (ISs) to receiver (Rx). DN is an added quantity that can be calculated by:

$$D_N = 1/\sqrt{2\pi} \cdot \int_{-\infty}^{x} \exp(-(x^2/2))dx \quad (2)$$

4.2 Microwave Ovens

Another interference comes from Microwave ovens. Microwave ovens produce electromagnetic radiation in which it's frequencies about several hundred gigahertzes down to several hundred megahertz. The wavelengths are approximately between 1 and 20 cm. Microwaves operate on higher frequencies and this allow it to carry more information rather than radio waves. Microwave oven operates on 2.45-GHz frequency range. Yet another potential source of noise and interference for IEEE 802.15.4 (ZigBee) communication. The microwave interference model (Packet Error Rate) on ZigBee communication can be expressed as [17]:

$$PER(x,y) = \left\{ \begin{array}{ll} \left|\alpha \cdot \ln\left(\frac{x}{2}\right)\right| + \beta \cdot D_N, & x = y \\ \gamma \cdot \ln\left(\frac{x}{y}\right)\Big| + \gamma \cdot D_N, & y \leq 2m \\ \varepsilon \cdot \ln\left(\frac{x}{y}\right)\Big| + \beta \cdot D_N, & Others \end{array} \right\} \quad (3)$$

where x: the distance between transceiver (Tx) and receiver (Rx), y: the distance from interference source (ISs) to receiver (Rx), $\alpha = 0.01$, $\beta = 0.02$, $\gamma = 0.03$, $\varepsilon = 0.05$, and DN is an added quantity that can be calculated by Eq. 2.

5 Multi-objective Modeling

We are interested in introducing a multi-objective model to address the problem of interference in wireless sensors. The important factor is Packet Error Rate (PER) and we found two models that rely on it.

5.1 Energy Efficiency Model

Energy and thus battery is an important issue in sensors. Energy consumption is affected by factors like packet error rate and packet payload length [10, 18]. We can define the Energy Efficiency as follows:

$$\eta = \frac{Ec \cdot \ell}{Ec \cdot (\ell + h) + Es} \cdot (1 - PER) \tag{4}$$

where:

Ec	energy consumption in communication.
E*s*	energy consumption in startu $\surd$.
ℓ	Packet payload length.
h	Packet header length.
PER	packet Error Rate

5.2 Network Throughput

Network throughput is attributed to the rate of successful received packet that is delivered over a medium. So, if this factor increases then the network efficiency will be increased. This factor is affected by two important factors like packet payload length and packet error rate. We can define the network output as follows [19]:

$$u_{tput} = \frac{\ell \cdot (1 - PER)}{T_{flow}} \tag{5}$$

where:

ℓ	Packet payload length.
PER	packet Error Rate.
Tflow	is the end-to-end latency which is a time spent between packets that is generated at a transmitter and received at the sink node through the multi-hop route.

6 Experimental Environment and Results

We assumed patients' home area about 20 m × 10 m. Wearable sensors are assumed to be use in our study to transmit patients' health status to sink node. These simulation parameters are mentioned in Table 2. We take in consideration

Table 2 Simulation parameters

#	Parameter	Values
1	Ec	30.5 mA
2	Es	15.5 mA
3	Payload size	50 bytes
4	Tflow	50 s
5	Packet header length	16 bytes
6	$D_{Tx\text{-}Rx}$	From 1 to 10 m
7	$D_{Is\text{-}Rx}$	From 1 to 10 m
8	$D_{Hop\text{-}Rx}$	10 m
9	$D_{Tx\text{-}hop}$	From 1 to 10 m

that a real time detection of Electrocardiogram (ECG) needs transmission rate about 10 packets/s with packet size about 50 byte [20].

NetBeans IDE 8.0.2. was used to compile the tool jMetal 4.5., while the test environment was Dual core CPU-T3200, 3 GB RAM and Windows 7 32-bit. We used three genetic algorithms in our tests (NSGA-II, SPEA-II and OMOPSO). The parameters of NSGA-II and SPEA-II algorithms were as follows: population size is 20, the number of generations is 250, the crossover probability is 0.9, and the mutation probability is to (1/s), where s = number of variables. Furthermore, the parameter of OMOPSO algorithm was as follows: swarm size was set to 20, the number of generations was about 250, and the probability was about (1/s).

6.1 The ISs Is Mobilized

In this test we make the interference source mobilized. However, the distance fixed between transmitter and receiver and it's about 10 m. The network is shown in Fig. 5. The tests are organized as following:

6.1.1 Wi-Fi Is the Interference Source

Figure 6 presents the comparison between NSGA-II, SPEAII and OMOPSO algorithms in terms of Network Throughput (in kbps) and Energy Efficiency and when the Wi-Fi is an interference source. The figure presents the minimum, maximum and average for each objective.

6.1.2 Microwave Oven Is the Interference Source

Figure 7 shows the comparison between NSGA-II, SPEAII and OMOPSO algorithms in terms of Network Throughput (in kbps) and Energy Efficiency and when the microwave oven is an interference source.

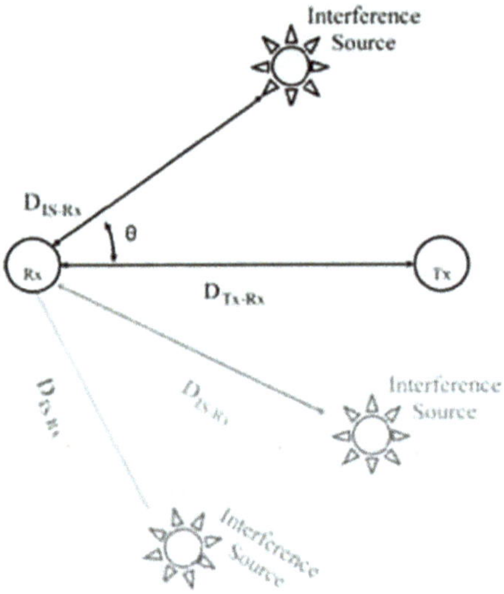

Fig. 5 ISs are mobilized

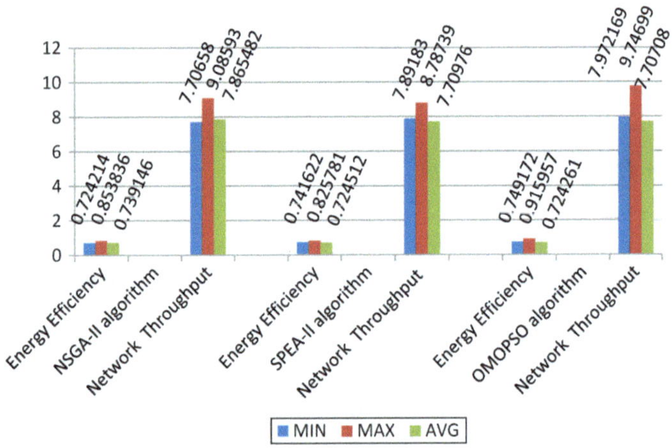

Fig. 6 Result of algorithms when Wi-Fi is mobilized

6.2 *The Transmitter Is Mobilized*

In this test we make the transmitter source mobilized. However, the interference source is fixed and the distance between it and receiver about 10 m. The network is shown in Fig. 8. The tests are organized as following:

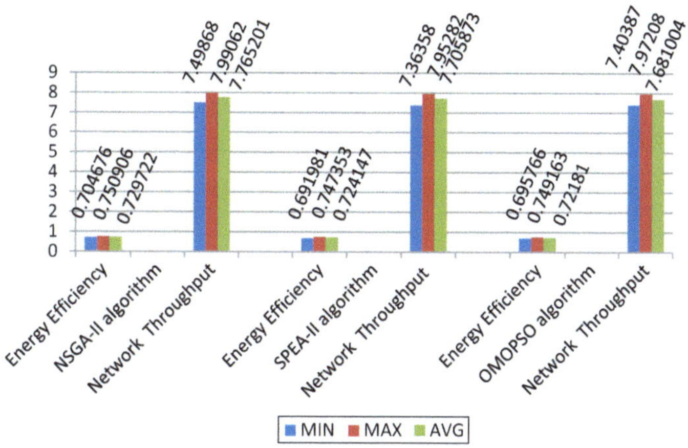

Fig. 7 Result of algorithms when microwave-oven is mobilized

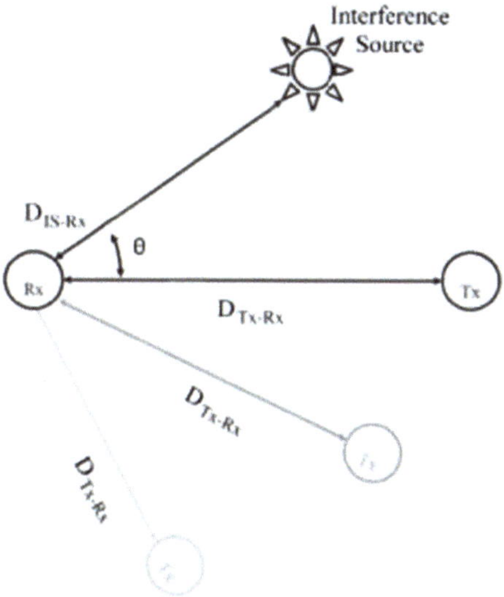

Fig. 8 The transmitter is mobilized

6.2.1 Wi-Fi Is the Interference Source

Figure 9 shows the comparison between NSGA-II, SPEAII and OMOPSO algorithms in terms of Network Throughput (in kbps) and Energy Efficiency and when the Wi-Fi is an interference source. The Figure presents the minimum, maximum and average for each objective.

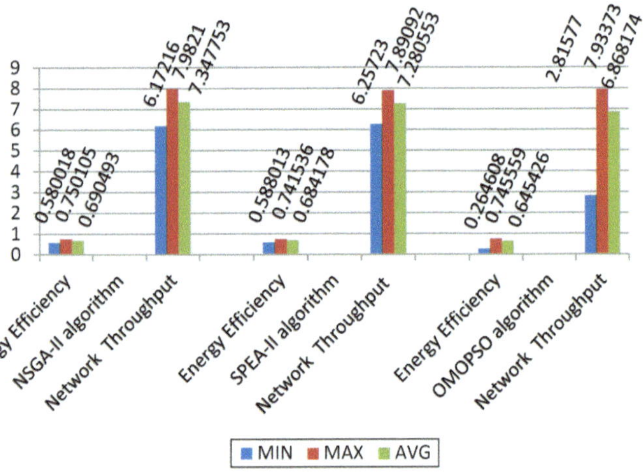

Fig. 9 Result of algorithms when Wi-Fi is an interference resource

6.2.2 Microwave Oven Is the Interference Source

Figure 10 shows the comparison between NSGA-II, SPEAII and OMOPSO algorithms in terms of Network Throughput (in kbps) and Energy Efficiency and when the microwave oven is an interference source. The figure presents the minimum, maximum and average for each objective.

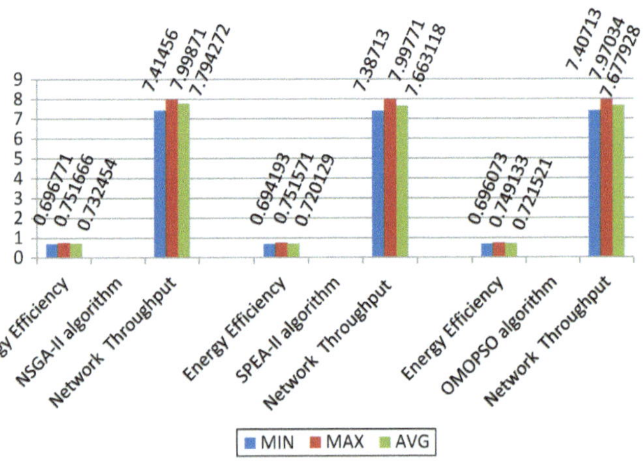

Fig. 10 Result of algorithms when microwave oven is an interference resource

6.3 The Transmitter Is Mobilized with Multi Hop Network

In this test we have used a multi hop network in which is used to cover a wide area like big houses and hospitals. In this test we make the distance fixed between interference source and receiver and it's about 10 m. However, the transmitter source is mobilized. The distance between receiver and next hop (repeater) about 10 m while the distance between the transmitter and the next hop is varied between 1 up to 10 m. The network is depicted in Fig. 11.

6.3.1 Wi-Fi Is the Interference Source

Figure 12 presents the comparison between NSGA-II, SPEAII and OMOPSO algorithms in terms of Network Throughput (in kbps) and Energy Efficiency and when the Wi-Fi is an interference source. The figure presents the average, maximum, and minimum for each objective.

6.3.2 Microwave Oven Is the Interference Source

Figure 13 shows the comparison between NSGA-II, SPEAII and OMOPSO algorithms in terms of Network Throughput (in kbps) and Energy Efficiency and when the microwave oven is an interference source. The figure presents the average, maximum, and minimum for each objective.

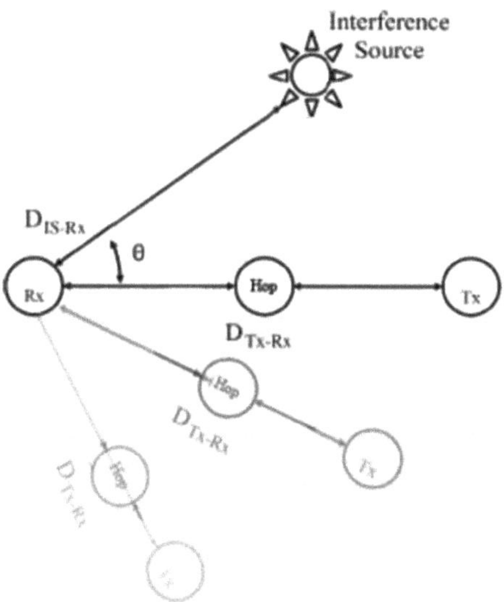

Fig. 11 The transmitter is mobilized with multi hop network

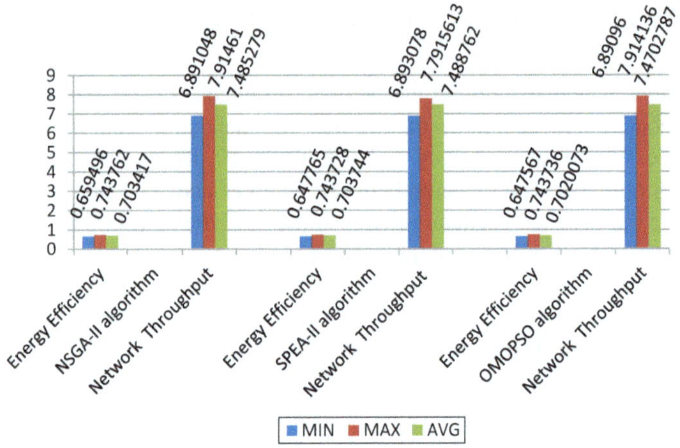

Fig. 12 Result of algorithms when Wi-Fi is an interference resource

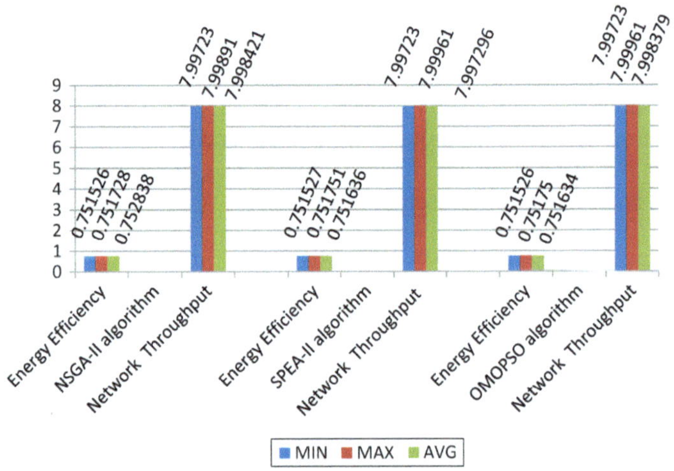

Fig. 13 Result of algorithms when microwave-oven is an interference resource

7 Conclusion

When Health sensors operate on the free 2.4-GHz ISM band then they are vulnerable to interference from other devices such as Wi-Fi and microwave oven. In this work we have modeled this issue s a multi-objective problem. The first objective is maximization of energy efficiency. The second objective is maximization of the Packet Throughput of the network. Both objectives are non conflicting as was seen by the results. In other words, both objectives are maximized when PER variables increases.

In future work, it will be interesting to have a live environment for testing but this requires resources and budget. Also, it would be interesting if we can extend the 2D model to a 3D model and find the interdependencies among variables.

References

1. Potdar V, Sharif A, Chang E (2009) Wireless sensor networks: a survey. In: International conference on advanced information networking and applications workshops, WAINA'09, IEEE, pp 636–641
2. Ilyas M (2013) Emerging applications of sensor networks. In: International conference on 2nd symposium on wireless sensors and cellular networks, pp 13–17
3. Zakrzewski M, Junnila S, Vehkaoja A, Kailanto H, Vainio A-M, Defee I, Lekkala J, Vanhala J, Hyttinen J (2009) Utilization of wireless sensor network for health monitoring in home environment. In: IEEE international symposium on industrial embedded systems, SIES'09, IEEE, pp 132–135
4. Kintner-Meyer M, Brambley MR (2002) Pros & cons of wireless. ASHRAE J 44(11):54
5. Ruiz J (2007) Going wireless. ASHRAE J 49(6):34–41
6. Guo W, Healy WM, Zhou M (2010) ZigBee-wireless mesh networks for building automation and control. In: 2010 international conference on networking, sensing and control (ICNSC), IEEE, pp 731–736
7. ZigBee Alliance (2010) ZigBee. http://www.zigbee.org/. Accessed 22 Feb 2016
8. IEEE 802.15.4 Working Group (2010) IEEE 802.15.4. http://www.ieee802.org/15
9. Anastasi G, Conti M, Di Francesco M, Passarella A (2009) Energy conservation in wireless sensor networks: a survey. Ad Hoc Netw 7(3):537–568
10. Hamdan M, Yassein MB, Shehadeh HA (2015) Multi-objective optimization modeling of interference in home health care sensors. In: 2015 11th international conference on innovations in information technology (IIT), IEEE, pp 219–224
11. Hamdan M, Shehadeh HA, Obeidat QY (2015) Multi-objective optimization of electrocardiogram monitoring network for elderly patient in home. Int J Open Probl Comput Math 8 (1):82–95
12. Berre ML, Hnaien F, Snoussi H (2011) Multi-objective optimization in wireless sensors networks. In: 2011 international conference on microelectronics (ICM), IEEE, pp 1–4
13. Abidin HZ, Din NM, Jalil YE (2013) Multi-objective optimization (MOO) approach for sensor node placement in WSN. In: 2013 7th international conference on signal processing and communication systems (ICSPCS), IEEE, pp 1–5
14. Yassein MB, Hamdan M, Shehadeh H (2015) Performance evaluation of health monitoring network for elderly patient in home. Asian J Math Comput Res 9(2):108–118
15. Pantelopoulos A, Bourbakis NG (2010) A survey on wearable sensor-based systems for health monitoring and prognosis. IEEE Trans Syst Man Cybern Part C Appl Rev 40(1):1–12
16. Malhi K, Mukhopadhyay SC, Schnepper J, Haefke M, Ewald H (2012) A zigbee-based wearable physiological parameters monitoring system. IEEE Sens J 12(3):423–430
17. Guo W, Healy WM, Zhou M (2012) Impacts of 2.4-GHz ISM band interference on IEEE 802.15. 4 wireless sensor network reliability in buildings. IEEE Trans Instrum Meas 61 (9):2533–2544
18. Joe I (2006) Energy efficiency maximization for wireless sensor networks. In: Mobile and wireless communication networks. Springer, pp 115–122

19. Vuran M, Akyildiz IF (2008) Cross-layer packet size optimization for wireless terrestrial, underwater, and underground sensor networks. In: INFOCOM 2008. The 27th conference on computer communications, IEEE, pp 780–788
20. Liang X, Balasingham I (2007) Performance analysis of the IEEE 802.15. 4 based ECG monitoring network. In: Proceedings of the 7th IASTED international conferences on wireless and optical communications, pp 99–104

Raspberry Pi Based Lightweight and Energy-Efficient Private Email Infrastructure for SMEs

Sufian Hameed and Muhammad Arsal Asif

1 Introduction

Electronic mail or Email is no doubt one of the biggest services being used over the Internet today and its importance in modern business communication is undeniable. Email has been around for more than 44 years and its transmission is based on Simple Mail Transfer Protocol (SMTP) [1]. With the wide-spread global usage of emails, the concept of email privacy is also under rapid threat. Email privacy is a very complicated issue that has to deal with unauthorized access and inspection of electronic mail. This unauthorized access can happen while an email is stored on untrusted email servers of the service providers (like Gmail, Yahoo, Hotmail, etc.) or due to broader government surveillance programs such as PRISM [2]. Third-party mail service is easier to use, but they require sacrifice of control and flexibility. Running a private mail server can be an ideal solution to protect email privacy against unauthorized storage access. Private mail server allows full control over both the server and the emails, complete access to mail server's logs, and access to raw email files in a user's mail directory. Nevertheless, this control and flexibility comes with an added cost [3].

A Smart City is a development vision that integrate multiple ICT solutions to enhance quality and performance of a working environment by reducing cost and resource consumption. Based on this vision, this article propose PiMail, an affordable, lightweight and energy-efficient private email system based on Raspberry Pi [4]. Raspberry Pi is low-cost, low-power and highly portable single board computer. It is one of the smallest, credit-card sized, single board computer available in the market that has the highest performance to cost ratio. Raspberry Pi makes it possible to create an affordable, energy-efficient and portable miniature

S. Hameed (✉) · M.A. Asif
National University of Computer and Emerging Sciences (NUCES),
Islamabad, Pakistan
e-mail: sufian.hameed@nu.edu.pk

L. Ismail and L. Zhang (eds.), *Information Innovation Technology in Smart Cities*,
https://doi.org/10.1007/978-981-10-1741-4_22

private mail server according to the need of individual users or small enterprise. In short, PiMail fulfils the following promises.

- A low cost infrastructure that would cost a onetime investment of around $35–$40 to purchase Raspberry Pi.
- Personalized email address like MrX@mydomain.com with an annual recurring cost of domain registration with a registrar like namecheap.com.
- Low electricity consumption with an email server that can run 24/7/365 for under $5 of electricity per year.
- The ability to connect from anywhere, and read and send email, using a secure IMAP connection on your phone, tablet or computer.
- Complete control over your personal communication. Emails are stored over PiMail server, and nobody scan them to sell adverts.
- Smart spam filtering with SpamAssassin [5].
- Efficient virus scanning with ClamAV [6].

We developed a test-bed implementation of PiMail using Raspbian OS, Postfix [7] mail transfer agent (MTA), ClamAV antivirus and SpamAssassin. We used different experiments to evaluate email processing latency, throughput and CPU/memory utilization of PiMail. The small size, low power consumption and performance benchmarks make Raspberry Pi an ideal candidate for personal email server in home and small organizations. An early version of this work appeared as a conference paper [8]. In this article, we provide new substantial results after porting PiMail on different Raspberry Pi models. The rest of the article is organized as follows. Section 2 summarize the basic properties of Raspberry Pi computer. Section 3 describes the software stack architecture of PiMail. In Sect. 4, we discuss the test-bed and demonstrate the performance of the proposed system under different experiments. Finally we discuss and conclude in Sects. 5 and 6.

2 The Raspberry Pi Computer

In early 2006, Eben Upton, a British engineer, brought together a group of teachers, academics and IT professional to address the lack of programmable hardware to teach computer science at school level. This lead to the official formation of Raspberry Pi foundation, a charitable organization, in 2009 [9]. In 2011, the Raspberry Pi foundation developed a credit card sized, single-board computer called Raspberry Pi. The first version of Pi i.e. Raspberry Pi 1 (model A) officially went on sale from 29th February 2012. The foundation has since then released couple of updates to Pi 1 and in February 2015 the Raspberry Pi 2 (model B) was introduced as the second generation of Raspberry Pi.

2.1 System Specification for Raspberry Pi

Raspberry Pi is based on an integrated circuit (system on chip) combining an ARM processor (CPU), a Broadcom VideoCore graphics processor (GPU), and RAM on a single chip [10]. Moreover, there is a microSD card slot for storage and I/O units such as USB, Ethernet, audio, RCA video, and HDMI. Power is provided via a 5 V micro-USB connector. Table 1 summarizes the hardware specification of Raspberry Pi 1 and Raspberry Pi 2 (see Fig. 1).

2.2 System Software

In terms of system software, the Raspberry Foundation has done a tiring effort in optimizing software to get the best out of Raspberry Pi. The recommended OS for Raspberry Pi is Raspbian [11], which is a port of the well-known Linux distribution, Debian. Raspbian is optimized for the ARMv6 and ARMv7 instruction set

Table 1 Hardware specification of Raspberry Pi 1 and 2 (model B)

	Raspberry Pi 1	Raspberry Pi 2
System of chip	Broadcom BCM2835	Broadcom BCM2836
CPU	700 MHz single core ARM v6	900 MHz quad-core ARM Cortex-A7
GPU	Broadcom VideoCore IV	Broadcom VideoCore IV
Memory	512 MB SDRAM	1 GB SDRAM (shared with GPU)
Ethernet	Onboard 10/100 Ethernet RJ45 jack	Onboard 10/100 Ethernet RJ45 jack
USB	2 USB 2.0 ports	4 USB 2.0 ports
Video output	HDMI (rev 1.3 & 1.4)	HDMI (rev 1.3 & 1.4)
Audio output	3.5 mm jack, HDMI	3.5 mm jack, HDMI
Storage	MicroSD slot	MicroSD slot

Fig. 1 Raspberry Pi 2 (Model B) board [4]

with hardware floating point support. Raspbian is optimized around 35,000 pre-built packages, for easy installation on Raspberry Pi. This includes WebKit, LibreOffice, Scratch, Pixman, XBMC/Kodi, libav and PyPy. With the introduction of ARMv7 core, Raspberry Pi 2 model can also run Ubuntu and Pi 2 compatible version of Windows 10 [12].

3 System Design

We have ported PiMail on two different models of Raspberry Pi i.e. Raspberry Pi 1 and Raspberry Pi 2. The PiMail software stack for an individual Raspberry Pi 2 is shown in Fig. 2. PiMail runs Raspbian [11] (a distribution of Debian optimized for the Raspberry Pi hardware) from a 16 GB microSD card storage. The PiMail server is placed on top of the Raspbian operating system. We have used Postfix, SpamAssassin, ClamAV and Davecot tools (briefly discussed below) to setup the mail server.

3.1 Postfix

Postfix [7] is a fast, easy to administer and secure Mail Transfer Agent (MTA) that can support LDAP, SMTP and AUTH (SASL). It was developed by Wietse Venema in 1997 as an alternative to SendMail.

Postfix performs a number of steps to deliver an email to any particular Inbox. Postfix receives an email via the Simple Mail Transfer Protocol Daemon (SMTPD) server. It then strips away the SMTP encapsulation, perform some sanity checks and alerts the cleanup server with the details on sender, recipients and content. The cleanup server inserts the details into the inbound mail queue and informs the queue

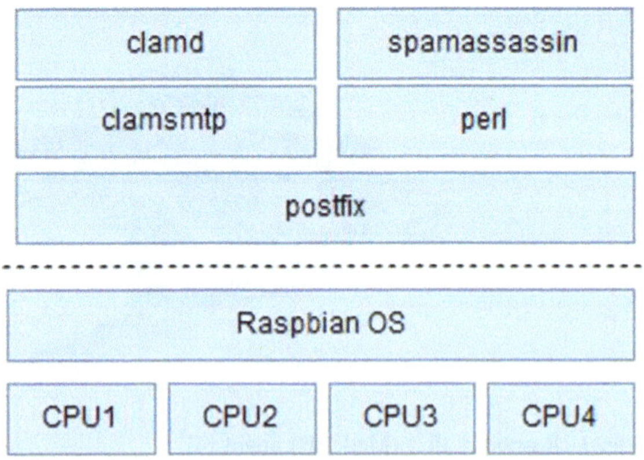

Fig. 2 PiMail software stack

manager that a new mail has arrived. This mail queue acts as a temporary buffer to the mail directory. As soon as the mail is inserted into the inbound queue a unique queue id is assigned to it. During our experiments we used this unique queue id to keep track of the mail.

3.2 SpamAssassin

Usman Sheikh SpamAssassin [5], which is one of the most widely used content-based filter, uses a variety of defense mechanisms to filter spam before it reaches the mailbox. The defense mechanisms mainly includes header tests, body phase tests, Bayesian filtering, automatic address whitelist/blacklist, automatic sender reputation system, manual address whitelist/blacklist, collaborative spam identification database, DNS blacklist and character sets. Due to these thorough tests it is difficult for the spammers to identify simple work around while creating spam. During the evaluations, we used SpamAssassin with its default settings to identify spam emails and separate them for legitimate ones. After completion of any single test, SpamAssassin assign cumulative score to the mail and check against the user defined threshold. If the calculated score is higher than the user defined threshold, the mail is marked as spam and sent to the spam box.

3.3 ClamAV

Clam AntiVirus [6] is an open source antivirus toolkit designed for the scanning of emails at the mail gateways. ClamAV analyze the mail from the inbound queue using shared libraries of the anti-virus engine. If the scan results in a positive ID the file/mail is moved to a quarantine folder, else the mail is queued back into the mailing inbound folder.

3.4 DoveCot

Dovecot [13] is a secure IMAP server that provide IMAP functionality to fetch the mail from the mail directory. It is also used to provide simple authentication and security layer (SASL) to validate the identity of a user before he can send or receive an email.

4 Test-Bed and Evaluations

For evaluating the system performance of email processing with PiMail, we augmented Raspberry Pi 1 and Pi 2 with Postfix MTA, SpamAssassin content filter, ClamAV antivirus and Dovecot IMAP server, and deployed it over the LAN. For all the experiments, we used a desktop machine connected via LAN to send mails with different size and frequencies to PiMail server. We conducted experiments in four different scenarios as follows.

- S1: In scenario 1 (S1), the SMTP server runs postfix without any spam filter.
- S2: In scenario 2 (S2), ClamAV is used as an anti-virus with Postfix.
- S3: In scenario 3 (S3), SpamAssassin is used as a content-based filter with Postfix.
- S4: In scenario 4 (S4), SpamAssassin is used as a content-based filter and ClamAV is used as an anti-virus with Postfix.

We run different experiments using these four scenarios on Raspberry Pi 1 and Pi 2 to study the impact of message size, processing delays, end-to-end throughput, CPU and memory utilizations (Fig. 3).

4.1 Processing Delays

We used two different modes to measure the email processing latency for the four scenarios listed above. First we bombard (as rapidly as possible) the PiMail server with 50 messages to saturate the mail server. After that we added a delay on 1 s between any two messages. For all the experiments we used fix message size of 8 KB, under the assumption of being the average size of email message [14].

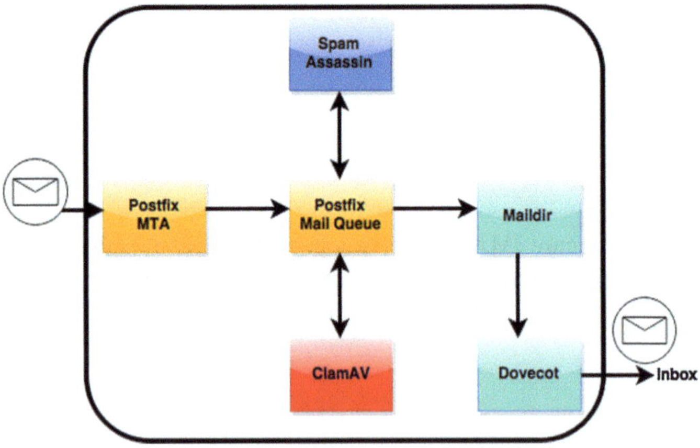

Fig. 3 PiMail testbed

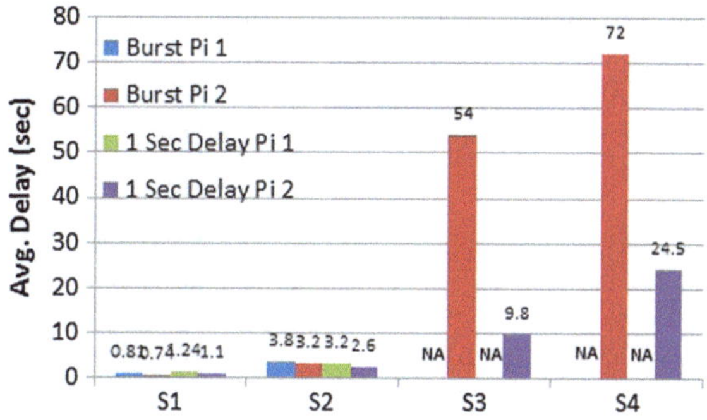

Fig. 4 Average processing delay

Figure 4 shows the average email processing delay of all the four scenarios. In the burst mode S1 was able to process an email with an average delay of 0.81 s with Pi 1 and 0.74 s with Pi 2. This increased to 3.8 and 3.2 s for Pi 1 and Pi 2 respectively in S2.

Content filtering with SpamAssassin is involving task on PiMail server which overwhelms Pi 1 device. Processing of emails in S3 and S4 takes somewhere between 30 and 60 min in burst mode on Pi 1. In Fig. 4 and all the subsequent evaluations involving SpamAssassin on Pi 1, we have labelled NA (not applicable). This however is not the case with quad core Pi 2, where S3 and S4 processing takes on average 54 and 72 s respectively. After introducing a delay of 1 s between two consecutive messages the average email processing delay was observed to be 1.1, 2.6, 9.8 and 24.5 s in S1, S2, S3 and S4 respectively on Pi 2. SpamAssassin processing with the burst of incoming messages creates a bottle neck. Delay of 1 s smooth out the delays by 66% in worst case scenario (S4).

4.2 Throughput

In order to measure the end-to-end throughput of the PiMail server, we used the same settings (sending of 50 email messages of 8 KB each in burst and with 1 s delay) discussed above. Figure 5 shows the throughput of all the four scenarios. Even with the bombardment of messages S1 was able to receive approximately 74 and 81 messages per minute in burst mode on Pi 1 and Pi 2 respectively. This reduced to 15.78 and 18.75 messages per minute for S2 on Pi 1 and Pi 2 respectively. After introducing a delay of 1 s between two consecutive messages the average end-to-end throughput came out to be 54.55, 23.08, 6.12 and 2.45 messages per minute in S1, S2, S3 and S4 respectively on Pi 2. PiMail can effectively handle simple email processing in burst mode. Virus scanning and Spam filtering introduce processing delays which directly effects the overall system throughput. PiMail cannot handle high frequency of email involving SpamAssassin content filtering with Pi 1.

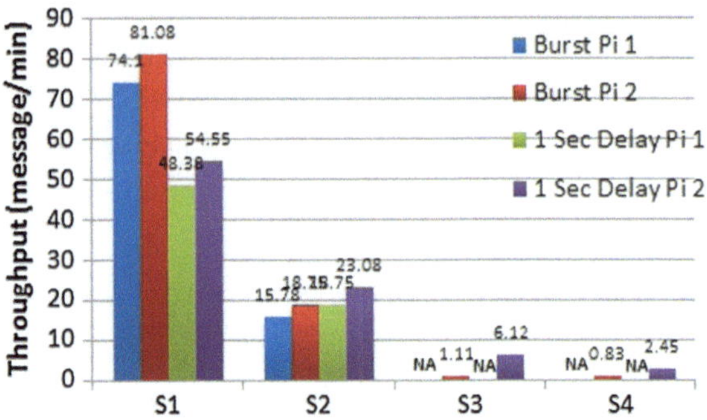

Fig. 5 Average throughput

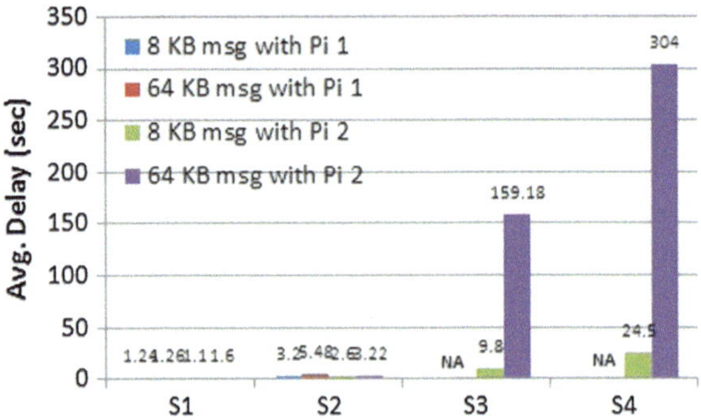

Fig. 6 Effect of size on processing delay with 1 s delay

4.3 *Effect of Message Size*

We measure how the size of the message effects the time required to process it and the end-to-end throughput. For this we sent 50 messages of varying sizes (8–64 KB) and 1 s delay with PiMail running on Pi 1 and Pi 2. Figure 6 shows the average processing delays of all the four scenarios. The processing delays of 64 KB message for S3 and S4 spikes exponentially to 159 and 304 s respectively on Pi 2 (as discussed above Pi 1 is not capable of handling content filtering with high frequency of emails). Message size also have a direct impact on end-to-end throughput. Figure 7 shows that with the increase in message size the throughput can dip to 0.2 message per minute. SpamAssassin utilize content based filtering, which is directly related to the size of the message. Stand-alone mail server deployment or addition of ClamAV have negligible effect on the size of the message.

4.4 *CPU and Memory Utilization*

We evaluated the CPU and Memory utilization of PiMail on Pi 2 (Pi 1 was excluded due to its incapability of handling S3 and S4) by continuously sending

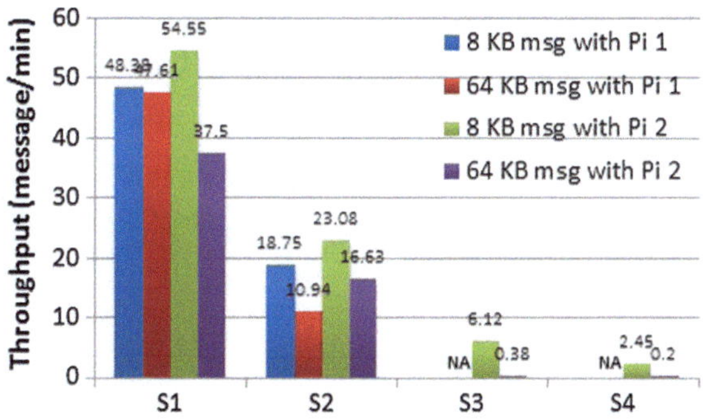

Fig. 7 Effect of size on throughput with 1 s delay

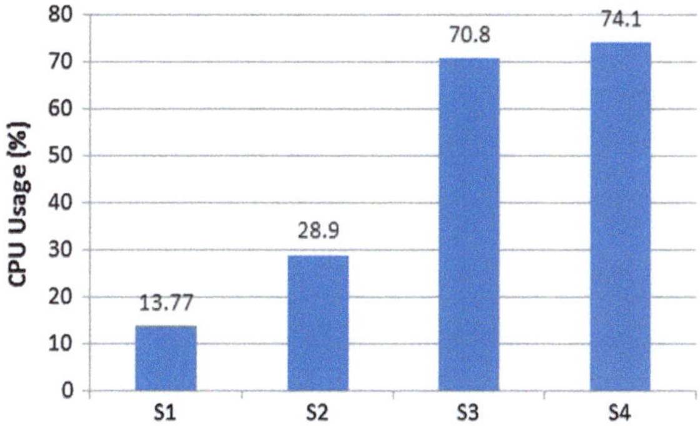

Fig. 8 Average CPU usage (%)

email message of 8 KB every 0.6 s for a total time of 8 min or 480 s. These experiment settings are based on email statistics from a large university discussed in [14]. The average CPU utilization results (Fig. 8) show that S3 and S4 are expensive in terms of CPU usage (70–75%). This is because content-based filtering has to apply different rules and filters. In contrast, the CPU usage of S1 and S2 remained as low as 13–29% on average.

As shown in Fig. 9, the memory usage in S2 and S4 remained constant to 53 and 63.52% respectively. Memory utilization is highest in virus filtering as compared to other scenarios. Surprisingly, there was not much difference in memory utilization when SpamAssassin (S3) was added to simple mail server (S1).

4.5 *Processing Delays with Low Email Volume*

In the end we measure the effect of low volume (without burst) of emails on the PiMail server running on Pi 2. We sent 50 email messages of 8 KB each with an interval of 1 min between any two messages. Figure 10 shows the average email processing delays for all the four scenarios. In S1, the PiMail server was able to

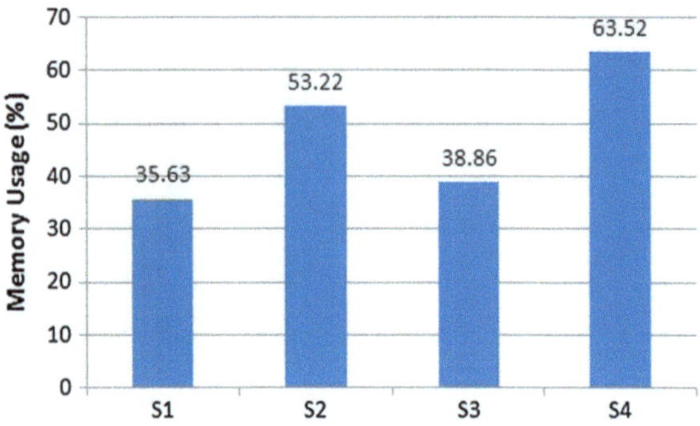

Fig. 9 Average memory usage (%)

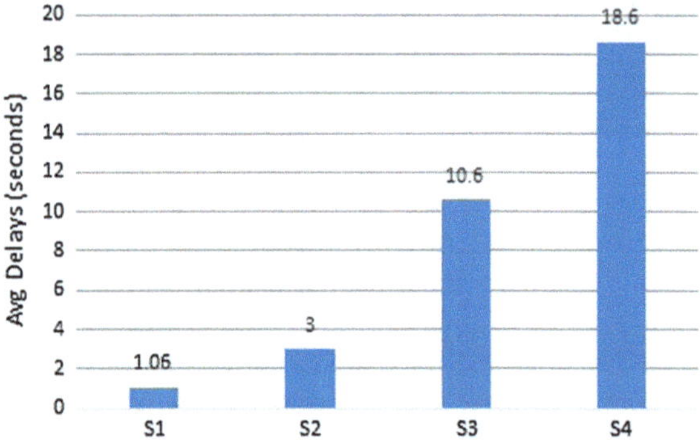

Fig. 10 Average processing delay with 1 email per 60 s on Raspberry Pi 2

process an email with an average delay of 1.06 s. This increased to 3.2, 10.6 and 18.6 s in S2, S3 and S4 respectively.

This indicates that PiMail server can handle low volume (without burst) of emails without creating any backlogs and it can fully cater for the needs of individual users and SMEs (small and medium-sized enterprises).

5 Discussion

5.1 Longer Interval Delays

The experiments in Sect. 4 were mainly designed to stress test the performance of PiMail under different settings. In Sect. 4.5. we explored longer interval delay time of 1 min between average sized emails. Figure 10 shows that with longer interval delays, PiMail server running on Pi 2 processed the emails well inside the interval delays without creating any backlogs. At this point we consider it un-necessary to explore any longer interval delay between emails.

5.2 *Email Attachments*

Reasonable proportion of emails today (10–20%) comes with an attachment i.e. graphics, spreadsheets, pdf, Word documents, etc. Emails with and without attachment(s) follow the same workflow in PiMail. The email processing delays and throughput mainly depends on the size of attachment/message. In Sect. 4.3. we measured the effect of message size on the processing delays and throughput. Based on the results we observed that stand alone mail server deployment or addition of ClamAV have negligible effect on the size of the message. On the other hand SpamAssassin utilize content based filtering, which is directly related to the size of the message. Thus, large attachment size will affect the PiMail performance in scenario S3 and S4.

5.3 *Target Users*

The main target users of PiMail infrastructure are individuals or SMEs with 10–250 users and aggregate email volume of 2500–4000 emails per day.

6 Conclusion

In this article, we propose PiMail, an affordable, lightweight and energy-efficient private email infrastructure based on Raspberry Pi. To the best of our knowledge, we performed the first extensive study that benchmarks the performance of Raspberry Pi used as a portable and private mail server.

The experiments were designed to stress test the performance of PiMail under different configurations. Based on the results, we observed that content-based spam filtering with SpamAssassin is the most resource hungry process. With high volume of emails, PiMail experienced performance bottleneck when configured to perform content filtering.

Having said that, if we are to focus on providing email services to individuals and SMEs, it would be unrealistic to have back to back emails or even with a short interval of just 1 s. With an interval of 20–30 s, even the most decorated configuration of PiMail (S4) will not exhaust the resources and there will be no backlogs. In the end, we can conclude that PiMail, running on Raspberry Pi 2, is capable handling a volume of 4000 emails with a frequency 3 emails per minute, which is more than enough for individuals and SMEs.

Acknowledgements We thank Imran Rauf for providing infrastructural support of Raspberry Pi. We also thank Farhan Kamal Khan for his helpful and interesting comments and suggestions.

References

1. Klensin J (2008) Simple mail transfer protocol. The internet society, rfc 5321. The Internet Society, RFC 5321
2. Prism surveillance program. https://en.wikipedia.org/wiki/PRISMsurveillance-program/
3. Cost of private mail server. http://jeffreifman.com/how-to-install-yourown-private-e-mail-server-in-the-amazon-cloud-aws/
4. Raspberry pi. https://www.raspberrypi.org/
5. Spamassassin. http://spamassassin.apache.org/
6. Clamav: Opensource antivirus engine for mail gateway. http://www.clamav.net/
7. Postfix mail transfer agent. http://www.postfix.org/
8. Hameed S, Asif MA, Khan FK (2015) PiMail: affordable, lightweight and energy-efficient private email infrastructure. In: Proceedings of 11th international conference on innovations in information technology (IIT)
9. T.R.P Foundation. About us. https://www.raspberrypi.org/about/
10. Raspberry pi 2 model b. https://www.raspberrypi.org/products/raspberrypi-2-model-b/
11. Raspbian os. http://www.raspbian.org/
12. Windows 10 for raspberry pi. http://www.WindowsOnDevices.com
13. Dovecot: secure imap server. http://www.dovecot.org/
14. Hameed Sufian, Xiaoming Fu, Nastry Nishanth, Hui Pan (2013) Fighting spam using social gatekeepers. Netw Sci 2(1–2):28–41

Piezoelectric-Based Energy Harvesting for Smart City Application

Mahmoud Al Ahmad and Areen Allataifeh

1 Introduction

One of the top processing issues in urban centers is energy management due to the complexity of their energy systems [1]. For that, researchers have dedicated huge effort to solve this issue. Moreover; digital technology has been integrated with physical city to form what is called a "smart city". The smart city improves life quality using technology in managing its resources to enhance economic, social, and environmental sustainability and efficiency [2]. Sophisticated and progressive systems present a primary part in smart cities in automating and improving processes inwards the cities; starting from intelligent systems that can monitor and control infrastructures independently, to buildings with smart designs that can collect rain water. Multi applications have been conducted and problems associated along with these applications have been resolved during the last decade [3]. Energy harvesting system is one of the applications that has been implemented in smart cities to enhance quality of life by generating green energy with fewer hazards on the environment. Harvesters have the ability to convert environmental energy to electrical energy. Researchers have developed various methods for energy harvesting from solar, wind, heat, and ambient vibration.

Various researches have focused on the transformation of vibration energy from the ambient; generated from human motion, air flow, and water current/wave as shown in Fig. 1. This paper presents a general review and overview of possible usage of a bias-less device which has the ability to generate electrical energy for remote applications. The harvester is composed of a self-biased piezoelectric oscillator that will be utilized in controlling switching frequency and amplitude of the transistor in the boost converter. With this device, controlling the DC output

M. Al Ahmad (✉) · A. Allataifeh
UAE University, Al Ain, UAE
e-mail: m.alahmad@uaeu.ac.ae

L. Ismail and L. Zhang (eds.), *Information Innovation Technology in Smart Cities*,
https://doi.org/10.1007/978-981-10-1741-4_23

Fig. 1 Vibration sources used for energy harvesting

voltage of the converter becomes easier and without the need of an external source for biasing.

Piezoelectric energy harvesters have been implemented to harvest energy from various vibrations. Wind energy is one of the sources that harvesters were used to convert it to electrical energy using a bimorph actuator. Combining the harvester with wireless transmission provided a power supply for communication and remote sensors [4]. Furthermore, they were planted in the sea to harvest electrical energy from longitudinal motion of sea waves. The harvester used in such an application composed of a cantilever attached with piezoelectric patches and a proof mass. It was found that the output power amount depends on the cantilever ratio of the width to the thickness, the ratio of proof mass to cantilever, sea depth, wave height, and the ratio between sea depth and wave height [5]. A prototype of windmill built with ten piezoelectric harvesters arranged circularly on it was tested under wind speed of 1–12 mph. Output power of such a harvester was about 7.5 mW when speed of the wind reached 10 mph [6]. The pressurized water flow was used as a source of mechanical energy for the shear mode harvester to convert it to electrical energy [7]. On the other hand, Spornraf et al. presented piezoelectric bend transducers that were excited using laminar flow [8].

In another field, piezoelectric harvesters have been used in powering aircraft electronic system. During the flight, the harvester converts mechanical energy from the inlet airflow to electrical energy. A prototype has been designed and tested by using an air cylinder to simulate the airflow. The results showed a linear relationship between the airflow velocity and sound pressure, and open circuit voltage. This proves the ability for this kind of harvesters to be implemented in aircraft application [9].

In the same context, the piezoelectric harvester could be mounted on a car damper to generate electrical energy from tire movements [10]. Pneumatic tire deflections under carried load have been used to harvest energy. This energy depends on tire geometry, vehicle speed, and air pressure. The piezoelectric stacks that are fabricated from lead zirconate titanate are embedded inside the tire. The harvester is modeled as the first mode vibration of a cantilever. Although it harvests small amounts of energy, it's enough to power wireless sensors [11]. Piezoelectric harvesters could be used also in wearable applications and can be integrated in watches, clothes and shoes [12]. Since 1990, many harvesters have been designed to generate electrical power from arm movements; an example in this respect is the Maestro brand, Swiss watches. The integrated micro generators in the wrist watch converts movement mechanism and stores it [13]. One of the ordinary activities of human is walking, thus piezoelectric materials have been invested in the generation of energy from walking by inserting them in the shoes and use their generated energy in charging phones. Kymissis et al. examined a prototype of a shoe with three harvesters planted in it. Another wearable application is to harvest from a backpack by replacing its strap with a PVDF strap [14]. Electrostatic self-assembly process was used to build electrodes on the surface of the strap to ensure that it can hold loadings and to increase its ratability. Results improved that this system could harvest about 45.6 mW [15].

In another field, a stairway was designed to harvest energy in the form of potential energy from human [16]. The excitation signals depend on the walking space as well as the characteristics of the person itself [17]. To take advantage from vibrations of high-rise buildings that occurs due to wind loadings, a coupled piezoelectric cantilever with a proof mass had been designed in 2013 by Xie et al. [18] It was optimized by studying the effect of the location, length, radius of the attached mass, and the ratio between the piezoelectric part and the beam. Two years later, they developed their model and presented a novel harvester. It's composed of two groups of generators connected in series by a common shaft. This harvester has the ability not only to harvest energy from harmonic movements, but also works as a damper to dissipate building vibrations [19].

Recently, Moure et al. have evaluated for the first time the integration of piezoelectric cymbal harvesters in asphalt with diameter of 29 cm. It was found that each cymbal harvests a power up to 16 μW due to passing of single heavy vehicle. When integrating 30,000 cymbals in 100 m of road, the induced energy densities were around 40–50 MW h/m2 which is about 65 MW in a year [20]. Under the same concept, traffic on the bridges creates vibrations that have been utilized to harvest energy by Sazonov et al. Results of the research showed the ability of such systems to harvest energy up to 12.5 mW. Experiments were conducted to operate wireless systems using the harvested energy on a low traffic volume of the rural highway bridge [21].

The potential of harvesting energy from the mobility of people in a Macquarie University's library has been studied by Li and Strezof. They examined the places in the building where the mobility is at its highest levels. They found that the best places to generate electricity from are in the main entrance, the areas of loopy meeting and the cafeteria. The high traffic places in the library have been defined in order to optimize the harvesting process and enhance its efficiency to reach the maximum power possible. The Pavegan tiles were used along the pathway to harvest energy from movements. Number of tiles depends on their size and the deployment method. The best deployment method were determined depending on the width and length of the pathways, as there is a lengthwise and widthwise arrangement of the tiles [22]. Different researches have been conducted during last decade to increase output power yield and enhance harvester performance. Three main methods have been used to develop piezoelectric generators: Choosing appropriate operation mode, changing materials and changing structure of the harvester.

Two operation modes exist for piezoelectric harvesters, 33 and 31 mode. B. Lee et al. studied the difference between the two operational modes of cantilever-based piezoelectric generator fabricated using silicon process. They found that 31 mode device has an output power higher than 33 mode, and open circuit voltage with less maximum value and resonance frequency comparing to 33 mode [23]. D. Kim et al. also studied the effect of each mode in the performance of the harvester and derived the Norton equivalent representation for each method. They concluded that 33 mode has the potential to offer modest enhancement compared to 31 mode [24].

Water vapor and oxygen in the ambient air could affect piezoelectric materials and their performance. Thus, different materials have been developed and utilized as piezoelectric harvesters; such as classical ceramic materials, polymers, and Aluminum nitride. Other researchers used lead-free piezoelectric films to design MEMS harvester [25–27]. The structure of the piezoelectric harvester plays an essential role in enhancing output power of the harvester. Different structures have been developed and studied during last decade to enhance harvester performance. Cantilever harvester is the conventional type of piezoelectric harvesters. To achieve flexibility and efficient energy harvesting, an array of cantilevers have been arranged. The alignment of the layers in an arrayed harvester as well as the thickness of the supporting layer plays a big role in optimizing voltage, current, and output power of the harvester. The arrayed harvester could respond at lower frequencies as compared to the single cantilever when fabricated from the same materials [28, 29]. Other structures have been designed for energy harvesting, such as ring shape [30, 31], cylinder [32], and a sandwich structure that are composed of two piezoelectric sheets with a metal shim sandwiched between them [33].

In 2012, a research was conducted to generate electric energy from free falling droplets which can be applied to generate power from raindrop energy. The harvester was designed using two layers of lead zirconate titanate films as piezoelectric material and a shim layer between them as shown in Fig. 2. The impact of the

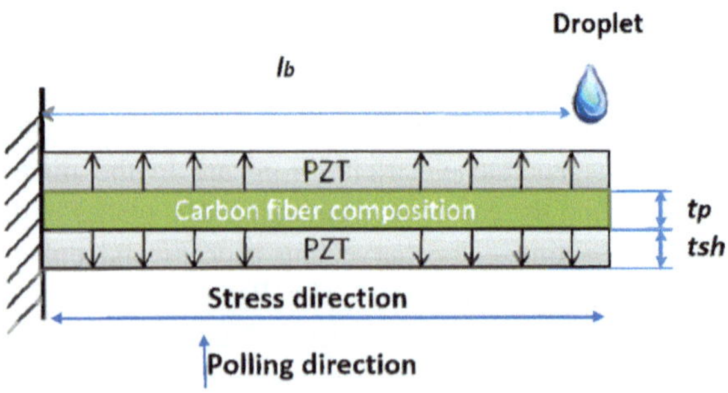

Fig. 2 Schematic diagram of droplet energy harvesting concept

falling of droplets on the top of the cantilever will transfer the kinetic energy from them to the cantilever causing it to vibrate due to mechanical stress. The generated output energy due to this impact was calculated using electromechanical relationships. The performance of this cantilever is affected by the resonance frequency as the output power increases with the increase at lower resonance frequencies. Experimental results showed that the power yield is 37 times higher than previous work done before [34]. Later that same year, an enhanced version of such cantilever for the same application was presented. Numbers of layers were increased to five layers composed from the same material, lead zirconate titanate, and succeeded to reach an output energy yield of 400 mJ. It is concluded that multilayers of PZT cantilevers enhances output power yield regardless of layers total quantity. Drop intensity and location also affects the output yield of the harvester. In the experiments, three intensities were tested: moderate, low and strong rain intensity, by varying number of drops per second [35].

In the same field, a research was carried out to study the induced strain in piezoelectric cantilever due to radio frequency (RF) propagation signal. When applying RF propagation field on a cantilever, it initiates an alternating current in the electrodes of the piezoelectric cantilever which in turn actuates the cantilever to mechanical movement when the frequency meets the resonance frequency of the structure. Figure 3 illustrates the cross-section of the piezoelectric cantilever that was used. It's composed of lead-zirconate-titanate with 1 mm width, 7 mm length, and 0.24 mm thickness. It has a gold metallized bottom, top, and intermediate electrode with thickness of 80 nm [36].

Lots of researchers have been interested in enhancing energy harvesters. That was the motivation to develop a new methodology that assists researchers in optimizing energy harvesters easily and quickly. This methodology significantly minimizes working time cycle needed to obtain a working prototype.The proposed methodology in estimating d33 electrical output power of a conventional cantilever shown in Fig. 4 relies on the analogy of electromechanical input impedance. It utilizes power system theories, electromagnetic theories, and the direct mechanical

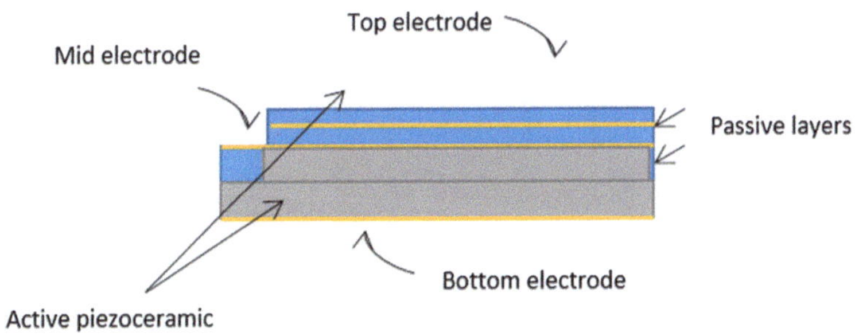

Fig. 3 Cross-sectional view of the Cantilever harvester

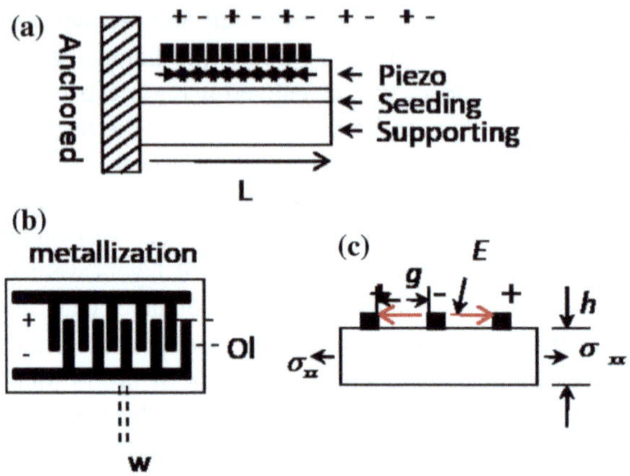

Fig. 4 Cantilever with d33-mode **a** cross section of the device, **b** electrode topology, **c** direction of electric field and mechanical stress

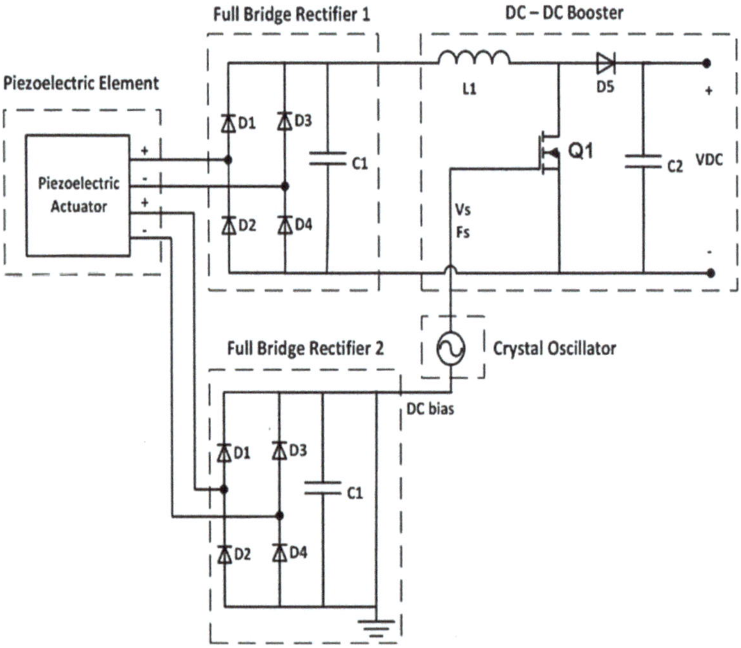

Fig. 5 Harvested Power conversion circuit

to electrical analogy to derive the models. These models compute output power taking in consideration material type of the harvester as well as its dimension. After proving the methodology experimentally, it was concluded that it gives crucial additional information that is used in developing harvester operation and design compared to the conventional methods [37].

The harvester design requires a power conversion circuit that can transform from one DC level to another DC level which is called a DC/DC converter, the DC/DC converter can be then used as step up voltage (boost) or step down voltage (buck) [38].

The novel idea in this research is in using a bias-less device capable to produce a reasonable amount of power used for battery-less operation of sensors for remote application. The harvester consists of a self-biased oscillator which is a piezoelectric vibrational oscillator that will be used for controlling voltage and frequency of the switch i.e. transistor in the DC/DC boost converter, so the output DC voltage can be controlled easily and the energy saved without external biasing source.

The basic structure for a piezoelectric energy harvester involves a piezoelectric layer, or beam, attached to a vibrating mechanical structure. The vibration causes bending (stress) in this piezoelectric layer and subsequently the bending induces electric charges. In many cases, a proof mass is attached to the end of the piezoelectric beam to covert the vibration, or equivalently acceleration, into an effective inertial force to further increase the bending of the beam. The mass also can be varied to tune the effective resonant frequency of the structure to the frequency of vibrations, thereby maximizing the output power as much as possible. Modeling piezoelectric harvester is being actively pursued in the literature to meet the requirements of design and development engineers [39–42]. A comprehensive model that effectively relates the power output to the piezoelectric harvester structure and materials involved is highly desired.

The piezoelectric harvester circuit which is shown in Fig. 5 contains two main sections for energy harvesting. In the first section the input from piezoelectric element enter to the AC/DC full wave bridge rectifier with smoothing capacitor C1 that transform the input AC signal to DC signal [43]. The second section transforms the converted DC signal to other level by step up the voltage using DC/DC boost converter, then the output DC signal will be transferred to the storage component then to the load which is a low power application operated without battery (Fig. 5).

2 Piezoelectric Energy

Piezoelectric bimorphs constitute of two piezoelectric beams that are separated by a shim. The three latter layers are usually fixed on one side and left free at the other. When a mechanical load is applied to the bimorph, the strain induced in the piezoelectric material generates a voltage, which represents an energy conversion from the mechanical domain to the electrical domain; a transformer with a special turn ratio accounts to this conversion. Modeling piezoelectric bimorph harvesters using an electric circuit based on the electromechanical analogy is shown in Fig. 6 [44–46].

In Fig. 6, the left side of the network model represents the mechanical domain, where σ_{in} models the input vibrations, Z_{in} models the input impedance of the network, R_b models the mechanical damping, Lm models the mass of the harvester,

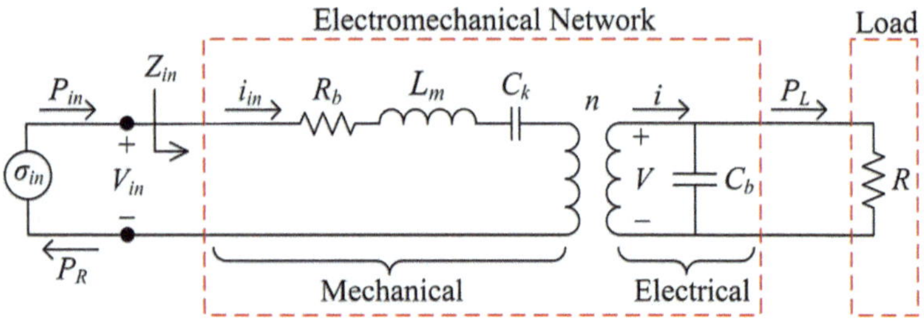

Fig. 6 Equivelent mechnical-electrical model circuit

and C_k models the inverse of the stiffness. At resonance, in which by definition the total impedance seen by the source is purely resistive, and using Kirchhoff's voltage law, the mesh voltage relation on the primary side of the network in Fig. 6 is:

$$\sigma_{in} = R_b\dot{S} + nV \tag{1}$$

where S is the strain, n is the turn ratio of the transformer, and V is the voltage, and the 'dot' represents the derivative. The voltage V is given simply by:

$$V = iR \tag{2}$$

where i is the current and R is the load resistance. Therefore

$$\sigma_{in} = R_b\dot{S} + niR \tag{3}$$

The current generated as a result of the mechanical stress evaluated at zero electric field is [4],

$$i = awl_e d_{31} c_p \dot{S} = \alpha\dot{S} \tag{4}$$

where a is a constant that is 1 or 2 depending on the wiring of the harvester, w is the width of the piezoelectric material, l_e is the length of the electrode in the piezoelectric harvester, d_{31} is the piezoelectric strain coefficient, and c_p is the Young's Modulus of the piezoelectric material, and α is the product of a, w, l_e, d_{31}, and c_p.

Hence, and after substituting Eq. (4) into (3),

$$\sigma_{in} = (R_b + n\alpha R)\dot{S} \tag{5}$$

Equation (5) provides important insight regarding the transfer of impedances in electromechanical analogy networks. By relying on power system theory, the equivalent load resistance should be carried to the primary side and hold a new value of n^2R. However, Eq. (5) explicitly indicates that the load resistance will be carried to the primary side with a scaling factor equal to $n\alpha$. This finding is one of the important contributions of this communication.

3 Harvester Design and Experiment

The novelty here is to use a bimorph piezoelectric vibrational oscillator output shown in Fig. 8 as a biasing source for the switch i.e. transistor Q1 that control the output DC voltage by charging and discharging the inductor L1. When the switch is ON the output from rectifier is fed to the inductor L1 so in the ON time the inductor store energy and diode D5 reversed biased isolating the output, When the switch in the OFF state the stored energy in the inductor will transfer through the diode to the storage element then to a load. The output DC voltage controlled by switching signal from piezoelectric oscillator applied to transistor Q1 [47]. The proposed harvester contains the stages of energy harvesting using an innovative design. The piezoelectric cantilever will vibrate and the signal is fed to AC/DC rectifier which will change the signal to DC level voltage, but the output voltage is low. The low output voltage is then stepped up using DC/DC converter, which boosts the DC voltage to a considerable level by external biasing circuit which will decrease the efficiency of this system.

The invented design provides a solution for biasing problem using a piezoelectric vibrational oscillator which generates a sine wave output of desired frequency and voltage required for biasing the converter switch that increase the efficiency of the harvesting power. The invented system is renewable so no waste in energy will happen and no external source needed which increase the efficiency of the system. The piezoelectric vibrational oscillator will be used as biasing for converter switch so we can control the switching voltage Vs and switching frequency Fs to have an optimum energy harvesting and this will be an extra

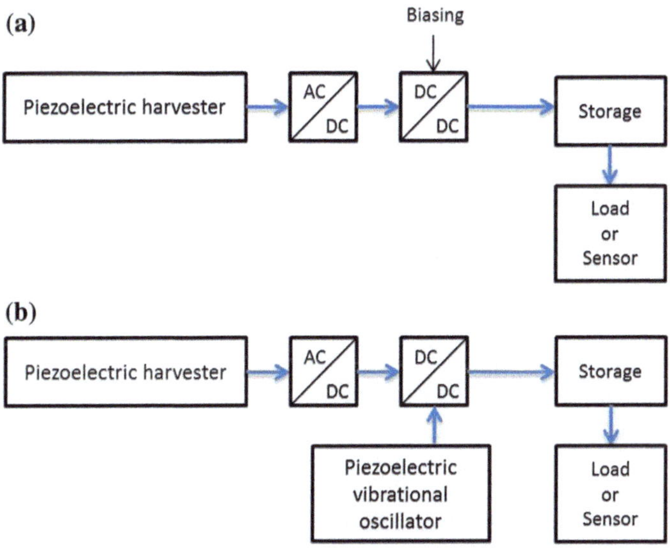

Fig. 7 Harvester design **a** Conventional design **b** Innovative design

advantage of the invented design, so you can control the output DC voltage VDC to the required level with a high efficiency compared with the conventional method that is limited to a constant values of biasing. The Innovative design enables us to match with optimum load condition for a desired application.

The proposed harvester design shown in Fig. 7 includes the stages of energy harvesting using a conventional design Fig. 7a and innovative design Fig. 7b. The piezoelectric cantilever will vibrate and the signal is fed to smoothing capacitor, but the output voltage is low. The low output voltage is then stepped up using DC/DC converter which boosts the DC voltage to a considerable level by external biasing circuit which will decrease the efficiency of this system as shown in Fig. 7a. The design in Fig. 7b invented a solution for biasing problem using a piezoelectric vibrational oscillator which generates a sine wave output of desired frequency and voltage required for biasing the converter switch that increase the efficiency of the harvesting method compared with the traditional one. The invented system is renewable so no waste in energy will happen and no external source needed which increase the efficiency of the system.

The piezoelectric vibrational oscillator will be used as biasing for converter switch so we can control the switching voltage V_s and switching frequency F_s to have an optimum energy harvesting and this will be an extra advantage of the invented design so you can control the output DC voltage V_{DC} to the required level with a high efficiency compared with the conventional method that is limited to a constant values of biasing. The Innovative design enables us to match with optimum load condition for a desired application.

There are four dominant factors affect the output Dc voltage VDC of the harvesting circuit are the input Piezoelectric voltage Vi (Piezo 1), input Piezoelectric frequency Fi, output voltage of the piezoelectric oscillator (Piezo 2) applied to transistor called switching voltage Vs, and switching frequency of the piezoelectric oscillator Fs. The DC output can be controlled by changing the duty cycle of the switch i.e. transistor, so if the duty cycle increase the output DC voltage will increase.

The fabricated circuit in Fig. 8 is the optimal design of piezoelectric harvesting circuit after choosing optimum values of the components C1, L1, and C2 as 1nF, 2mH, and 10nF respectively that give the maximum output DC voltage from the harvester. The circuit implemented practically in a printed circuit board (PCB) and experimental measurements were taken to the output DC voltage by sweeping the values of input voltage Vi, input frequency Fi, switching voltage Vs and switching frequency Fs. All DC measurements were taken using digital millimeters, and all waveforms were obtained via an oscilloscope.

The piezoelectric generator is connected to an electric shaker with integrated power amplifier driven with a function generator as shown in Fig. 9. When the resonant vibration has achieved, the generator has the maximum output power under the same load impedance, the shaker provides a variable mechanical excitation up to 31 N sine peak force with a 1/2 inch stroke in response to a sine wave input.

Fig. 8 Fabricated Piezoelectric harvester circuit

Fig. 9 Experimental setup

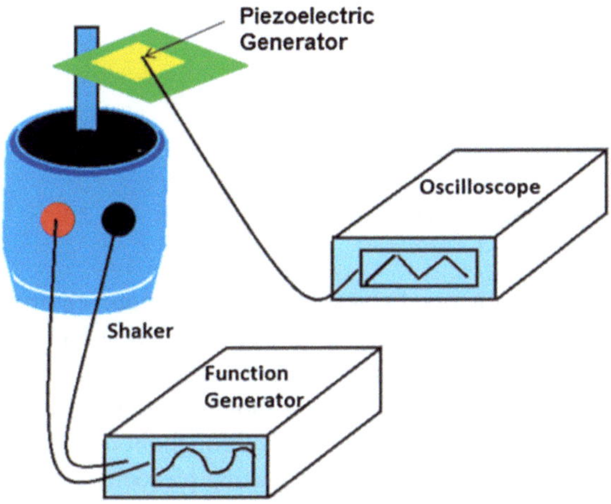

Piezoelectric element is a two-layered cantilever device that generates an ac voltage when a mechanical stress applied. The output power is critically determined by the mechanical deformation of the bender structure. The optimal design for piezoelectric element used in the experimental measurements is the DuraAct Piezoelectric energy transducer shown in Fig. 10a provided by Physik Instrument (PI) Ceramic [48]. The bender structure with the DuraAct P-876.A12 provides the greatest power output at different load resistance at the same excitation conditions (frequency: 1 kHz, displacement: 5 mm) as shown in Fig. 10b.

The presented piezoelectric self-biased energy harvester generates 5.4 VDC output under boundary conditions of (1 kHz, 2 Vpk). This invented design is a source covering several primary ultra-low power applications such as Wireless Sensor Network, precision agriculture, and biomedical uses for body area networks.

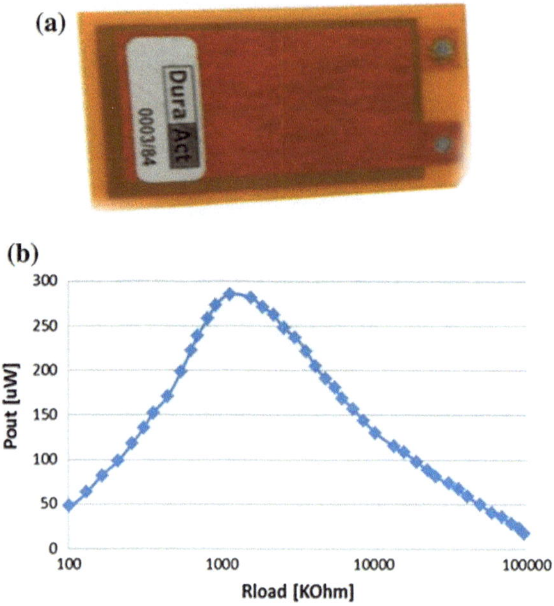

Fig. 10 Piezoelectric description **a** transducer, **b** power versus load. Measurements for DuraAct P-876.A12

4 Conclusion

Ambient energy harvesting sources in our environments has created significant interest as it offers a ultimate energy solution for small power applications including portable, flexible and wearable electronics; wireless sensor nodes, biomedical implants and environmental monitoring devices. The only way to power them is using ambient energy that lasts throughout the lifetime. This work presented a novel self-biased energy harvesting using piezoelectric bimorph design. The piezoelectric cantilever outputs are converted to DC voltages through two separate full bridge rectifiers. One rectifier output is then converted up through DC to DC circuitry, while the other have been used to provide a bias for an electronic circuit oscillator to switch on-off a transistor. The overall circuit efficiency is about 90% with an optimum output power of 300 μW at load of 1.2 MΩ.

References

1. Calvillo CF, Sánchez-Miralles A, Villar J (2016) Energy management and planning in smart cities. Ren and Sus Ene Reviews 55:273–328
2. Höller J, Tsiatsis V, Mulligan C et al (2014) Smart cities. From machine-to-machine to the internet of things. pp 281–294
3. Hancke G, Silva B, Hancke G (2013) The role of advanced sensing in smart cities. Sensors 13:393–425

4. Tan YK, Panda SK (2007) A novel piezoelectric based wind energy harvester for low-power autonomous wind speed sensor. Ind Elec Soc, 33rd Annual Conference of the IEEE, pp 2175–2180
5. Xie XD, Wang Q, Wub N (2014) Potential of a piezoelectric energy harvester from sea waves. J Sound Vib 421–1429
6. Priya Sh (2005) Modelling of electrical energy harvesting using piezoelectric windmill. Appl Phys Lett 87:184101
7. Wang D, Liu N (2011) A shear mode piezoelectric energy harvester based on a pressurized water flow. Sen Act 167:449–458
8. Spornraft M, Schwesinger N (2014) Flow Energy Harvester with nanoscale, piezoelectric material. En self-sufficient Sen 24:1–4
9. Zou H, Chen H, Zhu X (2015) Piezoelectric energy harvesting from vibrations induced by jet-resonator system. Mechatronics 26:29–35
10. Lafarge B, Delebarre C, Grondel S et al (2015) Analysis and optimization of a piezoelectric harvester on a car damper. 2015 International Congress on Ultrasonics, vol 70, pp 970–973
11. Khameneifar F, rzanpour S (2008) Energy harvesting from pneumatic tires using piezoelectric transducers. Smart Mat 1:331–337
12. Junga W, Leea M, Kangb M et al (2015) Powerful curved piezoelectric generator for wearable applications. Nano Eng 13:174–181
13. Dobrescua EM, Dobreb EM, Ionelac GP (2012) Technical means of preservation of renewable human energy's. Procedia Economics and Finance, vol 3, pp 463–468
14. Kymissis J, Kendall C, Paradiso J, Gershenfeld N (1998) Parasitic power harvesting in shoes. Second international symposium on wearable computers digest of papers, pp 132–139
15. Granstrom J, Feenstra J, Sodano H, Farinholt K (2007) Energy harvesting from backpack instrumented with piezoelectric shoulder straps. Smart Mater Struct 16:1810
16. Puspitarinia D, Suziantia A, Al Rasyida H (2016) Designing a sustainable energy—harvesting stairway: determining product specifications using TRIZ method. Procedia-Social and Behavioral Sciences, vol 21, pp 938–947
17. Kluger JM, Sapsis TP, Slocum AH (2015) Robust energy harvesting from walking vibrations by means of nonlinear cantilever beams. J Sound Vib 341:174–194
18. Xiea XD, Wub N, Yuenc KV, Wangb Q (2013) Energy harvesting from high-rise buildings by a piezoelectric coupled cantilever with a proof mass. Int J Eng Sci 72:98–106
19. Xie XD, Wang Q, Wang SJ (2015) Energy harvesting from high-rise buildings by a piezoelectric harvester device. Energy 93:1345–1352
20. Moure A, Izquierdo MA, Rueda S, Gonzalo A et al (2016) Feasible integration in asphalt of piezoelectric cymbals for vibration energy harvesting. Energy Convers Manage 112:246–253
21. Sazonov E, Haodong L, Curry D, Pillay P (2009) Self-powered sensors for monitoring of highway bridges. Sens J IEEE 9:1422–1429
22. Li X, Strezov V (2014) Modelling piezoelectric energy harvesting potential in an educational building. Energy Convers Manage 85:435–442
23. Lee B, Lin S, Wu W et al (2009) Piezoelectric MEMS generators fabricated with an aerosol deposition PZT thin film. J Micromech Microeng 19(6):065014
24. Donghwan K, Nishshank N, Hewa-Kasakarage N, Hall NA (2014) A theoretical and experimental comparison of 3-3 and 3-1 mode piezoelectric microelectromechanical systems (MEMS). Sens Actuators 219:112–122
25. Kima S, Leunga A, Koob Ch Y et al (2012) Lead-free (Na0.5K0.5)(Nb0.95Ta0.05)O3–BiFeO3 thin films for MEMS piezoelectric vibration energy harvesting devices. Mat Letters 69:24–26
26. Littrell R, Grosh K (2012) Modeling and characterization of cantilever-based MEMS piezoelectric sensors and actuators. J MEMS 21:406–413
27. Elfrink R, Kamel TM, Goedbloed M et al (2008) Vibration energy hatvesting with aluminum nitride–based piezoelectric devices. Proceedings of PowerMEMS 2008, microEMS2008, Sendai, Japan

28. Saadona S, Sidek O (2015) Micro-Electro-Mechanical System (MEMS)-based piezoelectric energy harvester for ambient vibrations. World conference on technology, innovation and entrepreneurship, vol 195, pp 2353–2362
29. Moonkeun K, Beomseok H, Yong-Hyun H et al (2012) Design, fabrication, and experimental demonstration of a piezoelectric cantilever for a low resonant frequency microelectromechanical system vibration energy harvester. J.Micro/Nanolithog Des 11(3):033009
30. Xie XD, Wang Q, Wu N (2014) A ring piezoelectric energy harvester excited by magnetic forces. Int J Eng Sci 77:71–78
31. Xiangdong X, Quan W (2015) A mathematical model for piezoelectric ring energy harvesting technology from vehicle tires. Int J Eng Sci 94:113–127
32. Mei J, Li L (2015) Double-wall piezoelectric cylindrical energy harvester. Sens Actuators 233:405–413
33. Dewana A, Ayb S, Karima M, Beyenal H (2014) Alternative power sources for remote sensors: a review. J Power Sources 245:129–143
34. Alkhaddeim T, Al Shujaa B, AlBeiey W et al (2012) Piezoelectric Energy Droplet Harvesting and modeling. Sensors 1–4
35. Al Ahmad M, Jabbour GE (2012) Electronically droplet energy harvesting using piezoelectric cantilevers. Electron lett 48:647–649
36. Alahmad M, Alshareef H (2012) Energy harvesting from radio frequency propagation using piezoelectric cantilevers. Solid-State Electron 68:13–17
37. Tawfiq Sh, Al ahmad M (2015) Electromechanical analogy for d33 piezoelectric harvester power calculations. European Conference on ECCTD, 2015, pp 1–3
38. Ferreira B, Van der Merwe W (2014) The principles of electronic and electromechanic power conversion: a systems approach. Wiley-IEEE Press, Hoboken, New Jersey, 1st ed, Jan 2014
39. Roundy Sh, Wright P, Rabaey J (2003) A study of low level vibrations as a power source for wireless sensor nodes. Comput Commun 26:1131–1144
40. Stephen N (2006) On energy harvesting from ambient vibration. J Sound Vib 293:409–425
41. El-Hami M et al (2001) Design and fabrication of a new vibration-based electromechanical power generator. Sens Actuators A 92:335–342
42. Williams C et al (2002) Development of an electromagnetic micro-generator. pp 337–342
43. Sedra AS, Smith KC (2013) Microelectronic circuits: theory and applications, 6th edn. Oxford University Press, Oxford
44. Beeby S, Tudor M, White N (2006) Energy harvesting vibration sources for microsystems applications. Meas Sci Technol 17:R175
45. Flynn A, Sanders S (2002) Fundamental limits on energy transfer and circuit considerations for piezoelectric transformers. Power Electron IEEE Trans 17:8–14
46. Tilmans HAC (1996) Equivalent circuit representation of electromechanical transducers: I. Lumped-parameter systems. J Micromech Microeng 6(1):157–176
47. Rashid HM (2014) Power electronics: circuits, devices & applications, 4th edn. Pearson, Bosto
48. PI ceramic: 'P–876 DuraAct patch transducer' (2014) Available at https://www.piceramic.com

MIX
Papier aus verantwortungsvollen Quellen
Paper from responsible sources
FSC® C105338

If you have any concerns about our products,
you can contact us on
ProductSafety@springernature.com

In case Publisher is established outside the EU,
the EU authorized representative is:
Springer Nature Customer Service Center GmbH
Europaplatz 3, 69115 Heidelberg, Germany

Printed by Libri Plureos GmbH
in Hamburg, Germany